Lecture Notes in Physics

Founding Editors

Wolf Beiglböck, Heidelberg, Germany

Jürgen Ehlers, Potsdam, Germany

Klaus Hepp, Zürich, Switzerland

Hans-Arwed Weidenmüller, Heidelberg, Germany

Volume 991

Series Editors

Roberta Citro, Salerno, Italy

Peter Hänggi, Augsburg, Germany

Morten Hjorth-Jensen, Oslo, Norway

Maciej Lewenstein, Barcelona, Spain

Angel Rubio, Hamburg, Germany

Wolfgang Schleich, Ulm, Germany

Stefan Theisen, Potsdam, Germany

James D. Wells, Ann Arbor, MI, USA

Gary P. Zank, Huntsville, AL, USA

The Lecture Notes in Physics

The series Lecture Notes in Physics (LNP), founded in 1969, reports new developments in physics research and teaching - quickly and informally, but with a high quality and the explicit aim to summarize and communicate current knowledge in an accessible way. Books published in this series are conceived as bridging material between advanced graduate textbooks and the forefront of research and to serve three purposes:

- to be a compact and modern up-to-date source of reference on a well-defined topic;
- to serve as an accessible introduction to the field to postgraduate students and non-specialist researchers from related areas;
- to be a source of advanced teaching material for specialized seminars, courses and schools.

Both monographs and multi-author volumes will be considered for publication. Edited volumes should however consist of a very limited number of contributions only. Proceedings will not be considered for LNP.

Volumes published in LNP are disseminated both in print and in electronic formats, the electronic archive being available at springerlink.com. The series content is indexed, abstracted and referenced by many abstracting and information services, bibliographic networks, subscription agencies, library networks, and consortia.

Proposals should be sent to a member of the Editorial Board, or directly to the responsible editor at Springer:

Dr Lisa Scalone
Springer Nature
Physics
Tiergartenstrasse 17
69121 Heidelberg, Germany
lisa.scalone@springernature.com

More information about this series at http://www.springer.com/series/5304

Gianguido Dall'Agata • Marco Zagermann

Supergravity

From First Principles to Modern Applications

 Springer

Gianguido Dall'Agata
Physics and Astronomy
University of Padua
Padua, Italy

Marco Zagermann
Hamburg, Germany

ISSN 0075-8450   ISSN 1616-6361   (electronic)
Lecture Notes in Physics
ISBN 978-3-662-63978-8   ISBN 978-3-662-63980-1   (eBook)
https://doi.org/10.1007/978-3-662-63980-1

This Springer imprint is published by the registered company Springer-Verlag GmbH, DE part of Springer Nature.
The registered company address is: Heidelberger Platz 3, 14197 Berlin, Germany

Preface

The idea of supersymmetries between bosons and fermions continues to be a major driving force behind many interesting developments in contemporary high-energy and mathematical physics. This statement applies in particular to *supergravity*, which may be described either as the result of supersymmetrizing Einstein's general theory of relativity or, equivalently, as the result of turning supersymmetry into a local gauge symmetry.

Given the success of gauge symmetries in the Standard Model, it may seem rather natural to treat also supersymmetry as a local gauge symmetry. On the other hand, gravitational interactions between particles are notoriously irrelevant at accessible collider energies. One might therefore be concerned that the difference between global supersymmetry and supergravity could only affect Planckian energy scales and hence be without measurable implications for present-day phenomenology. This naive expectation, however, is in general not true, because many supergravity effects also involve the scale of supersymmetry breaking, which may be much lower than the Planck scale. In fact, supergravity may be directly relevant for issues such as supersymmetry breaking, dark matter, dark energy, Big Bang Nucleosynthesis and inflation, i.e. for topics of strong present interest.

From the viewpoint of unified frameworks of all interactions such as string theory, the inclusion of gravity is of course not just an option, but a necessity. In fact, the low-energy limits of all supersymmetric string theories are supergravity theories. Due to their inherently non-linear structure and the protective properties of supersymmetry, the supergravity approximations to string theory are often the only available tools for probing the non-perturbative structure of string theory.

In spite of all these applications, many researchers who use aspects of supergravity in their daily work never really studied the theoretical origins of these aspects or the derivations of equations they have encountered many times. This is largely due to supergravity's reputation of being an extremely complicated subject with only very few pedagogically suitable references.

Part of this perception is not completely wrong: compared to the literature on global supersymmetry or string theory, there are much fewer introductory texts on supergravity that treat the subject in a sufficiently deep but comprehensible manner. Admittedly, some supergravity Lagrangians may at first sight also look quite intimidating, and browsing through this book will certainly reveal a few examples. But the reader should be assured that this intimidation goes away rather quickly once one has learned where to focus on in these equations.

Another part of the frustration many students experience when they try to learn supergravity is that many approaches use advanced techniques such as superfields on curved superspace or superconformal tensor calculus, which are powerful computational tools, but require a dedicated study of these techniques before the really interesting stuff can be extracted. It is also fair to say that the usefulness of some of these techniques is often greater for global supersymmetry than for supergravity, or may, in fact, not even be practically available, as is the case for many extended or higher-dimensional supergravity theories.

Especially for particle phenomenologists, an additional obstacle is that many texts on supersymmetry work with the two-component spinor notation instead of the more familiar four-component spinor formalism.

Lowering these obstacles towards a deeper understanding of supergravity and explaining also those things that are usually not explained or even overlooked were precisely the central motivations for writing this book.

In order to reach this goal, we completely avoid the introduction of a technical machinery that involves unphysical fields, auxiliary symmetries or other artificial objects. Instead, we work directly with the physical on-shell field content and use straightforward and explicit computations as well as simple geometrical reasoning to arrive at the results. Two major themes of this book are also the differences between supergravity and global supersymmetry and the physical and mathematical consequences of these differences.

Throughout this book, we use four-component spinors instead of two-component spinors. This formalism should make the text more accessible to phenomenologically oriented readers and has the advantage that it can be easily extended to other spacetime dimensions.

This book is suited for beginning graduate or advanced undergraduate students in high-energy physics or mathematical physics, as well as for researchers working in these or related areas. It assumes familiarity with basic notions of general relativity, differential geometry and global supersymmetry.

The book is divided into three parts:

Part I discusses foundational material such as our spinor conventions in Chap. 1, the transition from global to local supersymmetry in Chap. 2, as well as spinors in curved spacetime in Chap. 3. It culminates in the detailed discussion of pure supergravity with and without a cosmological constant in Chap. 4.

Part II is devoted to the couplings of matter fields in global supersymmetry (Chap. 5) and supergravity (Chap. 6) and discusses the phenomenological consequences for particle physics and cosmology in Chap. 7.

In Part III, three more formal topics are introduced: extended supergravity in Chap. 8, gauged supergravity in Chap. 9 and supergravity in higher spacetime dimensions in Chap. 10.

Most chapters contain some exercises to illustrate special cases, discuss some extensions or prove statements from the main text. Moreover, an incomplete collection of references is given at the end of each chapter, where we refer to some original papers, review articles or textbooks we found useful for preparing this text.

Padova, Italy
Hamburg, Germany
29 June 2021

Gianguido Dall'Agata
Marco Zagermann

Acknowledgements

Over the years, we have benefitted much from the interaction with many colleagues on subjects related to this book, and it is a great pleasure to thank them all for the numerous fruitful discussions and collaborations. We feel guilty in not being able to list them all, but we do have to give special thanks to those colleagues who played a pivotal role in shaping our understanding of supergravity: Anna Ceresole, Riccardo D'Auria, Sergio Ferrara, Murat Günaydin, Antoine Van Proeyen and Fabio Zwirner. Without their support and their willingness to share with us so many of their deep insights into this subject, this book would have never been written.

This book grew out of introductory lectures on supergravity and closely related topics presented by G. Dall'Agata at SISSA (a.y. 2006/07–2009/10), Scuola Normale Superiore di Pisa (2011, 2017–2021), Padova Uni. (2008/09,2011/12–2012/13, 2016/17–2017/18), EPFL Lausanne (2009), GGI, Florence (LACES 2009) and at SEPnet (Queen Mary U. London, Univ. of Southampton and Royal Holloway U. London, 2010) and by M. Zagermann at LMU Munich (2006/07), Sogang University, Seoul (2007), MPI for Gravitational Physics, Potsdam ("String Steilkurs" 2005 and 2008), MPI for Physics, Munich (2009), Leibniz Universität Hannover (2009/10), DESY, Hamburg (2012) and GGI, Florence (LACES 2019). We thank C. Angelantonj, M. Bertolini, D. Cassani, O. Lechtenfeld, J. Louis, D. Lüst, J.-H. Park, R. Russo, A. Sagnotti, C. Scrucca, F. Steffen and S. We thank Theisen for inviting us to these lectures and for encouraging us to give such courses. It is a pleasure to thank all the participants of these courses for their countless insightful comments and questions that greatly helped us to improve the presentation, especially Johannes Brödel, Daniel Junghans, Stefano Lanza, Enrico Pajer, Dario Partipilo, Erik Plauschinn, Timm Wrase and Matteo Zatti. Moreover, we thank Vincent Bettaque, Niccolò Cribiori, Gianluca Inverso and Karl Kortum for their careful reading of parts of the manuscript and their valuable feedback.

We are indebted to Christian Caron and Lisa Scalone from Springer for their helpful support over the years and their remarkable patience.

Finally, we thank Luis Álvarez Gaumé, who first suggested to turn our notes into a book. We hope that the published version embodies the hands-on and physics first approach that he liked in our original presentation.

Contents

Part I

Foundations and Pure Supergravity

Introduction

The quest for a fundamental theory of all elementary particles and their interactions has been one of the most fascinating scientific endeavors during the past century. One of the main guiding principles in the construction of the ever more refined theories of high energy physics has been the systematic use of symmetry principles. They form the basic language in terms of which the Poincaré invariant quantum field theory based on the gauge group $SU(3) \times SU(2) \times U(1)$ we now call the Standard Model (SM) is formulated.

Among the many interesting ideas for physics beyond the Standard Model, the most fruitful one has arguably been the introduction of the concept of *supersymmetry*, i.e., a symmetry between bosonic and fermionic degrees of freedom. Indeed, supersymmetric extensions of the Standard Model have been put forward as possible solutions to the hierarchy problem, though the simplest models are challenged by current Large Hadron Collider results; they improve the unification of the three Standard Model gauge couplings at the Grand Unified Theory (GUT) scale, $M_{GUT} \cong 2 \times 10^{16} \text{GeV}$; and they might provide interesting dark matter candidates.

The one missing major player in these constructions, however, is the gravitational interaction with its elegant geometric description in terms of Einstein's general theory of relativity. Combining the principles of supersymmetry with gravity defines what is called *supergravity*, the topic of these lecture notes. As we now briefly explain, the idea of supergravity has many other intriguing applications in various areas of particle physics, cosmology, string theory, and mathematics that go far beyond the simple desire to marry supersymmetry with gravity and can serve as separate motivations to study supergravity theories.

© Springer-Verlag GmbH Germany, part of Springer Nature 2021
G. Dall'Agata, M. Zagermann, *Supergravity*, Lecture Notes in Physics 991,
https://doi.org/10.1007/978-3-662-63980-1_1

1.1   The Many Facets of Supergravity

Supergravity as a Gauge Theory of Supersymmetry
Supersymmetric field theories in rigid Minkowski spacetime feature supersymmetry as a global (or "rigid") symmetry. In view of the success of local gauge invariances in the Standard Model, it is natural to try to promote supersymmetry likewise to a local gauge symmetry. This leads directly to supergravity, and in fact this was the way supergravity was first constructed. Supergravity theories can hence equivalently be defined as the *gauge theories of supersymmetry*.

Just as for an ordinary Yang–Mills symmetry, the gauging of supersymmetry requires the introduction of a suitable gauge field that transforms into the spacetime derivative of the infinitesimal symmetry parameter:

$$\delta_{gauge}(\text{gauge field}) = \partial_\mu(\text{gauge parameter}) + \dots \qquad (1.1)$$

In supersymmetry, the symmetry parameter is a spinorial quantity, ϵ_α, with α being a spinor index. The supersymmetry gauge field is therefore not a vector field as in ordinary Yang–Mills theories but a vector-spinor field, $\psi_{\mu\alpha}$. On-shell (and for unbroken supersymmetry), this field propagates two helicity $\pm 3/2$ states, whereas off-shell it also contains states with helicity $\pm 1/2$ (just as an ordinary gauge field, A_μ, contains helicities ± 1 on-shell, but off-shell also two states with helicity 0).

We will see later that supersymmetry further implies that the graviton, $g_{\mu\nu}$, and the supersymmetry gauge field, $\psi_{\mu\alpha}$, sit in the same supersymmetry multiplet, the supergravity multiplet, so that local realizations of supersymmetry necessarily need to include gravity, which explains the name supergravity. Being the superpartner of the graviton, $\psi_{\mu\alpha}$ is called the *gravitino*.

Extended Supergravity and Unification
In extended supergravity theories with $\mathcal{N} \geq 2$ supersymmetries, the supermultiplet of the graviton contains $\mathcal{N}$ gravitini but also fields with spin ≤ 1. Extended supersymmetry can thus interpolate between the graviton and ordinary gauge fields, thereby leading to some sort of unification of the interactions mediated by these fields. For $\mathcal{N} > 2$, the supergravity multiplet also contains spin $1/2$ particles, and one may wonder whether sufficiently extended supergravity could even provide a unified theory of all interactions and matter particles. The attempts in this direction culminated in the construction of the maximally extended $\mathcal{N} = 8$ supergravity with the maximally possible compact gauge group SO(8) by de Wit and Nicolai in 1982 [1]. Unfortunately, this and other extended supergravity theories generally suffer from the problem that the corresponding gauge interactions are non-chiral, in contrast to what we see in the electroweak sector of the Standard Model. Moreover, the SO(8) gauge group studied in [1] is too small to accommodate the Standard Model gauge group SU(3) $\times$ SU(2) $\times$ U(1) as a subgroup, and the theory also has a very large negative cosmological constant. Although unification via extended

supergravity theories has not met with phenomenological success, the study of many extended supergravity theories in the early 1980s has proven to be a very valuable resource for many modern theoretical developments, notably in string theory; see below.

Better Behavior of Ultraviolet Divergencies

Another interesting motivation for studying supergravity is the *better behavior of ultraviolet divergencies* of its corresponding quantum theory in comparison with ordinary general relativity (GR). In the quantization of GR, one can remove the potential on-shell one-loop divergencies by field redefinitions. However, there is an ultraviolet (UV) divergence at two loops which cannot be removed [2, 3], and if matter fields are added, divergencies can appear already at one loop. However, if the matter content is consistent with supersymmetry, the situation improves again. It is even still under debate at the time of this writing whether maximally supersymmetric gravity in four dimensions is divergent or not. Recent calculations show that divergencies in $\mathcal{N} = 8$ supergravity do not show up before five loops [4] and current consensus is that the first counterterm appears at seven loops, though it is still possible that its coefficient is vanishing [5]. The study of the UV properties of $\mathcal{N} = 8$ supergravity and its connections with amplitudes in $\mathcal{N} = 4$ super Yang–Mills theory is presently a very active research area, and new techniques developed for their computation are now used in a much wider context.

Phenomenology

Whereas the lack of chiral gauge interactions precludes any direct use of extended supersymmetry and supergravity for phenomenological applications, $\mathcal{N} = 1$ supergravity theories are phenomenologically very interesting and could resolve some issues in globally supersymmetric extensions of the Standard Model. For instance:

- Supergravity can remove the large tree-level cosmological constant of spontaneously broken rigid supersymmetry.
- Supergravity suggests new mechanisms of supersymmetry breaking and their transmission to the Standard Model sector.
- When supersymmetry is broken in supergravity, the goldstino is "eaten" by the gravitino, which provides one way of explaining why the goldstino has not been seen.

Supergravity can also have important implications for the dark matter sector or early universe cosmology, as we will also explain later.

Supergravity as an Effective Theory of String Theory

While supergravity theories are in general not renormalizable, they do arise as the *low-energy effective actions of (super)string theory*, the best viable candidate for a consistent unified quantum theory of gravity and all gauge interactions. From this point of view, supergravity forms the interface between string theory and most of

its potentially observable low-energy phenomena, with its non-renormalizability
no longer being an issue. Moreover, even as the infrared limit of string theory,
supergravity with all its non-linear interactions still captures some of the non-
perturbative properties of string theory, which would hardly be accessible via the
conventional world sheet conformal field theory approach. For instance, string
theory is not only a theory of strings but also contains other extended objects
such as D-branes or other solitonic p-branes. Some of these objects first arose as
solutions of supergravity models, and the supergravity perspective has often given
interesting insights into their physics. Similarly, various duality symmetries have
been first, or better, understood by looking at the effective supergravity theories and
their solutions. Often when different compactifications lead to the same low-energy
spectra and vacua, one may uncover some new underlying symmetry of the higher
dimensional theories.

The Gauge/Gravity Correspondence
One of the biggest revolutions in our understanding of string theory came from
the observation that certain string models on curved spaces can be dual to non-
gravitating (often conformal) gauge field theories: this is the gauge/gravity correspondence or anti-de Sitter/conformal field theory (AdS/CFT) correspondence
[6, 7]. The weak coupling limit of string models, i.e., supergravity, is then an
important tool to compute quantities that are related to the strong coupling regime
in the dual field theory. A special role in this correspondence is played by
gauged supergravity,[1] because global symmetries in the CFT become local in the
corresponding supergravity model.

Geometry
Supergravity solutions that preserve some of the supersymmetry of the action are
particularly interesting also from a mathematical point of view. For one thing
this is due to the circumstance that supersymmetric solutions often satisfy *first-
order* differential equations, which in general are much easier to solve than the
usual second-order field equations of non-supersymmetric solutions. These first-
order differential equations define Killing spinors, which prove to be a powerful
concept in different areas of mathematics. The supersymmetric compactification
backgrounds of string or M-theory, for example, define compactification manifolds
with interesting mathematical structures such as restricted holonomy groups or special types of gauge bundles, which are consequences of the corresponding Killing
spinor equations. But also the spaces of fluctuations about such compactification
backgrounds carry non-trivial geometric structures that result in various interesting
scalar field geometries or "moduli spaces" in the dimensionally reduced field
theories, such as, e.g., Kähler, special Kähler, or quaternionic Kähler geometries, or
various types of coset spaces (see later chapters). Various cascades of dimensional

---

[1] Gauged supergravity refers to supergravity theories that contain also non-trivial conventional
gauge interactions, as we will see in Chap. 9.

reductions then reveal unexpected mathematical relations between different classes of restricted geometries that would have been difficult to obtain or even suspect otherwise.

Fake Supergravity

As a final motivation, we would like to mention the concept of *fake supergravity* [8, 9]. Fake supergravity describes classes of solutions (e.g., domain walls or black holes) of non-supersymmetric gravity theories for which the second-order field equations can be rewritten in a first-order form that resembles the Killing spinor equations of genuine supergravity theories. This is especially useful for the discussion of the stability of these gravity solutions, as it allows the use of the Nester–Witten argument [10] in a large class of theories without the usual consistency requirements and limitations imposed by actual supersymmetry. In this same framework, also cosmological solutions may be under better control [11].

1.2   Plan of the Lectures

Clearly, the present lecture notes cannot cover all the above topics and applications in full detail. Instead, their purpose is to give a survey of the basic ingredients needed to construct a supergravity action and to discuss its physical implications (mainly for particle physics), without introducing too much technical formalism or spending too much time on the many possible applications. Another focus will be on the differences between globally supersymmetric theories and supergravity, so that the reader may better understand what really needs supergravity and what can already be implemented in a globally supersymmetric theory without gravity.

More explicitly, the outline of the lecture notes is as follows: In the first part, which consists of Chaps. 1–4, we work our way toward the construction of the simplest possible supergravity theory in four dimensions, namely, pure 4D, $\mathcal{N} = 1$ supergravity. To this end, we first introduce our spinor conventions in the remainder of Chap. 1 and then explain, in Chap. 2, how the requirement of local supersymmetry naturally leads to the inclusion of the graviton supermultiplet. In Chap. 2, we also motivate the form of the supergravity action and the supersymmetry transformation laws. In order to write these down properly, we review spinors in curved spacetime and the vierbein formulation of general relativity in Chap. 3. Chapter 4 then is devoted to the detailed discussion of pure 4D, $\mathcal{N} = 1$ supergravity with and without a cosmological constant and gives the complete proof of its invariance under local supersymmetry. In these first chapters, we work out all the details and often show pedantically all the steps necessary for this construction. The serious reader is recommended to go through this material in full detail, as it greatly contributes to developing a good intuition that will also be helpful for the remaining chapters.

In the second part of the book, we describe how the minimal supergravity sector discussed in Part I can be coupled to matter multiplets and what the implications of these matter couplings for phenomenology are. To this end, we first discuss matter couplings in global supersymmetry in Chap. 5 and then explain,

in Chap. 6, the differences that arise when supergravity and local supersymmetry
are introduced. Various consequences of supergravity for the phenomenology of
particle physics and cosmology are explored in Chap. 7. In these chapters, we
emphasize in particular the differences between rigid and local supersymmetry, both
at a formal/mathematical and at a physical level.

In the third part of the book, the presentation will be more formal, and we will
discuss a number of more advanced topics. In Chap. 8, we introduce the concept of
electric-magnetic duality and show how its interplay with the R-symmetry group
essentially fixes the geometrical structures encountered in models with extended
supersymmetry, detailing especially the case of $\mathcal{N} = 2$ theories. We then give, in
Chap. 9, a brief but modern introduction to gauged supergravity models, which are
playing a prominent role in many interesting recent developments in string theory.
Special emphasis will be given here to the case of $\mathcal{N} = 8$ gauged supergravity.
Finally, we conclude with some remarks on higher-dimensional theories and the
relations between these models and four-dimensional gauged supergravities in
Chap. 10.

We used a number of references throughout the lectures, mainly to point to
some additional sources of information on the discussed topics. There are obviously
already many very good reviews on various aspects of supergravity theories, which
we used as inspiration to prepare these lectures; some of them are [12–29]. Also,
very good complementary recent references on supergravity are [30, 31].

1.3   A Quick Guide Through Our Spinor Conventions

Before we embark on our journey through the world of supergravity theories, we
briefly summarize the 4D spinor conventions used in this book.[2] For a generic
treatment of spinors in any spacetime dimension, see appendix 10.A. Very good
references are also [18, 32].

Assuming an orthonormal set of basis vectors, e_a, of Minkowski space
$(a, b, \ldots = 0, 1, 2, 3)$, the Minkowski metric with our choice of signature is

$$\eta_{ab} = \mathrm{diag}(-1, +1, +1, +1). \tag{1.2}$$

The generators of Lorentz transformations are denoted by $M_{ab} = -M_{ba}$ when they
are taken as anti-Hermitian generators or as $\mathcal{M}_{ab} = -i M_{ab}$ when they are taken as
Hermitian generators and satisfy the Lorentz algebra $\mathfrak{so}(1, 3)$,

$$[M_{ab}, M_{cd}] = -2\, \eta_{c[a} M_{b]d} + 2\, \eta_{d[a} M_{b]c} \tag{1.3}$$

---

[2] Throughout the book, it is understood that $c = 1$ and $\hbar = 1$.

> In these notes, symmetrization () and antisymmetrization [] are always taken with weight one, i.e., $(ab) = \frac{1}{2}(ab + ba)$, $[ab] = \frac{1}{2}(ab - ba)$, etc.

A spinor representation is a representation of the above Lorentz algebra that does not integrate to an ordinary (i.e., "single-valued") representation of the corresponding Lorentz *group*. Instead, it gives rise only to a "double-valued" representation of the Lorentz group in the sense that spatial rotations by 2π give minus the identity.

Mathematically, this is possible because the Lorentz group[3] is not simply connected but contains closed loops that cannot be continuously contracted to a point. The universal covering group of the Lorentz group is a group that is locally isomorphic to the Lorentz group but with a different global structure such that all closed curves are fully contractible. Spinor representations are then equivalently described as single-valued representations of this universal covering group that project to double-valued representations of the Lorentz group itself. The universal covering group of the 4D Lorentz group happens to be isomorphic to $SL(2, \mathbb{C})$, the group of unimodular complex (2×2) matrices.

In four dimensions, there are two commonly used notations for spinor representations: the two-component spinor notation and the four-component spinor notation. The two-component spinor notation is based on the abovementioned accidental isomorphism between the universal cover of the 4D Lorentz group and $SL(2, \mathbb{C})$, which does not have a direct analogue in a generic spacetime dimension. The four-component spinor notation, by contrast, readily generalizes to any spacetime dimension and is based on the notion of Clifford algebras and gamma matrices.

In this book, we will use the four-component spinor formalism, because it is more common in the supergravity literature. The two-component spinor notation, on the other hand, is frequently used in texts on global supersymmetry, so we briefly include it here as well to facilitate the translation of one formalism into the other.

1.3.1 Two-Component Spinors

The group $SL(2, \mathbb{C})$ has two equivalence classes of irreducible two-dimensional complex representations, corresponding to the defining representation of $SL(2, \mathbb{C})$ and its complex conjugate representation. These two representations are the minimal spinor representations of the 4D Lorentz group and are often denoted by $(1/2, 0)$ and $(0, 1/2)$ or by dotted and undotted two-component spinors, λ_A and $\chi_{\dot{A}}^*$. The Lorentz

---

[3] The full isometry group $O(1, 3)$ of 4D Minkowski spacetime decomposes into four disconnected components. In this book, we mean by Lorentz group only the component, $SO(1, 3)_0$, of Lorentz transformations that are continuously connected to the identity element. This subgroup is often called the "proper orthochronous Lorentz group".

group generators, M_{ab}, act on these via representation matrices that can be chosen as

$$(1/2, 0) : \qquad \rho(M_{ab}) = -\frac{1}{4}(\sigma_a \overline{\sigma}_b - \sigma_b \overline{\sigma}_a) \tag{1.4}$$

$$(0, 1/2) : \qquad \rho^*(M_{ab}) = e \left[-\frac{1}{4}(\overline{\sigma}_a \sigma_b - \overline{\sigma}_b \sigma_a) \right] e^{-1}, \tag{1.5}$$

where $\sigma_i = -\overline{\sigma}_i$ ($i = 1, 2, 3$) are the Pauli matrices, $\sigma_0 = \overline{\sigma}_0 = \mathbb{1}_2$, and

$$e \equiv \begin{pmatrix} 0 & 1 \\ -1 & 0 \end{pmatrix}. \tag{1.6}$$

That (1.5) is indeed the complex conjugate of (1.4) follows from the identity $e^{-1}\sigma^\mu e = \overline{\sigma}^{\mu*} = \overline{\sigma}^{\mu T}$.

The matrix e is an invariant of $SL(2, \mathbb{C})$ in the sense that

$$A^T e A = e \quad \forall A \in SL(2, \mathbb{C}), \tag{1.7}$$

which implies that quantities such as $\lambda^T e \omega$ are Lorentz invariant products of two spinors λ_A and ω_A. This is often written as $\lambda_A \omega^A$ or $\lambda^A \omega_A$ using suitable raising and lowering conventions for spinor indices with the matrix e.

All finite-dimensional irreducible representations of $SL(2, \mathbb{C})$ can be obtained from symmetrized tensor products of these elementary building blocks, often denoted by $(n/2, m/2)$, where n and m count the $(1/2, 0)$ and $(0, 1/2)$ factors, respectively. For $(m + n) =$ even/odd, these describe single-/double-valued representations of the Lorentz group, corresponding to bosons/fermions.

1.3.2 Four-Component Spinors

The above two-component spinor formalism has many nice features and, as already mentioned, is frequently used in the literature on global supersymmetry. In these lectures, however, we will use four-component spinors instead, as these are more standard in the four-dimensional supergravity and particle phenomenology literature and generalize easily to other spacetime dimensions.

More precisely, we will use four-component *Majorana* spinors to describe all fermions in 4D. Majorana spinors satisfy a reality condition (see below) and naturally describe fermions that are gauge invariant or transform in real representations of the gauge group. This is true in particular for the gauge-invariant gravitino,[4] the superpartner of the graviton, but also for the gaugini, the superpartners of the

---

[4] As we will discuss later, an exception to this gauge invariance of the gravitino arises in the presence of Fayet–Iliopoulos terms in $\mathcal{N} = 1$ supergravity. These terms require a gauging of

gauge bosons, which (before spontaneous symmetry breaking) always transform in the adjoint and hence a real representation of the gauge group. Fermions that sit in $\mathcal{N} = 1$ chiral multiplets, on the other hand, may have chiral gauge interactions and hence may sit in complex gauge group representations. While these fermions can be nicely described by two-component spinors, they may equally well be described in terms of the chiral components of different sets of four-component Majorana spinors, as we will do in this book.

1.3.2.1 Dirac Spinors

A general four-component Dirac spinor, ψ_α ($\alpha = 1, \ldots, 4$), has four complex components and corresponds to a direct sum of two two-component spinors of the form $(1/2, 0) \oplus (0, 1/2)$, i.e., it combines one undotted two-component spinor, λ_A, and one dotted spinor, $\chi_{\dot{A}}^*$. A very convenient way to do so is to define

$$\psi = \begin{pmatrix} e \cdot \chi^* \\ \lambda \end{pmatrix}, \tag{1.8}$$

so that the Lorentz generators are represented by the manifestly reducible matrices

$$\Sigma_{ab} = \begin{pmatrix} e\rho^*(M_{ab})e^{-1} & 0 \\ 0 & \rho(M_{ab}) \end{pmatrix}. \tag{1.9}$$

The (4×4) matrices Σ_{ab} can conveniently be expressed in terms of Dirac gamma matrices,

$$\gamma_a = \begin{pmatrix} 0 & i\bar{\sigma}_a \\ i\sigma_a & 0 \end{pmatrix}, \tag{1.10}$$

so that

$$\Sigma_{ab} = \frac{1}{4}[\gamma_a, \gamma_b]. \tag{1.11}$$

The gamma matrices defined above satisfy the Clifford algebra

$$\{\gamma_a, \gamma_b\} = 2\,\eta_{ab}\,\mathbb{1}_4. \tag{1.12}$$

As one easily verifies, these Clifford algebra relations already imply the Lorentz algebra commutation relations for the matrices Σ_{ab} if one uses (1.11) as their definition. From this point of view, the particular expression (1.9) is just a special case that corresponds to the particular representation (1.10) of the Clifford algebra

---

the R-symmetry group, under which also the gravitino is charged, but these theories are often anomalous.

(1.12). This representation (1.10) is called the *Weyl representation*, and its main virtue is that the Lorentz generators assume the block diagonal form (1.9), making their reducibility manifest.[5] However, there are infinitely many other irreducible representations of (1.12) that one could also use; apart from the Weyl representation, the most common representations are the Dirac representation and the Majorana representation.[6] Fortunately, all these representations are equivalent to the Weyl representation, and we rarely will make use of particular representations. For convenience, however, we will always assume "friendly" representations whose defining properties are

$$\gamma_0^\dagger = -\gamma_0 \tag{1.13}$$

$$\gamma_i^\dagger = +\gamma_i \tag{1.14}$$

$$\gamma_a^T = \pm\gamma_a, \tag{1.15}$$

where the last equation means that each gamma matrix is either symmetric or antisymmetric. The Weyl, Dirac, and Majorana representations are obviously all friendly representations.

An important object in the following will be the completely antisymmetrized products of several gamma matrices,

$$\gamma_{a_1 \dots a_p} \equiv \gamma_{[a_1} \gamma_{a_2} \cdots \gamma_{a_p]}, \tag{1.16}$$

where, as usual, the antisymmetrization involves a prefactor $1/p!$, so that, e.g., $\gamma_{ab} = 1/2[\gamma_a, \gamma_b]$, etc. Note that the Clifford relation (1.12) implies

$$\gamma_{a_1 a_2 \dots a_p} = \begin{cases} \gamma_{a_1} \gamma_{a_2} \cdots \gamma_{a_p} & \text{if all } a_i \text{ are different} \\ 0 & \text{otherwise} \end{cases} \tag{1.17}$$

The 16 matrices $\gamma_M = \{\mathbb{1}_4, \gamma_a, \gamma_{ab}, \gamma_{abc}, \gamma_{abcd}\}$ are linearly independent and form a basis of the complex (4×4) matrices.

As we have seen, an irreducible representation of the Clifford algebra (1.12) induces a spinor representation (1.11) of the Lorentz group. This is true in any spacetime dimension, which makes the Clifford algebra approach to spinor representations so useful for higher-dimensional supergravity theories. As was stressed in footnote 5, however, the Lorentz algebra representations obtained in

---

[5] Note that the Weyl representation is *irreducible* as a representation of the *Clifford algebra* (1.12). It is only the induced representation Σ_{ab} of the *Lorentz algebra* that is reducible.

[6] The Dirac representation corresponds to $\gamma_0 = i\sigma_3 \otimes \mathbb{1}_2 \equiv \begin{pmatrix} i\mathbb{1}_2 & 0 \\ 0 & -i\mathbb{1}_2 \end{pmatrix}$, $\gamma_i = \sigma_2 \otimes \sigma_i$, and the Majorana representation is given by $\gamma_0 = i\sigma_2 \otimes \sigma_3$, $\gamma_1 = -\sigma_1 \otimes \mathbb{1}_2$, $\gamma_2 = \sigma_2 \otimes \sigma_2$, $\gamma_3 = \sigma_3 \otimes \mathbb{1}_2$, but they are not really needed for this book.

this way are in general not irreducible, even if the underlying Clifford algebra representation is irreducible.

For supersymmetry, it is convenient to work with minimal representations of the Lorentz group, as these typically correspond also to the minimal amount of supersymmetry one can have in the respective spacetime dimension. There are essentially two ways to reduce the number of degrees of freedom of a Clifford algebra spinor so as to obtain irreducible representations (irreps) of the corresponding Lorentz algebra. One possibility is to impose a chirality condition, which leads to Weyl spinors, and the other one is to impose a reality condition, which leads to Majorana spinors.[7] This is not always possible in every spacetime dimension: The Weyl condition can only be imposed in even dimensions, whereas the possibility to impose a Majorana condition shows a somewhat more complicated dependence on the spacetime dimension, as we will see in Chap. 10. Moreover, the Weyl and Majorana condition can often not be imposed simultaneously. For the moment, we restrict ourselves to four spacetime dimensions, where one can impose a Weyl *or* a Majorana condition, but not both of them at the same time.

1.3.2.2 The Weyl Condition
The Weyl condition projects out the part of a spinor that has a particular handedness. It is imposed with the γ_5 matrix:

$$\gamma_5 \equiv \gamma^5 \equiv -i\gamma^0\gamma^1\gamma^2\gamma^3 = +i\gamma_0\gamma_1\gamma_2\gamma_3. \tag{1.18}$$

The Clifford algebra (1.12) implies

$$(\gamma_5)^2 = \mathbb{1}_4 \tag{1.19}$$

$$\{\gamma_5, \gamma_a\} = 0 \Rightarrow [\gamma_5, \Sigma_{ab}] = 0 \tag{1.20}$$

so that the chirality projectors

$$P_L \equiv \frac{1}{2}\left(1+\gamma^5\right), \quad P_R \equiv \frac{1}{2}\left(1-\gamma^5\right), \tag{1.21}$$

can be used to define left- and right-handed spinors.

$$\psi_L \equiv P_L\psi, \qquad \psi_R \equiv P_R\psi. \quad \text{(Weyl condition)} \tag{1.22}$$

---

[7] Note that imposing a Weyl or Majorana condition in 4D does not necessitate the use of the Weyl or the Majorana representation of the gamma matrices. The Weyl condition just takes on a particularly simple form in the Weyl representation, and the Majorana condition leads to a particularly simple result in the Majorana representation. We usually do not make use of these simplified forms and write down the conditions in a covariant way.

Because of (1.20), this projection is consistent with Lorentz covariance, and left- and right-handed spinors form separate representations of the Lorentz group. In the Weyl representation, $\gamma_5 = \sigma_3 \otimes \mathbb{1}_2$, i.e., the left- and right-handed spinors, are simply the upper or lower two components of ψ as in (1.8). Note that $\gamma^5 \psi_L = \psi_L$, while $\gamma^5 \psi_R = -\psi_R$.

In a friendly representation, $\gamma_0 \gamma_a^\dagger \gamma_0 = \gamma_a \Rightarrow \Sigma_{ab}^\dagger \gamma_0 = -\gamma_0 \Sigma_{ab}$, and the *Dirac conjugate* of a general four-component spinor is defined by

$$\overline{\psi} \equiv i \psi^\dagger \gamma^0 = -i \psi^\dagger \gamma_0 \tag{1.23}$$

so that bilinears such as $\overline{\psi} \chi$ are Lorentz invariant.

In a friendly representation, we also have that $\gamma_5^\dagger = \gamma_5$, and hence $P_{L,R}^\dagger = P_{L,R}$. Moreover $P_L \gamma_0 = \gamma_0 P_R$ and $P_R \gamma_0 = \gamma_0 P_L$, and this implies

$$\overline{\psi_R} = \overline{\psi} P_L, \text{ and } \overline{\psi_L} = \overline{\psi} P_R, \tag{1.24}$$

where

$$\overline{\psi_R} \equiv \overline{(P_R \psi)} = -i(P_R \psi)^\dagger \gamma_0. \tag{1.25}$$

Because of this we will often write

$$\overline{\psi}_L \equiv \overline{\psi} P_L = \overline{\psi_R}, \qquad \overline{\psi}_R \equiv \overline{\psi} P_R = \overline{\psi_L}. \tag{1.26}$$

The matrix γ_5 also enters a number of very useful duality relations between the antisymmetrized products of gamma matrices, as the reader is asked to verify in Exercise 1.1.

1.3.2.3 The Majorana Condition
The Majorana condition is a reality condition that can be written as

$$\psi^* = B\psi, \quad \text{(Majorana condition)} \tag{1.27}$$

with some matrix B. This condition is self-consistent (i.e., $\psi^{**} = \psi$) and Lorentz covariant if B satisfies

$$B^* B = \mathbb{1}_4, \qquad \gamma_a^* = B \gamma_a B^{-1} \Rightarrow \Sigma_{ab}^* B = B \Sigma_{ab}. \tag{1.28}$$

In the Weyl basis (1.10), one can choose

$$B = \begin{pmatrix} 0 & e \\ -e & 0 \end{pmatrix}, \tag{1.29}$$

so that a Majorana spinor would be of the form (1.8), but with $\lambda = \chi$. From this we see that a Majorana spinor describes the same number of independent degrees of freedom as a Weyl spinor.

Another equivalent, but for many purposes more convenient, way to write the Majorana condition is via the charge conjugation matrix, C, which satisfies

$$C^T = -C, \qquad \gamma_a^T = -C\gamma_a C^{-1}. \tag{1.30}$$

In a friendly representation, one can moreover choose C such that it also satisfies

$$C^{-1} = -C = C^\dagger. \tag{1.31}$$

In terms of C, the charge conjugate spinor is defined as

$$\psi^c = C\overline{\psi}^T = iC\gamma^{0T}\psi^* \tag{1.32}$$

and a *Majorana spinor* is defined as

$$\psi^c = \psi. \qquad \text{(alternative form of Majorana condition)} \tag{1.33}$$

This is equivalent to (1.27) if we identify

$$B = (i\, C\gamma_0^T)^{-1}, \tag{1.34}$$

so that, in terms of B, charge conjugation reads

$$\psi^c = B^{-1}\psi^*. \tag{1.35}$$

The advantage of C is that for a Majorana spinor the Dirac conjugate can be written as

$$\overline{\psi} = \psi^T C. \tag{1.36}$$

Notice that using (1.30) one finds the symmetry properties

$$C^T = -C, \qquad (C\gamma^{abc})^T = -(C\gamma^{abc}), \quad (C\gamma^{abcd})^T = -(C\gamma^{abcd}),$$
$$(C\gamma^a)^T = (C\gamma^a), \quad (C\gamma^{ab})^T = (C\gamma^{ab}). \tag{1.37}$$

For *anti*-commuting Majorana spinors, this then implies

$$\overline{\psi}_1 M \psi_2 = \begin{cases} +\overline{\psi}_2 M \psi_1 & \text{for } M = \mathbb{1}_4, \gamma_{abc}, \gamma_{abcd} \\[2mm] -\overline{\psi}_2 M \psi_1 & \text{for } M = \gamma_a, \gamma_{ab} \end{cases} \tag{1.38}$$

*Unless stated otherwise, we will, in the following, always use anti-commuting Majorana spinors but often also take in addition the chiral projections ψ_L and ψ_R of these Majorana spinors, which therefore are **not** independent.*

More specifically, we have, in our conventions,

$$(\psi_L)^c = \psi_R, \qquad (\psi_R)^c = \psi_L. \tag{1.39}$$

To show this, we use

$$B^{-1}\gamma_5^* = -\gamma_5 B^{-1} \Leftrightarrow B^{-1}P_L^* = P_R B^{-1} \tag{1.40}$$

so that

$$(\psi_L)^c = (P_L\psi)^c = B^{-1}P_L^*\psi^* = P_R B^{-1}\psi^* = P_R\psi^c = P_R\psi = \psi_R. \tag{1.41}$$

Note that (1.39) implies that ψ_R is no longer a Majorana spinor, because that would require $(\psi_R)^c$ being equal to ψ_R. Thus, in 4D, a four-component spinor cannot be simultaneously chiral and Majorana. Nevertheless, it makes sense to talk about the projection ψ_R or ψ_L of a given Majorana spinor ψ.

From the definition of γ_5 and (1.30), one also gets

$$C\gamma_5 = \gamma_5^T C \Leftrightarrow C P_L = P_L^T C \tag{1.42}$$

so that for a Majorana spinor ψ,

$$\overline{\psi}_L = \overline{\psi}P_L = \psi^T C P_L = \psi^T P_L^T C = (\psi_L)^T C \tag{1.43}$$

even though ψ_L is not Majorana. From this we can obtain more symmetry properties for the chiral projections that are very similar to those for the Majorana spinors themselves,

$$\overline{\chi}_L\psi_L = \chi_L^T C\psi_L = -\psi_L^T C^T \chi_L = \overline{\psi}_L\chi_L,$$
$$\overline{\chi}_L\gamma^a\psi_R = -\overline{\psi}_R\gamma^a\chi_L, \quad \overline{\chi}_L\gamma^{ab}\psi_L = -\overline{\psi}_L\gamma^{ab}\chi_L \tag{1.44}$$
$$\overline{\chi}_L\gamma^{abc}\psi_R = \overline{\psi}_R\gamma^{abc}\chi_L.$$

We also note that under charge conjugation

$$(\gamma_a)^c = \gamma_a, \quad (\gamma_5)^c = -\gamma_5, \tag{1.45}$$

in the sense that $(\gamma^a\psi)^c = \gamma^a\psi^c$, etc.

In all the subsequent formulae, the Hermitian conjugate, $+h.c.$, of a field operator is denoted with a superscript *, whereas the superscript † is reserved for matrix expressions when in addition to Hermitian conjugation also a transposition of the matrix is involved. On ordinary complex numbers and classical fields, the Hermitian conjugation acts as complex conjugation, where, however, the order of anti-commuting spinor fields is exchanged to mimic the effect of Hermitian conjugation on the corresponding quantum fields. This results in a minus sign when the original spinor order is restored. Fortunately, the effect of this Hermitian conjugation can simply be obtained by writing down the charge conjugate expression with all the rules obtained so far, including (1.45), but without exchanging the order of the spinors.

As an example, we show $(\overline{\psi}_L \gamma^a \chi_R)^* = (\overline{\psi}_L \gamma^a \chi_R)^c = \overline{\psi}_R \gamma^a \chi_L$:

$$(\overline{\psi}_L \gamma^a \chi_R)^* = (\psi_L^T C \gamma^a \chi_R)^* = -\psi_L^\dagger C^* \gamma^{a*} \chi_R^* = -\psi_L^\dagger C^* \gamma^{a*} B (\chi_R)^c$$

$$= -\psi_L^\dagger C^* B \gamma^a \chi_L. \tag{1.46}$$

Inserting (1.34), $C^* = C$, and $(\gamma^{0T})^{-1} = -C(\gamma^0)^{-1} C^{-1} = -C^{-1}(\gamma^0)^{-1} C$, this becomes

$$(\overline{\psi}_L \gamma^a \chi_R)^* = -i\psi_L^\dagger (\gamma^0)^{-1} \gamma^a \chi_L = \overline{\psi}_L \, \gamma^a \chi_L = \overline{\psi}_R \gamma^a \chi_L = (\overline{\psi}_L)^c (\gamma^a)^c (\chi_R)^c$$

$$= (\overline{\psi}_L \gamma^a \chi_R)^c, \tag{1.47}$$

where in the second equation we used $(\gamma^0)^{-1} = -\gamma^0$, which follows from $(\gamma^0)^2 = -\mathbb{1}_4$.

In the following, we will often need to rewrite three or four Fermi terms to complete our analysis of the supersymmetry properties of an action, and hence Fierz identities will be extremely useful. We list here the main ones for two spinors:

$$\psi_R \overline{\chi}_R = -\frac{1}{2} \overline{\chi}_R \psi_R \, P_R + \frac{1}{8} \overline{\chi}_R \gamma_{ab} \psi_R \, \gamma^{ab} \, P_R, \tag{1.48}$$

$$\psi_R \overline{\chi}_L = -\frac{1}{2} \overline{\chi}_L \gamma^a \psi_R \, \gamma_a \, P_L, \tag{1.49}$$

where for the sake of clarity we explicitly left the projectors on the right hand side.

In the rest of these lectures, we will also often make use of spinor one-forms $\psi = dx^\mu \psi_\mu$. Exchanging such spinor one-forms then leads to an additional minus sign from the anti-commutativity of the wedge product, and hence we have

$$\psi_R \wedge \overline{\psi}_R = -\frac{1}{8} \overline{\psi}_R \wedge \gamma_{ab} \psi_R \, \gamma^{ab} \, P_R, \tag{1.50}$$

$$\psi_R \wedge \overline{\psi}_L = \frac{1}{2} \overline{\psi}_L \wedge \gamma^a \psi_R \, \gamma_a \, P_L. \tag{1.51}$$

where now $\overline{\psi}_L \wedge \psi_L = 0$ because of (1.44) and the wedge product. A crucial consequence is the *cyclic identity*:

$$\gamma^a \psi_L \wedge \overline{\psi}_L \wedge \gamma_a \psi_R = 0. \tag{1.52}$$

1.3.3   Susy Algebra in Four Dimensions

Using the conventions described so far, the $\mathcal{N} = 1$ supersymmetry algebra in four dimensions has the following form:

$$\{Q, \bar{Q}\} = -2i\gamma^a \mathcal{P}_a,$$
$$[\mathcal{P}_a, Q] = 0,$$
$$[\mathcal{M}_{ab}, Q] = \frac{i}{2} \gamma_{ab} Q,$$
$$[R, Q] = i \gamma_5 Q, \tag{1.53}$$
$$[\mathcal{P}_a, \mathcal{P}_b] = 0,$$
$$[\mathcal{P}_a, \mathcal{M}_{bc}] = -2i \, \eta_{a[b} \mathcal{P}_{c]},$$
$$[\mathcal{M}_{ab}, \mathcal{M}_{cd}] = 2i \, \eta_{c[a} \mathcal{M}_{b]d} - 2i \, \eta_{d[a} \mathcal{M}_{b]c}.$$

Here, Q is the supersymmetry generator described by a Majorana spinor; $\mathcal{M}_{ab}$ and $\mathcal{P}_a$ denote, respectively, the usual generators of Lorentz transformations and spacetime translations; and R is the U(1) internal R-symmetry generator. In (1.53), we used Hermitian generators $\mathcal{P}_a$ and $\mathcal{M}_{ab}$. Just as in (1.3), we will sometimes also use their anti-Hermitian counterparts,

$$P_a = i \, \mathcal{P}_a, \qquad M_{ab} = i \, \mathcal{M}_{ab} \tag{1.54}$$

when this is more convenient. Note that, for simplicity, we have not included internal bosonic symmetry generators other than the R-symmetry, as they commute with the above generators.

Exercises

1.1. Using the totally antisymmetric epsilon tensor with

$$\epsilon_{0123} = 1, \tag{1.55}$$

check the duality relations

$$\gamma^{abc} = i\,\epsilon^{abcd}\gamma_d\gamma_5, \qquad i\,\gamma_a\gamma_5 = \frac{1}{3!}\epsilon_{abcd}\gamma^{bcd},$$

$$\gamma^{abcd} = -i\,\epsilon^{abcd}\gamma_5, \qquad i\,\gamma_5 = \frac{1}{4!}\epsilon_{abcd}\gamma^{abcd}, \qquad (1.56)$$

$$\gamma^{ab} = \frac{i}{2}\epsilon^{abcd}\gamma_{cd}\gamma_5.$$

1.2. Verify that $M_{ab} = \frac{1}{2}\gamma_{ab}$ satisfies $[M_{ab}, M_{cd}] = -2\,\eta_{c[a}M_{b]d} + 2\,\eta_{d[a}M_{b]c}$.

1.3. Using just the Clifford algebra (1.12) in an *arbitrary* representation and the definition (1.11) of Σ_{ab}, compute the rotation matrix $R(\theta) = e^{\theta\Sigma^{12}}$ for a rotation in the $(1, 2)$-plane by a finite angle θ and read off from your result that $R(2\pi) = -\mathbb{1}$, i.e., that Σ_{ab} is indeed a spinor representation.

References

1. B. de Wit, H. Nicolai, N = 8 Supergravity. Nucl. Phys. **B208**, 323 (1982)
2. M.H. Goroff, A. Sagnotti, The ultraviolet behavior of Einstein gravity. Nucl. Phys. **B266**, 709 (1986)
3. A.E.M. van de Ven, Two loop quantum gravity. Nucl. Phys. **B378**, 309–366 (1992)
4. Z. Bern, L.J. Dixon, R. Roiban, Is N = 8 supergravity ultraviolet finite?. Phys. Lett. **B644**, 265–271 (2007) [hep-th/0611086]
5. J. Carrasco, *Generic Multiloop Methods for Gauge and Gravity Scattering Amplitudes, a Guided Tour with Pedagogic Aspiration* (Strings, Munich, 2012). http://wwwth.mpp.mpg.de/members/strings/strings2012/strings_files/program/Talks/Tuesday/Carrasco.pdf
6. J.M. Maldacena, The large N limit of superconformal field theories and supergravity. Int. J. Theor. Phys. **38**, 1113–1133 (1999) [arXiv:hep-th/9711200 [hep-th]]
7. E. Witten, Anti-de Sitter space and holography. Adv. Theor. Math. Phys. **2**, 253–291 (1998) [arXiv:hep-th/9802150 [hep-th]]
8. D.Z. Freedman, C. Nunez, M. Schnabl, K. Skenderis, Fake supergravity and domain wall stability. Phys. Rev. **D69**, 104027 (2004) [hep-th/0312055]
9. A. Celi, A. Ceresole, G. Dall'Agata, A. Van Proeyen, M. Zagermann, On the fakeness of fake supergravity. Phys. Rev. **D71**, 045009 (2005) [hep-th/0410126]
10. K. Skenderis, P.K. Townsend, Gravitational stability and renormalization group flow. Phys. Lett. B **468**, 46 (1999) [hep-th/9909070]
11. K. Skenderis, P. Townsend, Pseudo-supersymmetry and the domain-wall/cosmology correspondence. J. Phys. A **A40**, 6733–6742 (2007) [hep-th/0610253]
12. P. Van Nieuwenhuizen, Supergravity. Phys. Rept. **68**, 189–398 (1981)
13. H.P. Nilles, Supersymmetry, supergravity and particle physics. Phys. Rept. **110**, 1 (1984)
14. B. de Wit, D.Z. Freedman, Supergravity: the basics and beyond. Bonn Superym. ASI 0135 (1984). MIT-CTP-1238
15. L. Castellani, R. D'Auria, P. Fre, *Supergravity and Superstrings: A Geometric Perspective*, vol. 1. Mathematical Foundations (World Scientific, Singapore, 1991), pp. 1–603
16. L. Castellani, R. D'Auria, P. Fre, *Supergravity and Superstrings: A Geometric Perspective*, vol. 2. Supergravity (World Scientific, Singapore, 1991), pp. 607–1371

17. J. Wess, J. Bagger, *Supersymmetry and Supergravity* (Princeton University Press, Princeton, 1992)
18. A. Van Proeyen, Tools for supersymmetry. hep-th/9910030
19. J.-P. Derendinger, *Introduction to Supergravity*. Lectures at the Summer School "Gif 2000" in Paris
20. B. de Wit, Supergravity. hep-th/0212245
21. A. Van Proeyen, Structure of supergravity theories. hep-th/0301005. To be published in the series Publications of the Royal Spanish Mathematical Society
22. P. van Nieuwenhuizen, Supergravity as a Yang-Mills theory. hep-th/0408137
23. F. Zwirner, *Supersymmetry Breaking in Four and More Dimensions*. Lectures at the 2005 Parma School of Theoretical Physics
24. H. Samtleben, Lectures on gauged supergravity and flux compactifications. Class. Quant. Grav. **25**, 214002 (2008) [arXiv:0808.4076]
25. R. Kallosh, L. Kofman, A.D. Linde, A. Van Proeyen, Superconformal symmetry, supergravity and cosmology. Class. Quant. Grav. **17**, 4269–4338 (2000) [Erratum: Class. Quant. Grav. **21**, 5017 (2004)] [arXiv:hep-th/0006179 [hep-th]]
26. L. Andrianopoli, M. Bertolini, A. Ceresole, R. D'Auria, S. Ferrara, P. Fre, T. Magri, N = 2 supergravity and N = 2 super Yang-Mills theory on general scalar manifolds: symplectic covariance, gaugings and the momentum map. J. Geom. Phys. **23**, 111–189 (1997) [arXiv:hep-th/9605032 [hep-th]]
27. B. de Wit, H. Samtleben, M. Trigiante, On Lagrangians and gaugings of maximal supergravities. Nucl. Phys. B **655**, 93–126 (2003) [arXiv:hep-th/0212239 [hep-th]]
28. B. de Wit, H. Samtleben, M. Trigiante, The maximal D = 4 supergravities. JHEP **06**, 049 (2007) [arXiv:0705.2101 [hep-th]]
29. S. Weinberg, *The Quantum Theory of Fields*, vol. 3. Supersymmetry (Cambridge University Press, Cambridge, 2000), 419 p.
30. D.Z. Freedman, A. Van Proeyen, *Supergravity* (Cambridge University Press, Cambridge, 2012)
31. E. Lauria, A. Van Proeyen, $\mathcal{N} = 2$ Supergravity in $D = 4, 5, 6$ Dimensions. Lect. Notes Phys. **966** (2020) [arXiv:2004.11433 [hep-th]]
32. P.C. West, Supergravity, brane dynamics and string duality [arXiv:hep-th/9811101 [hep-th]]

From Global to Local Supersymmetry

2

In this chapter, we revisit the simplest *globally* supersymmetric field theory in four dimensions, the free massless Wess–Zumino model for one chiral multiplet, and discuss how this theory has to be changed when supersymmetry is turned into a local symmetry. As we will see, making supersymmetry local by a simple iterative procedure (the "Noether method") directly exhibits the need for the gravitino field and its superpartner, the graviton, and suggests the supersymmetry transformation laws of these fields. We end this chapter by a discussion of the basic properties of the gravitino in Sect. 2.2.

2.1 Promoting Supersymmetry to a Local Symmetry

Consider the free massless Wess–Zumino model for one chiral multiplet (ϕ, χ), where $\phi(x)$ is a complex scalar and $\chi(x)$ a Majorana spinor field with Lagrangian

$$\mathscr{L} = -\partial_\mu \phi \partial^\mu \phi^* - \left(\overline{\chi}_R \slashed{\partial} \chi_L + \overline{\chi}_L \slashed{\partial} \chi_R \right). \tag{2.1}$$

We recall that the mass dimensions of the fields are $D[\phi] = 1$ and $D[\chi] = 3/2$.

Before we continue, we should make a short remark on our index conventions. In this book, we generally use Greek indices $\mu, \nu, \ldots = 0, 1, 2, 3$ to denote the local coordinates, x^μ, of 4D spacetime manifolds. For consistency, we also do this for 4D Minkowski spacetime when, as in the above Lagrangians, it is considered as a special example of a differentiable manifold. On the other hand, when we consider Minkowski space as a vector space (e.g., a tangent space at a point of a differentiable manifold) with orthonormal basis vectors, e_a, we use the Latin indices $a, b, \ldots = 0, 1, 2, 3$, as we did in Sect. 1.3. For Minkowski spacetime, this distinction is of course not really necessary if one works with coordinates x^μ that correspond to a global inertial frame, as we also do here, because then the coordinate-induced tangent vectors, ∂_μ, are orthonormal and could be identified with the orthonormal

© Springer-Verlag GmbH Germany, part of Springer Nature 2021
G. Dall'Agata, M. Zagermann, *Supergravity*, Lecture Notes in Physics 991,
https://doi.org/10.1007/978-3-662-63980-1_2

basis vectors e_a. The gamma matrices γ^μ in Minkowski spacetime are then also simply the same constant matrices γ^a described in Sect. 1.3. We will see later, however, that on a general curved spacetime manifold, this distinction between local coordinate indices μ, ν and the indices a, b of an orthonormal basis of the tangent spaces is in fact crucial for a proper definition of spinor fields on curved spacetimes.

The Lagrangian (2.1) is supersymmetric under the variations[1]

$$\delta_\epsilon \phi = \bar\epsilon_L \chi_L \qquad \Longleftrightarrow \qquad \delta_\epsilon \phi^* = \bar\epsilon_R \chi_R \tag{2.2}$$

$$\delta_\epsilon \chi_L = \frac{1}{2}\slashed\partial\phi\epsilon_R \qquad \Longleftrightarrow \qquad \delta_\epsilon \chi_R = \frac{1}{2}\slashed\partial\phi^*\epsilon_L, \tag{2.3}$$

with $D[\epsilon] = -1/2$. In these conventions, it follows that

$$\delta_\epsilon \overline\chi_L = -\frac{1}{2}\bar\epsilon_R\slashed\partial\phi \qquad \Longleftrightarrow \qquad \delta_\epsilon \overline\chi_R = -\frac{1}{2}\bar\epsilon_L\slashed\partial\phi^*, \tag{2.4}$$

as one can easily check by using (1.43) as well as $\epsilon_R^T(\gamma^\mu)^T C = -\epsilon_R^T(C\gamma^\mu)$, or by simply taking the charge conjugate of both sides. The Lagrangian (2.1) is invariant under the supersymmetry transformations (2.2)–(2.3) up to a total derivative.

To check this explicitly, we first use (1.44) to write the fermionic term of the Lagrangian (2.1) as $\mathscr{L}_{\text{fer}} = -\overline\chi_R\slashed\partial\chi_L + \partial_\mu(\overline\chi_R)\gamma^\mu\chi_L$. For the supersymmetry variation of the Lagrangian, it is sufficient to trace the terms involving ϵ_L, which come only from the variation of ϕ and $\overline\chi_R$:

$$\delta\mathscr{L} = -\partial_\mu(\delta\phi)\partial^\mu\phi^* - \delta\overline\chi_R\slashed\partial\chi_L + \partial_\mu(\delta\overline\chi_R)\gamma^\mu\chi_L + \text{h.c.} \tag{2.5}$$

Integrating by parts now the first and the second term gives

$$\delta\mathscr{L} \quad = \quad \delta\phi\,\square\phi^* + 2\partial_\mu(\delta\overline\chi_R)\gamma^\mu\chi_L + \partial_\mu\underbrace{\left(-\delta\phi\partial^\mu\phi^* - \delta\overline\chi_R\gamma^\mu\chi_L\right)}_{\equiv\mathscr{K}^\mu} + \text{h.c.}$$

$$\overset{(2.2),(2.4)}{=} \quad \bar\epsilon_L\chi_L\square\phi^* - \partial_\mu(\bar\epsilon_L\slashed\partial\phi^*)\gamma^\mu\chi_L + \partial_\mu\mathscr{K}^\mu + \text{h.c.}$$

---

[1] Note that, since we are not using auxiliary fields, the supersymmetry algebra closes only on-shell:

$$[\delta_{\epsilon_2}, \delta_{\epsilon_1}]\phi = \frac{1}{2}(\bar\epsilon_1\gamma^\mu\epsilon_2)\partial_\mu\phi$$

$$[\delta_{\epsilon_2}, \delta_{\epsilon_1}]\chi_L = \frac{1}{2}(\bar\epsilon_1\gamma^\mu\epsilon_2)\partial_\mu\chi_L + [\dots]\slashed\partial\chi_L,$$

where $[\dots]$ denotes a non-vanishing expression of the fields and supersymmetry parameters. The last term then vanishes due to the field equation $\slashed\partial\chi_L = 0$, and one obtains the usual susy algebra $[\delta_{\epsilon_2}, \delta_{\epsilon_1}] = \frac{1}{2}(\bar\epsilon_1\gamma^\mu\epsilon_2)\partial_\mu$ on all fields.

$$\begin{aligned}
&= &&\bar{\epsilon}_L \chi_L \Box \phi^* - \partial_\mu(\bar{\epsilon}_L)\slashed{\partial}\phi^* \gamma^\mu \chi_L - \bar{\epsilon}_L \underbrace{\partial_\mu \partial_\nu \phi^* \gamma^\nu \gamma^\mu}_{\Box \phi^* \mathbb{1}_4} \chi_L + \partial_\mu \mathscr{K}^\mu + \text{h.c.} \\
&= &&-\partial_\mu(\bar{\epsilon}_L)\slashed{\partial}\phi^* \gamma^\mu \chi_L + \partial_\mu \mathscr{K}^\mu + \text{h.c.}
\end{aligned} \tag{2.6}$$

As promised, the result is that under global supersymmetry, where the supersymmetry parameter is constant, $\partial_\mu \epsilon = 0$, the Lagrangian transforms into a total derivative:

$$\delta_\epsilon \mathscr{L} = \partial_\mu(\mathscr{K}^\mu + \mathscr{K}^{\mu*}) \equiv \partial_\mu K^\mu . \tag{2.7}$$

When dealing with local supersymmetry, however, the parameter ϵ becomes a local function of the coordinates, $\epsilon = \epsilon(x)$, and the Lagrangian is no longer invariant up to a total derivative. The new non-invariant part of the Lagrangian reads

$$\delta_\epsilon \mathscr{L} = (\partial_\mu \bar{\epsilon})j^\mu = (\partial_\mu \bar{\epsilon}_L)j_L^\mu + (\partial_\mu \bar{\epsilon}_R)j_R^\mu, \tag{2.8}$$

where

$$j_L^\mu \equiv -\slashed{\partial}\phi^* \gamma^\mu \chi_L, \qquad j_R^\mu \equiv -\slashed{\partial}\phi \gamma^\mu \chi_R \tag{2.9}$$

give the super-Noether current $j^\mu = j_L^\mu + j_R^\mu$. In fact, it can be easily checked that this supercurrent is a conserved current, namely, that $\partial_\mu j^\mu = 0$, upon using the equations of motion for the fields ϕ and χ. It should also be noted that the dimension of these currents is $D[j_{L,R}^\mu] = 7/2$.

We can now apply Noether's method and associate to the supercurrent (2.9) a gauge field that compensates the non-invariance of the Lagrangian (2.8). This gauge field, $\psi_{\mu\alpha}$, has to have a spinorial index (i.e., the index $\alpha = 1, 2, 3, 4$, which we will suppress again in the following), such that

$$\delta_\epsilon \psi_{\mu L,R} = M_P \partial_\mu \epsilon_{L,R}, \qquad \delta_\epsilon \bar{\psi}_{\mu L,R} = M_P \partial_\mu \bar{\epsilon}_{L,R}, \tag{2.10}$$

where M_P is a mass parameter that is needed to relate the mass dimension $3/2$ of the fermionic field ψ_μ and the dimension of the supersymmetry parameter $D[\epsilon] = -1/2$. As suggested by the notation, M_P will later be identified with the (reduced) Planck mass.

The Noether procedure now tells us that we need to add a new piece to the Lagrangian:

$$\mathscr{L}'_{\text{WZ}} = -\frac{1}{M_P}\left(\bar{\psi}_{\mu L} j_L^\mu + \bar{\psi}_{\mu R} j_R^\mu\right). \tag{2.11}$$

Again M_P is needed to get a Lagrangian density whose total mass dimension is 4, and this dimensionful coupling in the action can be viewed as a first sign that we eventually need gravity in local supersymmetry.

Using (2.10) in the variation of (2.11), we now precisely compensate the variation of the original Wess–Zumino multiplet, but now there is a new piece to compensate in the variation of (2.11) from $\delta_\epsilon j^\mu_{R,L}$, which is in general non-vanishing. To see this, it suffices to consider the variation of the term $\overline{\psi}_{\mu L} j^\mu_L$ that is quadratic in the scalar fields. This term comes from the variation of χ_L inside j^μ_L:

$$\overline{\psi}_{\mu L} \slashed{\partial} \phi^* \gamma^\mu \delta_\epsilon \chi_L = \frac{1}{2} \overline{\psi}_{\mu L} \gamma^\nu \gamma^\mu \gamma^\rho \epsilon_R \partial_\nu \phi^* \partial_\rho \phi = \overline{\psi}_{\mu L} \gamma_\nu \epsilon_R T^{\mu\nu} + \dots \qquad (2.12)$$

where, using some gamma matrix algebra,

$$T^{\mu\nu} = \partial^{(\mu} \phi \partial^{\nu)} \phi^* - \frac{1}{2} \eta^{\mu\nu} (\partial_\sigma \phi \partial^\sigma \phi^*), \qquad (2.13)$$

and the dots stand for terms involving $\gamma^{\nu\mu\rho}$. One can show that variations bilinear in χ likewise give the energy momentum tensor for the field χ. So,

$$\delta \mathscr{L}'_{WZ} \sim \frac{1}{M_P} \overline{\epsilon} \gamma_\mu \psi_\nu T^{\mu\nu} + \dots \qquad (2.14)$$

In order to cancel this term, we now introduce a new current which is a symmetric tensor $g_{\mu\nu}$ with transformation rule

$$\delta g_{\mu\nu} \sim \frac{1}{M_P} \overline{\epsilon} \gamma_{(\mu} \psi_{\nu)} \qquad (2.15)$$

and add a new piece to the Lagrangian with a coupling between the tensor field $g_{\mu\nu}$ and energy momentum tensor:

$$\mathscr{L}''_{WZ} \sim -g_{\mu\nu} T^{\mu\nu}. \qquad (2.16)$$

As only the spacetime metric can couple to the energy momentum tensor, local supersymmetry requires the coupling of the Wess–Zumino multiplet to gravity described by a dynamical spacetime metric, $g_{\mu\nu}$, and ψ_μ must be its superpartner, the gravitino, as follows from the transformation law (2.15). As in ordinary gauge theories, one also adds kinetic terms for these new "gauge" fields, and we thus expect a final result of the form

$$\mathscr{L} = \underbrace{\mathscr{L}_{kin}(\phi) + \mathscr{L}_{kin}(\chi)}_{\mathscr{L}_{WZ}} + \underbrace{\mathscr{L}_{int}(\phi, \chi, g_{\mu\nu}, \psi_\mu)}_{\mathscr{L}'_{WZ} + \mathscr{L}''_{WZ} + \dots} \qquad (2.17)$$
$$+ \mathscr{L}_{kin}(g_{\mu\nu}) + \mathscr{L}_{kin}(\psi_\mu)$$

where the dots indicate possible further interaction terms.

We used the chiral multiplet to guess the supersymmetry transformation rules of the supergravity multiplet. These rules, however, should hold also in the absence of the chiral multiplet, and we thus arrive at a motivated guess for the Lagrangian and transformation laws of pure $\mathcal{N} = 1$ supergravity:

$$\mathscr{L}_{\text{pure sugra}} = \underbrace{\mathscr{L}_{\text{kin}}(g_{\mu\nu})}_{\frac{M_P^2}{2}\sqrt{-g}R} + \underbrace{\mathscr{L}_{\text{kin}}(\psi_\mu)}_{-\frac{1}{2}\bar{\psi}_\mu \gamma^{\mu\nu\rho}\partial_\nu\psi_\rho|\text{cov}} + \mathscr{L}_{\text{int}}(g_{\mu\nu},\psi_\mu), \tag{2.18}$$

using

$$\delta g_{\mu\nu} \simeq \frac{1}{M_P}\bar{\epsilon}\gamma_{(\mu}\psi_{\nu)}|\text{cov}, \tag{2.19}$$

and

$$\delta\psi_\mu \simeq M_P \partial_\mu\epsilon|\text{cov}, \tag{2.20}$$

where *cov* stands for a proper spacetime covariantization, and $\mathscr{L}_{\text{int}}$ denotes possible interaction terms that are not contained in the covariantizations of the kinetic terms (e.g., four Fermion terms). In Chaps. 3 and 4, we will discuss this spacetime covariantization, which requires an appropriate description of spinors in curved spacetimes. As we will see in Chap. 4, the additional interaction terms not related to spacetime covariantization can, in fact, elegantly be absorbed into the (covariantized) kinetic terms by working with covariant derivatives with non-trivial torsion. Before we come to this, however, let us briefly pause and take a quick look at some properties of the gravitino.

2.2 The Gravitino

As we have seen from the previous discussion, a new fundamental ingredient in the construction of locally supersymmetric actions is the gravitino, which acts as the gauge field associated to the supersymmetry gauge transformation. In this section we want to discuss two important points regarding the gravitino field: its action and the role of the gravitino multiplet.

2.2.1 The Gravitino Action

The gravitino field should propagate spin $3/2$ degrees of freedom, because it is the superpartner of the metric field, which is known to propagate spin 2 degrees of freedom.[2] The correct Lorentz representation for describing such degrees of freedom is $(1, 1/2) \oplus (1/2, 1)$, where we labeled the representations of the Lorentz group as described in Sect. 1.3, which means that one should use $\psi_{(AB)\dot{A}}$ and $\psi_{(\dot{A}\dot{B})A}$ fields. These fields generically contain both spin $3/2$ and spin $1/2$ degrees of freedom with respect to the spatial rotation group appropriate for the description of a massive particle:

$$1 \otimes 1/2 = 3/2 \oplus 1/2. \tag{2.21}$$

Unfortunately, as shown by Fierz and Pauli [1, 2], it is impossible to write down a local, Lorentz invariant action with such fields only. One needs at least an auxiliary spin $1/2$ field. As discussed by Rarita and Schwinger [3], one can build a consistent action by employing a gravitino field that is a vector-spinor, ψ_μ, sitting in a reducible representation of the Lorentz group. It does indeed contain both the gravitino and an auxiliary spinor degree of freedom as follows from the Clebsch–Gordan decomposition of its index structure

$$(1/2, 1/2) \otimes [(1/2, 0) \oplus (0, 1/2)] = (1, 1/2) \oplus (1/2, 1) \oplus (1/2, 0) \oplus (0, 1/2). \tag{2.22}$$

The action for the free field is of the form

$$\mathcal{L}_{3/2} = -\frac{1}{2}\overline{\psi}_\mu \gamma^{\mu\nu\rho} \partial_\nu \psi_\rho + \frac{1}{2}m_{3/2}\overline{\psi}_\mu \gamma^{\mu\nu} \psi_\nu, \tag{2.23}$$

where $m_{3/2}$ is a (real) mass parameter that we will identify below with the physical mass of the gravitino in Minkowski spacetime.

We now show that the equation of motion following from (2.23) propagates 2 degrees of freedom in the massless case and 4 degrees of freedom in the massive case. Let us first focus on the massless case, where $m_{3/2} = 0$. In this case, the Lagrangian $\mathcal{L}_{3/2}$ is invariant, up to a total derivative, under the gauge transformations

$$\delta\psi_\mu = \partial_\mu \Lambda, \tag{2.24}$$

where $\Lambda(x)$ is an arbitrary Majorana spinor-valued function. For $m_{3/2} = 0$, the equation of motion following from $\mathcal{L}_{3/2}$ is

$$\gamma^{\mu\nu\rho} \Psi_{\nu\rho} = 0, \tag{2.25}$$

---

[2] For the sake of readability, we do not distinguish carefully here between spin and helicity, i.e., "spin s" should be understood as "helicity $\pm s$" in the massless case.

where we introduced the gravitino field strength, $\Psi_{\mu\nu} = 2\partial_{[\mu}\psi_{\nu]}$, which is invariant under (2.24). This equation is also equivalent to

$$\gamma_\mu \gamma_\sigma \gamma^{\mu\nu\rho} \Psi_{\nu\rho} = 4\gamma^\mu \Psi_{\mu\sigma} = 0. \tag{2.26}$$

In order to see that this is the correct equation that propagates only spin 3/2 degrees of freedom, we can look at it in momentum space. Assuming that we have plane-wave solutions with momentum k^μ, the Fourier transform of $\Psi_{\mu\nu}$ can be decomposed as

$$\Psi_{\mu\nu}(k) = a_i(k) \, k_{[\mu}\epsilon^i_{\nu]} + b_i(k) \, \tilde{k}_{[\mu}\epsilon^i_{\nu]} + c(k) \, k_{[\mu}\tilde{k}_{\nu]} + d_{ij}(k) \, \epsilon^i_{[\mu}\epsilon^j_{\nu]}, \tag{2.27}$$

where we introduced a complete set of longitudinal and transverse vectors

$$k^\mu = (k^0, \vec{k}), \quad \tilde{k}^\mu = (k^0, -\vec{k}), \quad \epsilon^{\mu i}, \tag{2.28}$$

with $i = 1, 2$, $k^0 = |\vec{k}| \neq 0$, and

$$k_\mu \tilde{k}^\mu < 0, \quad k_\mu \epsilon^{\mu i} = \tilde{k}_\mu \epsilon^{\mu i} = 0, \quad \epsilon^i_\mu \epsilon^{\mu j} = \delta^{ij}. \tag{2.29}$$

It is straightforward to see that $b_i = d_{ij} = 0$ because of the Bianchi identity $\partial_{[\mu}\Psi_{\nu\rho]} = 0$, which in momentum space is

$$b_i(k) \, k_{[\mu}\tilde{k}_\nu\epsilon^i_{\rho]} + d_{ij}(k) \, k_{[\mu}\epsilon^i_\nu\epsilon^j_{\rho]} = 0. \tag{2.30}$$

Hence off-shell the independent degrees of freedom are contained in the spinors a_i and c, which are Majorana, and therefore sum up to 12 independent real components in four dimensions. In momentum space, the equation of motion (2.26) now becomes

$$\epsilon^i_\mu k_\nu \gamma^\nu a_i + \tilde{k}_\mu k_\nu \gamma^\nu c - k_\mu(\epsilon^i_\nu \gamma^\nu a_i + \tilde{k}_\nu \gamma^\nu c) = 0. \tag{2.31}$$

Contracting with $\epsilon^{\mu j}$, $\tilde{k}^\mu$, and k^μ, this implies $k_\mu\gamma^\mu a_i = 0$, $k_\mu\gamma^\mu c = 0$, and $\epsilon^i_\mu\gamma^\mu a_i + \tilde{k}_\mu\gamma^\mu c = 0$, which further brings us to

$$c = 0. \tag{2.32}$$

Thus, all the on-shell degrees of freedom are contained in the two spinors $a_i(k)$, subject to the constraints

$$k_\mu\gamma^\mu a_i = 0, \quad \epsilon^i_\mu\gamma^\mu a_i = 0. \tag{2.33}$$

Overall we therefore have only two degrees of freedom, because the first equation is a projector halving the number of degrees of freedom contained in a_i and the second shows that they are not independent.

In the massive case, the Lagrangian (2.23) leads to the equations of motion

$$\gamma^{\mu\nu\rho}\partial_\nu\psi_\rho = m_{3/2}\gamma^{\mu\nu}\psi_\nu, \tag{2.34}$$

which are also not in the standard form proposed by Dirac for a fermion field. This, however, was to be expected, because the gravitino field is a sum of representation of the Lorentz group of different dimension.

We can obtain the Dirac equation of motion for a spin $3/2$ field by looking at two different contractions of (2.34). First, take its divergence:

$$0 = \partial_\mu\left(\gamma^{\mu\nu\rho}\partial_\nu\psi_\rho\right) = m_{3/2}\,\partial_\mu\left(\gamma^{\mu\nu}\psi_\nu\right), \tag{2.35}$$

where the first term vanishes because of symmetry properties. Then contract (2.34) with γ_μ:

$$0 = 2\gamma^{\nu\rho}\partial_\nu\psi_\rho = 3\,m_{3/2}\,\gamma^\nu\psi_\nu, \tag{2.36}$$

which is vanishing because of (2.35). The result is that we removed the auxiliary spin $1/2$ component ψ. Finally, by using some gamma matrix algebra and by applying the constraints (2.35) and (2.36) to the original equations of motion, we get

$$\gamma^{\mu\nu\rho}\partial_\nu\psi_\rho = \gamma^\mu\gamma^{\nu\rho}\partial_\nu\psi_\rho - \gamma^\rho\partial^\mu\psi_\rho + \partial\!\!\!/\psi^\mu = \partial\!\!\!/\psi^\mu, \tag{2.37}$$

for the left-hand side of (2.34), and

$$\gamma^{\mu\nu}\psi_\nu = \gamma^\mu\gamma^\nu\psi_\nu - \psi^\mu = -\psi^\mu \tag{2.38}$$

for the right-hand side of (2.34). We thus proved that the massive Rarita–Schwinger action leads to equations of motion equivalent to

$$\left(\partial\!\!\!/\psi_\mu + m_{3/2}\,\psi_\mu\right) = 0, \qquad \partial_\mu\psi^\mu = 0, \qquad \gamma^\mu\psi_\mu = 0, \tag{2.39}$$

where the second equation follows from the divergence of (2.38), using (2.36). The final result is the standard Dirac form for the equations of motion of the irreducible component of the gravitino field. We can thus interpret $m_{3/2}$ as the physical mass of the gravitino in Minkowski space.[3] A counting of the degrees of freedom, similarly to the massless case, would reveal four independent states [1, 2].

---

[3] We will see in Sect. 4.2.2 that in the presence of a (negative) cosmological constant, the concept of mass will be slightly different from the one we are used to in the flat case.

2.2.2 The Gravitino Multiplet

In the previous section, we mentioned that the gravitino is associated to the graviton by a supersymmetry transformation and that therefore these two fields will sit in the same supermultiplet. However, from the representations of the supersymmetry algebra, one can see that, for minimal supersymmetry, there is also a supermultiplet that pairs the gravitino with a vector field, namely,

$$\text{the gravitino multiplet:} \quad \{\psi_\mu, A_\mu\}. \tag{2.40}$$

One may expect this multiplet to play a role in a supergravity theory, too. However, this is not the case.

In fact, it has been shown that trying to add interactions to the free gravitino multiplet or to use a massless gravitino multiplet as a matter multiplet in a supergravity theory leads to undesired pathologies. For instance, minimal couplings to an external $U(1)$ vector field are already problematic. In fact, once one introduces covariant derivatives in the gravitino kinetic term $\partial_\mu \rightarrow D_\mu = \partial_\mu - e A_\mu$, the divergence of the massless gravitino equation of motion imposes $\gamma^{\mu\nu\rho} F_{\mu\nu} \psi_\rho = 0$ and therefore improperly sets to zero some of the degrees of freedom. On the other hand, the minimal coupling of a massive gravitino to external electromagnetic fields results in equations of motion which exhibit faster-than-light propagation of signals [4]. The only known way outs, so far, are given either by Vasiliev's higher spin theories [5], where one introduces an infinite number of non-normalizable couplings in a fixed spacetime with a non-vanishing cosmological constant, or by the promotion of the gravitino's gauge invariance to an additional supersymmetry transformation relating it to the metric field.

As we will see later on, in Chap. 8, this important fact leads also to one of the main striking differences between global and local supersymmetry: In the case of globally supersymmetric theories, the models with $\mathcal{N} > 1$ supersymmetry can always be rewritten in terms of the $\mathcal{N} = 1$ language. More precisely, even highly (globally) supersymmetric models can always be considered as special models in the class of minimally supersymmetric theories in the sense that their couplings are fully compatible with all $\mathcal{N} = 1$ requirements but just sit at a special point in parameter space that allows the existence of additional global supersymmetries. In other words, several global supersymmetries do not interfere with each other's individual consistency conditions and can peacefully coexist.

In supergravity this is not the case. The graviton multiplet with more than minimal supersymmetry has to be rewritten in terms of an $\mathcal{N} = 1$ graviton multiplet coupled to $\mathcal{N} - 1$ massless gravitino multiplets. However, as we just saw, this is not consistent unless one introduces new scales and couplings which lead to higher spin models or when the gauge invariance of the additional gravitini is realized as an additional supersymmetry. A local supersymmetry, however, always introduces various non-renormalizable and non-linear couplings of the scalar fields, as we will discuss in detail in later chapters. Due to their non-linear nature, these additional scalar field couplings cannot simply be "superimposed" for the different local supersymmetries, but they distort each other. The resulting scalar field couplings therefore satisfy distorted consistency conditions compared to the situation when only one local supersymmetry is present, and, generically, the consistency conditions for precisely one local supersymmetry are no longer satisfied in extended supergravity. Thus, extended supergravity models can in general not be described in terms of $\mathcal{N} = 1$ language. We will see examples of this when we discuss the scalar manifolds of, e.g., hypermultiplets in $\mathcal{N} = 2$ supergravity or of the scalars in $\mathcal{N} = 8$ supergravity.

We close this chapter by noting that despite the abovementioned difficulties for gravitino multiplet couplings, it is nevertheless possible to construct a (globally) supersymmetric action for this multiplet, with the gravitino and vector field satisfying free equations of motion and both being massless fields:

$$\mathcal{L} = -\frac{1}{2}\overline{\psi}_\mu \gamma^{\mu\nu\rho}\partial_\nu \psi_\rho - \frac{1}{4}F^{\mu\nu}F_{\mu\nu}. \tag{2.41}$$

The gravitino has the expected gauge invariance $\delta\psi_\mu = \partial_\mu \Lambda$, like the gauge boson $\delta A_\mu = \partial_\mu \Sigma$, but this is unrelated to the supersymmmetry transformation rule, which, on the other hand, relates the gravitino to the vector field strength [6]:

$$\delta\psi_\mu = \frac{1}{4}\gamma_{\mu\nu\rho}\epsilon \, F^{\nu\rho} - \frac{1}{2}\gamma^\rho \epsilon \, F_{\mu\rho}, \quad \delta A_\mu = \bar{\epsilon}\psi_\mu. \tag{2.42}$$

Exercises

2.1. Verify the invariance of (2.41) under the supersymmetry transformations (2.42).

References

1. M. Fierz, On the relativistic theory of force-free particles with any spin. Helvetica Physica Acta **12**(I), 3–37 (1939)
2. M. Fierz, W. Pauli, On relativistic wave equations for particles of arbitrary spin in an electromagnetic field. Proc. Roy. Soc. Lond. A **173**, 211 (1939)
3. W. Rarita, J. Schwinger, On a theory of particles with half integral spin. Phys. Rev. **60**, 61 (1941)
4. G. Velo, D. Zwanziger, Propagation and quantization of Rarita-Schwinger waves in an external electromagnetic potential. Phys. Rev. **186**, 1337–1341 (1969)
5. M.A. Vasiliev, Higher spin gauge theories in four-dimensions, three-dimensions, and two-dimensions. Int. J. Mod. Phys. D **5**, 763–797 (1996) [arXiv:hep-th/9611024 [hep-th]]. Higher spin gauge theories: star product and AdS space [arXiv:hep-th/9910096 [hep-th]]
6. B. de Wit, J.W. van Holten, Multiplets of linearized SO(2) supergravity. Nucl. Phys. **B155**, 530 (1979)

Gravity and Spinors

3

Supergravity is a supersymmetric theory of gravity based on the principles of general relativity. In this subsection we set up our notation and review some elementary material on general relativity with particular emphasis on those aspects that are relevant for the formulation of supergravity theories.

3.1 The Standard Metric Formulation

General relativity is concerned with the dynamics of Lorentzian metric tensors, $g = g_{\mu\nu}(x)dx^{\mu} \otimes dx^{\nu}$, on differentiable manifolds, $\mathcal{M}$. In these lectures, we always use the metric signature $(-, +, \ldots, +)$, and local coordinate bases of tangent vectors are denoted as $\partial_{\mu} \equiv \frac{\partial}{\partial x^{\mu}}$, i.e., $g_{\mu\nu}(x) = g(\partial_{\mu}, \partial_{\nu})$, (See Fig. 3.1).

A connection, ∇, on a differentiable manifold $\mathcal{M}$ is a prescription for a covariant derivative of the tensor fields of $\mathcal{M}$. It is fixed by its action on the coordinate basis vectors ∂_{μ} (or the dual covectors dx^{μ}), i.e., by the (generically x-dependent) connection coefficients $\Gamma^{\rho}_{\mu\nu}$,

$$\nabla_{\mu}(\partial_{\nu}) \equiv \nabla_{\partial_{\mu}}(\partial_{\nu}) = \Gamma^{\rho}_{\mu\nu}\partial_{\rho} \quad \Longleftrightarrow \quad \nabla_{\mu}(dx^{\rho}) = -\Gamma^{\rho}_{\mu\nu}dx^{\nu}. \tag{3.1}$$

Once the action on the basis vectors is defined, we can compute the action of ∇ on any vector (and covector):

$$\nabla_V W = V^{\mu}\nabla_{\partial_{\mu}}\left(W^{\nu}\partial_{\nu}\right) = V^{\mu}\left(\partial_{\mu}W^{\nu} + \Gamma^{\nu}_{\mu\rho}W^{\rho}\right)\partial_{\nu}. \tag{3.2}$$

With an obvious abuse of notation, we therefore define

$$\nabla_{\mu}W^{\nu} \equiv \partial_{\mu}W^{\nu} + \Gamma^{\nu}_{\mu\rho}W^{\rho}, \tag{3.3}$$

and analogously for more general tensor fields.

© Springer-Verlag GmbH Germany, part of Springer Nature 2021
G. Dall'Agata, M. Zagermann, *Supergravity*, Lecture Notes in Physics 991,
https://doi.org/10.1007/978-3-662-63980-1_3

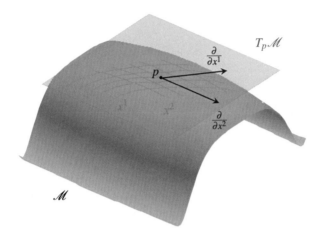

Fig. 3.1 The tangent space $T_p \mathcal{M}$ to a point $p \in \mathcal{M}$ with tangent vectors in a coordinate basis

On a manifold with a metric, there is a natural connection, the *Levi–Civita connection*, which is uniquely specified by the following requirements

(i) $\nabla_\mu g_{\nu\rho} = 0$ (metric compatibility)
(ii) $\nabla_\mu \partial_\nu - \nabla_\nu \partial_\mu = 0 \Leftrightarrow \Gamma^\rho_{\mu\nu} = \Gamma^\rho_{\nu\mu}$ (vanishing torsion)

The metric compatibility (i) means that the inner product of two vectors is unchanged under parallel transport, whereas the vanishing torsion[1] condition (ii) can, e.g., be interpreted as requiring that two covariant derivatives commute when they act on a scalar function or that $\Gamma^\rho_{\mu\nu}$ vanishes at the origin of a Riemannian normal coordinate system. For later reference we note that, using the one-form $\Gamma^\rho{}_\nu \equiv \Gamma^\rho_{\mu\nu} dx^\mu$, the vanishing torsion condition (ii) can alternatively be written in terms of the dual basis of one-forms, dx^ρ, as

(ii)' $\nabla dx^\rho \equiv (d + \Gamma^\bullet{}_\bullet \wedge) dx^\rho = \Gamma^\rho{}_\nu \wedge dx^\nu \overset{!}{=} 0.$

Either way, the two requirements of metric compatibility and vanishing torsion completely fix the Levi–Civita connection in terms of the metric via the standard Christoffel symbols,

$$\Gamma^\rho_{\mu\nu} = \Gamma^\rho_{\mu\nu}(g) = \frac{1}{2} g^{\rho\sigma} \left(\partial_\mu g_{\nu\sigma} + \partial_\nu g_{\mu\sigma} - \partial_\sigma g_{\mu\nu} \right). \tag{3.4}$$

---

[1] The torsion tensor, T, maps two vector fields, V, W, to another vector field defined by $T(V, W) = \nabla_V W - \nabla_W V - [V, W]$, implying $T_{\mu\nu}{}^\rho = \Gamma^\rho_{\mu\nu} - \Gamma^\rho_{\nu\mu}$.

From the connection, one obtains the Riemann tensor

$$R_{\rho\sigma}{}^{\mu}{}_{\nu} \equiv 2\,\partial_{[\rho}\Gamma^{\mu}_{\sigma]\nu} + 2\,\Gamma^{\mu}_{\tau[\rho}\Gamma^{\tau}_{\sigma]\nu},\tag{3.5}$$

the Ricci tensor

$$R_{\mu\nu} \equiv R_{\rho\mu}{}^{\rho}{}_{\nu},\tag{3.6}$$

and the Ricci scalar

$$R \equiv R_{\rho\mu}{}^{\rho}{}_{\nu}\,g^{\mu\nu} = R_{\mu\nu}\,g^{\mu\nu}.\tag{3.7}$$

The Einstein–Hilbert action then is

$$S_{EH} = \frac{1}{16\pi\,G_N}\int d^4x\,\sqrt{-g}\,R,\tag{3.8}$$

where $g \equiv \det(g_{\mu\nu})$, and G_N denotes Newton's constant. It is customary to introduce the reduced Planck mass, M_P, related to Newton's constant by[2]

$$M_P = \frac{1}{\sqrt{8\pi\,G_N}} = \frac{M_P^{old}}{\sqrt{8\pi}} = 2.44 \cdot 10^{18}\,\text{GeV},\tag{3.9}$$

so that the Einstein–Hilbert action can also be written as

$$S_{EH} = \frac{M_P^2}{2}\int d^4x\,\sqrt{-g}\,R.\tag{3.10}$$

3.2   The Vielbein Basis and Cartan's Formalism

In D-dimensional Minkowski spacetime, a vector field, V, and a spinor field, ψ, transform, respectively, in the fundamental and in the spinor representation of the Lorentz group,

$$V^{\mu} \to \Lambda^{\mu}{}_{\nu}V^{\nu}\tag{3.11}$$

$$\psi_{\alpha} \to \rho(\Lambda)_{\alpha\beta}\psi_{\beta},\tag{3.12}$$

---

[2] The difference between the "old" Planck mass value $M_P^{old} = 1.22 \cdot 10^{19}\,GeV$ and the reduced Planck mass (3.9) used in supergravity computations should be kept in mind when quantitative predictions are required.

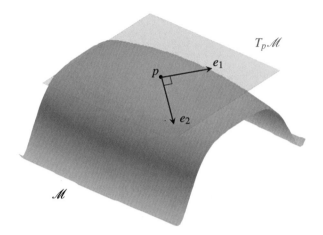

Fig. 3.2 Orthonormal non-coordinate basis for $T_p \mathcal{M}$

where $\Lambda^\mu{}_\nu \in SO(1, D - 1)$, and $\rho(\Lambda)_{\alpha\beta}$ is a spinor representation matrix (i.e., a double-valued representation matrix of $SO(1, D - 1)$; see Sect. 1.3) with α, β denoting suitable spinor indices.

On a curved manifold with arbitrary local coordinate transformations $x \to x'(x)$, the transformation (3.11) is naturally generalized by replacing $\Lambda^\mu{}_\nu$ by the Jacobi matrix $(\partial x'^\mu / \partial x^\nu)$, which in general is an (x-dependent) element of $GL(D, \mathbb{R})$. A spinor representation $\rho(\Lambda)_{\alpha\beta}$, by contrast, does *not* have an analogous extension to the full group $GL(D, \mathbb{R})$.

In order to define spinor fields in curved spacetimes, one therefore makes use of the equivalence principle and switches to a *local* inertial frame *at every point*. More precisely, one chooses as tangent space basis a set of local vector fields, e_a ($a = 0, 1, \ldots, (D - 1)$), that are mutually orthonormal at each point (see Fig. 3.2),

$$g(e_a, e_b) = \eta_{ab}. \tag{3.13}$$

Obviously, this condition does not fix the vectors e_a uniquely; rather they can be rotated by arbitrary local (i.e., x-dependent) Lorentz transformations, $e_a \to e_b \Lambda^b{}_a(x)$, without violating the orthonormality (3.13). It should also be noted that these orthonormal vector fields can in general only be defined on local patches unless $\mathcal{M}$ is parallelizable and that they can in general not be written as coordinate vectors $\frac{\partial}{\partial x^a}$ for some local coordinates x^a unless $\mathcal{M}$ happens to be flat. The non-coordinate basis e_a is often called an orthonormal frame, vielbein, or D-bein (in $D = 4$: vierbein or tetrad). In the following we will follow common practice and use the word vielbein also for the matrix $e^a_\mu(x)$ that mediates between the local, e_a, and the coordinate, ∂_μ, bases,

$$e_a = e^\mu_a(x)\partial_\mu, \qquad \partial_\mu = e^a_\mu(x)e_a \tag{3.14}$$

with $e^a_\mu e^\nu_a = \delta^\nu_\mu$, $e^\mu_a e^b_\mu = \delta^b_a$. This vielbein can be used to convert the "curved" (or "world") indices $\mu, \nu, \ldots$ of any tensor field to "flat" (or "local Lorentz") indices $a, b, \ldots$, e.g., $V^a = e^a_\mu V^\mu$, etc. In particular, they interpolate between the curved metric $g_{\mu\nu}(x)$ and the flat metric η_{ab}:

$$g_{\mu\nu}(x) = e^a_\mu(x) e^b_\nu(x) \eta_{ab}. \tag{3.15}$$

This statement is clearly invariant under the abovementioned local Lorentz transformations of the vielbein, $e^{a\,\prime}_\mu = e^b_\mu \Lambda^a{}_b(x)$, as $\Lambda^T \eta \Lambda = \eta$, and we see that the vielbein contain the same amount of information as the metric tensor. Equation (3.15) also implies

$$\sqrt{-g} = e \equiv \det e^a_\mu. \tag{3.16}$$

A general relativistic theory written in terms of a vielbein thus has two local invariances: general coordinate transformations acting with Jacobian matrices on all curved indices $\mu, \nu, \ldots$, and local Lorentz transformations acting with $\Lambda^b{}_a(x)$ on all flat indices $a, b, \ldots$. One can therefore think of the vielbein e^a_μ also as elements parameterizing the coset space

$$\frac{GL(D, \mathbb{R})}{SO(1, D-1)}.$$

In terms of the orthonormal basis e_a, a connection is now specified by its action on these basis vectors,

$$\nabla_\mu e_a = \omega_\mu{}^b{}_a(x)\, e_b \tag{3.17}$$

where we introduced new connection coefficients $\omega_\mu{}^b{}_a(x)$. A connection $\omega_\mu{}^b{}_a$ is equivalent to a given connection $\Gamma^\rho_{\mu\nu}$ in a curved basis if $\nabla_\mu e_a = e^\nu_a \nabla_\mu \partial_\nu$, which is equivalent to the condition

$$\nabla_\mu e^a_\nu \equiv \partial_\mu e^a_\nu + \omega_\mu{}^a{}_b e^b_\nu - \Gamma^\rho_{\mu\nu} e^a_\rho = 0, \tag{3.18}$$

where, as suggested by the notation, the left-hand side can be viewed as a total covariant derivative acting on curved and flat indices. Equation (3.18) is sometimes referred to as the "vielbein postulate," but here it is really just the statement that the two connections $\omega_\mu{}^b{}_a$ and $\Gamma^\rho_{\mu\nu}$ are actually equivalent connections expressed in two different basis systems.

Just as for the curved indices, one can now demand that $\omega_\mu{}^b{}_a$ preserve the metric tensor and be torsion-free so as to arrive at the flat index equivalent of the Levi–Civita connection. The corresponding connection coefficients $\omega_\mu{}^b{}_a = \omega_\mu{}^b{}_a(e)$ are then uniquely determined by the vielbein e^a_μ and define the (torsion-free) spin connection. We derive the expression for $\omega_\mu{}^b{}_a$ further below. In the following, it will be useful to define also a derivative operator, D_μ, that is covariant only with respect to local Lorentz transformations, but not necessarily with respect to general coordinate transformations, i.e., all local Lorentz indices will be contracted with spin connections, but there are no Christoffel symbols that contract any world index. As an example, consider this Lorentz covariant derivative acting on a vielbein e_ν^a:

$$D_\mu e_\nu^a \equiv \partial_\mu e_\nu^a + \omega_\mu{}^a{}_b\, e_\nu^b. \tag{3.19}$$

In view of (3.18), this expression does not vanish but is equal to $\Gamma^\rho_{\mu\nu} e_\rho^a$, where $\Gamma^\rho_{\mu\nu}$ is the curved connection equivalent to $\omega_\mu{}^b{}_a$. Recalling that the antisymmetrization $\Gamma^\rho_{[\mu\nu]} = \frac{1}{2}T_{\mu\nu}{}^\rho$ is just the torsion tensor of this connection, we can thus write

$$D_{[\mu} e_{\nu]}^a = \frac{1}{2}T_{\mu\nu}{}^a = \frac{1}{2}T_{\mu\nu}{}^\rho e_\rho^a. \tag{3.20}$$

Note that, even though the Lorentz covariant derivative, D_μ, is not covariant with respect to general coordinate transformations, the above *antisymmetrized* derivative still forms a proper tensor field. In supergravity, all equations can similarly be expressed in terms of antisymmetrized Lorentz covariant derivatives only. This allows one to use compact differential form notation, which, as we will see in Chap. 4, can substantially reduce the index clutter in some computations. To this end, we introduce the co-frame of one-forms, e^a, dual to the vectors e_b:

$$e^a \equiv e_\mu^a dx^\mu. \tag{3.21}$$

These one-forms are at the basis of Cartan's formulation of general relativity, as we briefly review in the remainder of this subsection.

This box summarizes our conventions regarding differential forms. The wedge product between basis one-forms is defined as

$$dx^{i_1} \wedge \ldots \wedge dx^{i_p} = \sum_{\sigma \in S_p} \mathrm{sign}(\sigma)\, dx^{\sigma(i_1)} \otimes \ldots \otimes dx^{\sigma(i_p)},$$

where S_p is the permutation group of p elements. The components of a p-form $F_p \in \Lambda^p(\mathcal{M})$ are normalized as

$$F_p = \frac{1}{p!}dx^{\mu_1} \wedge \ldots \wedge dx^{\mu_p} F_{\mu_1 \ldots \mu_p}.$$

(continued)

The exterior differential acts from the left as

$$dF_p = \frac{1}{p!}dx^\rho \wedge dx^{\mu_1} \wedge \ldots \wedge dx^{\mu_p} \, \partial_\rho F_{\mu_1 \ldots \mu_p}.$$

For instance, for a connection one-form A, the components of the curvature $F = dA$ are defined as $F_{\mu\nu} = 2 \, \partial_{[\mu} A_{\nu]}$. The inner product also starts from the left,

$$\iota_V(dx^{\mu_1} \otimes \ldots \otimes dx^{\mu_p} T_{\mu_1 \ldots \mu_p}) = dx^{\mu_2} \otimes \ldots \otimes dx^{\mu_p} V^{\mu_1} T_{\mu_1 \mu_2 \ldots \mu_p},$$

so that

$$\iota_V F_p = \frac{1}{(p-1)!}dx^{\mu_2} \wedge \ldots \wedge dx^{\mu_p} V^{\mu_1} F_{\mu_1 \mu_2 \ldots \mu_p}.$$

Another useful tool is the Hodge star dual, relating elements in $\Lambda^p(\mathcal{M})$ with those in $\Lambda^{(D-p)}(\mathcal{M})$, where D is the dimension of the manifold $\mathcal{M}$. Its definition on a basis of $\Lambda^p(\mathcal{M})$ is defined by:

$$\star dx^{\mu_1} \wedge \ldots \wedge dx^{\mu_p} = \frac{1}{(D-p)!} dx^{\mu_{p+1}} \wedge \ldots \wedge dx^{\mu_D} \epsilon_{\mu_{p+1} \ldots \mu_D}{}^{\mu_1 \ldots \mu_p},$$

where indices are raised and lowered with the metric $g_{\mu\nu}$, and $\epsilon_{\mu_1 \ldots \mu_D} = e_{\mu_1}{}^{a_1} \ldots e_{\mu_D}{}^{a_D} \epsilon_{a_1 \ldots a_D}$ with $\epsilon_{012 \ldots (D-1)} = 1$, so that

$$\star F_p = \frac{1}{p!(D-p)!} dx^{\mu_{p+1}} \wedge \ldots \wedge dx^{\mu_D} \epsilon_{\mu_{p+1} \ldots \mu_D}{}^{\mu_1 \ldots \mu_p} F_{\mu_1 \ldots \mu_p}.$$

This also implies that

$$\star^2 F_p = -(-1)^{p(D-p)} F_p.$$

In terms of the e^a, the expression (3.20) for the torsion tensor reads

$$De^a \equiv de^a + \omega^a{}_b \wedge e^b = T^a = \frac{1}{2}dx^\mu \wedge dx^\nu T_{\mu\nu}{}^a, \tag{3.22}$$

whereas the curvature tensor of the connection $\omega_\mu{}^b{}_a$ is given by the two-form

$$R^a{}_b = d\omega^a{}_b + \omega^a{}_c \wedge \omega^c{}_b. \tag{3.23}$$

The conditions for the connection $\omega_\mu{}^b{}_a$ to be equivalent to the Levi–Civita connection (i.e., for $\omega_\mu{}^b{}_a$ to be the torsion-free spin connection) can now be written as

- $D\eta_{ab} = 0$ (metric compatibility)
- $T^a = 0$ (vanishing torsion)

The first condition implies that the spin connection is antisymmetric in its indices when both are raised or lowered with η: $D\eta_{ab} \equiv d\eta_{ab} + \omega_a{}^c\eta_{cb} + \omega_b{}^c\eta_{ac}$, and $d\eta_{ab} = 0$, so that $\omega_{(ab)} = 0$. The second equation (in combination with (3.22)) has an obvious similarity with the alternative way (ii)' of writing the vanishing torsion condition for the curved indices,

$$De^a = 0. \tag{3.24}$$

Just as for the Levi–Civita connection in curved indices, the vanishing of the torsion tensor can be used to deduce an expression for the spin connection in terms of the vielbein, $\omega_\mu^{ab} = \omega_\mu^{ab}(e)$. It is instructive to do this calculation once, as it will also be useful later, and the same trick can be used for $T^a \neq 0$. To this end, consider

$$t^{dc,a} \equiv e^{\nu[d}e^{\mu c]}\left(\partial_\mu e_\nu^a + \omega_\mu{}^a{}_b e_\nu^b\right).$$

This tensor is zero because of the assumed vanishing torsion. Hence we can consider the sum

$$t^{dc,a} - t^{ca,d} - t^{ad,c} = 0. \tag{3.25}$$

From this sum there is only one term in the spin connection that survives, $e^{\rho a}\omega_\rho^{cd}$, and, multiplying by $e_{\mu a}$, we obtain

$$\omega_\mu^{cd}(e) = 2e^{\nu[c}\partial_{[\mu}e_{\nu]}^{d]} - e_{a\mu}e^{\nu[c}e^{\sigma d]}\partial_\nu e_\sigma^a. \qquad \text{(torsion-free spin connection)} \tag{3.26}$$

It can be checked that the expression (3.4) of the Levi–Civita connection implies the spin connection (3.26) and vice versa.

In Cartan's formalism the Einstein–Hilbert action now becomes

$$S_{EH} = \frac{M_P^2}{4}\int R^{ab} \wedge e^c \wedge e^d \epsilon_{abcd}, \tag{3.27}$$

where $\epsilon_{0123} = 1$. Recovering the standard form (3.10) is rather straightforward if we use the identification of the four-dimensional measure

$$dx^\mu \wedge dx^\nu \wedge dx^\rho \wedge dx^\sigma = -d^4x\, e\, \epsilon^{\mu\nu\rho\sigma}, \tag{3.28}$$

where $\epsilon_{\mu\nu\rho\sigma} = e^a_\mu e^b_\nu e^c_\rho e^d_\sigma \, \epsilon_{abcd}$, and express the forms through their components:

$$
\begin{aligned}
S_{EH} &= \frac{M_P^2}{4} \int \left(\frac{1}{2} dx^\mu \wedge dx^\nu R_{\mu\nu}{}^{ab} \right) \wedge \left(dx^\rho e^c_\rho \right) \wedge \left(dx^\sigma e^d_\sigma \right) \epsilon_{abcd} \\
&= \frac{M_P^2}{8} \int dx^\mu \wedge dx^\nu \wedge dx^\rho \wedge dx^\sigma \left(R_{\mu\nu}{}^{\tau\upsilon} e^a_\tau e^b_\upsilon e^c_\rho e^d_\sigma \epsilon_{abcd} \right) \\
&= -\frac{M_P^2}{8} \int d^4x \, e \, \epsilon^{\mu\nu\rho\sigma} \, \epsilon_{\tau\upsilon\rho\sigma} \, R_{\mu\nu}{}^{\tau\upsilon} \\
&= \frac{M_P^2}{2} \int d^4x \, e \, R.
\end{aligned}
\tag{3.29}
$$

In order to derive the equations of motion coming from the Einstein–Hilbert action, it is useful not to assume (3.26) from the outset, but rather to consider the vielbein and spin connection as independent fields, which then leads to the two field equations

$$
\frac{\delta S}{\delta e^a_\mu} = 0,
\tag{3.30}
$$

$$
\frac{\delta S}{\delta \omega^{ab}_\mu} = 0.
\tag{3.31}
$$

Given the Einstein tensor,

$$
G_{\mu\nu} \equiv R_{\mu\nu} - \frac{1}{2} g_{\mu\nu} R,
\tag{3.32}
$$

one finds

$$
\int d^4x \frac{\delta S}{\delta e^a_\mu} \delta e^a_\mu = -M_P^2 \int d^4x \, e \, e^\sigma_a \, G_\sigma{}^\mu \, \delta e^a_\mu
\tag{3.33}
$$

and then the vacuum Einstein equation for the torsion-free spin (i.e., the Levi–Civita) connection,

$$
\frac{\delta S}{\delta e^a_\mu} = 0 \Leftrightarrow G_{\mu\nu} = 0,
\tag{3.34}
$$

$$
\frac{\delta S}{\delta \omega^{ab}_\mu} = 0 \Leftrightarrow T^a = 0,
\tag{3.35}
$$

as the reader is urged to verify in the exercises at the end of this chapter.

Apart from encoding the Einstein equation, these equations are especially important also for the discussion of the invariance of the action under a symmetry. A general symmetry acting on the vielbein and the spin connection transforms the action as

$$\delta S = \int d^4 x \left(\frac{\delta S}{\delta e^a_\mu} \delta e^a_\mu + \frac{\delta S}{\delta \omega^{ab}_\mu} \delta \omega^{ab}_\mu \right), \qquad (3.36)$$

where the functional derivative with respect to the vielbein is meant to include only the explicit appearance of e^a_μ.

Depending on how we interpret the variation with respect to the spin connection, we can now distinguish the following formalisms:

- *Second-order formalism:* Here the spin connection is meant to be the explicit functional of the vielbein given in (3.26), and, therefore, the variation of the action has to be understood as

$$\delta S = \int d^4 x \left(\frac{\delta S}{\delta e^a_\mu} + \frac{\delta S}{\delta \omega^{cd}_v} \frac{\delta \omega^{cd}_v}{\delta e^a_\mu} \right) \delta e^a_\mu.$$

  The name "second order" comes from the fact that the equations of motion for the vielbein (or the metric) in this formalism involve second-order derivatives.
- *First-order formalism:* Here the spin connection and the vielbein are considered as independent fields, just as we did when we derived the Einstein equation above. This is also called the Palatini formalism, and in this formalism we have a priori $De^a \neq 0$. As we have seen before, it is the variation with respect to the spin connection that leads to the torsion constraint, while the variation with respect to the vielbein gives the Einstein equation. The action invariance under a symmetry is established by requiring a suitable variation for the spin connection $\delta \omega^{ab}$. We will see an example of this later in Chap. 4.
- *1.5-order formalism:* In this mixed formalism, one considers the spin connection and the vielbein again as independent but uses the fact that the variation of the action with respect to the spin connection is the spin connection equation of motion (or torsion constraint (3.35)) and hence vanishes on-shell. The invariance of the action then needs to be checked only by considering the variation with respect to the vielbein, and the spin connection only plays the role of an auxiliary field whose variation does not have to be considered in this formalism. Though this is not very useful in an ordinary gravity theory, it will simplify a lot the calculations when we check the supersymmetry invariance of the supergravity action.

Of course, these three formalisms must reduce to the same set of transformation rules on-shell. This happens as long as the spin connection ω^{ab} appears in the action at most quadratically, as in (3.27). This is not always the case for actions including higher order terms in the Riemann tensor like $R^{ab} \wedge R_{ab}$ or R^4 terms, etc., which however we do not consider in these lecture notes. Also, while classically equivalent, these formalisms may not be equivalent at the quantum level.

All these results can be extended easily to cases with non-vanishing (but fixed) torsion. We just need to solve (3.22) for a given T^a. This gives a spin connection which is the sum of (3.26) and a contorsion tensor, κ_μ^{ab}, defined as

$$\kappa_\mu^{ab} = e_\mu^c \left(T^{ab}{}_c - T_c{}^{ab} - T^b{}_c{}^a \right), \tag{3.37}$$

where indices are made flat by the use of vielbeins and raised and lowered with the flat metric. This solution will follow again from the variation of the action with respect to the spin connection, when it is considered as an independent field. When the action has more terms than just the Einstein–Hilbert term, and especially when there are new couplings of the spin connection to other fields, the variation of the action with respect to the spin connection will lead to a spin connection with non-vanishing torsion, as we will see Chap. 4.

We close this subsection by mentioning that the treatment of the vielbein and the spin connection as independent quantities can have a conceptual meaning that goes beyond the mere simplification of some mathematical computations. As we explain in Appendix 4.A, an interesting perspective on gravity can be gained by considering it as a gauge theory of the Poincaré group. This analogy will work only to a certain extent, but it will be very useful in understanding many specific new features that have to be introduced when one wants to promote supersymmetry to a local symmetry of nature. In fact supergravity *is* the gauge theory of supersymmetry, and therefore there must be a way to describe it as a theory where the gauge group is the Poincaré supergroup (or some other supergroup).

3.3   Spinors in Curved Spacetime

As we already mentioned, spinors transform in double-valued representations of the Lorentz group and cannot be extended to analogous representations of the general linear group. To describe spinors in curved spacetime, we therefore use a vierbein basis for the tangent space, so that local Lorentz transformations are still meaningful, and a spinor field now transforms as

$$\psi_\alpha \to \rho(\Lambda(x))_{\alpha\beta} \psi_\beta. \tag{3.38}$$

Here, as indicated, the Lorentz transformation $\Lambda(x)$ in general is allowed to be spacetime dependent. Meaningful derivatives acting on spinors thus have to be covariant with respect to such local Lorentz transformations. As the spin connection is the connection for local Lorentz transformations, and the latter are generated by $\frac{1}{2}\gamma_{ab}$ in the spinor representation, the correct Lorentz covariant derivative is

$$D_\mu \psi = \partial_\mu \psi + \frac{1}{4}\omega_\mu{}^{ab}\gamma_{ab}\psi, \tag{3.39}$$

where we suppressed the spinor indices, and the additional factor of $1/2$ comes from the sum over a double index. This derivative satisfies $D_\mu\left[\rho(\Lambda(x))\psi\right] = \rho(\Lambda(x))D_\mu\psi$.

We finally mention that gamma matrices with a curved index μ are obtained from the constant γ_a via contraction with a vierbein:

$$\gamma_\mu \equiv e_\mu^a \gamma_a. \tag{3.40}$$

The γ_μ are then in general no longer constant, and they transform non-trivially under a variation of the vierbein.

Exercises

3.1. Prove the relations (3.4) and (3.26).

3.2. Prove (3.34)–(3.35).

3.3. Check that

$$\{\gamma_{\mu\nu\rho}, \gamma^{ab}\} = -12\,\gamma_{[\mu}e_\nu^a e_{\rho]}^b.$$

3.4. Using the definition of the Riemann curvature from the spin connection (3.23) and the definition of the covariant derivative on a spinor (3.39), prove that

$$[D_\mu, D_\nu]\psi = \frac{1}{4}R_{\mu\nu}{}^{ab}\gamma_{ab}\psi.$$

Pure $\mathcal{N} = 1$ Supergravity in Four Dimensions

4

In this chapter, we discuss pure supergravity in four dimensions with minimal supersymmetry and show explicitly the construction of its action, also proving the invariance under supersymmetry transformations. We then extend this construction to the case of a non-trivial cosmological constant and finally discuss the concept of mass in spacetimes with negative curvature.

4.1   Pure Supergravity: The Action and SUSY Rules

In this section, we will work out in detail the supersymmetry transformation of the pure supergravity Lagrangian, so that the origin and inevitability of the various terms in the action should become as clear as possible. In order to simplify several computations, we often use the language of differential forms. For instance, the starting supergravity action is the sum of the Einstein–Hilbert and Rarita–Schwinger actions,

$$S = \int d^4x \, (\mathcal{L}_{EH} + \mathcal{L}_{RS}) = \int d^4x \, e \left(\frac{M_P^2}{2} R - \frac{1}{2} \overline{\psi}_\mu \gamma^{\mu\nu\rho} D_\nu \psi_\rho \right), \tag{4.1}$$

which, in the language of differential forms, becomes

$$S = \frac{M_P^2}{4} \int R^{ab} \wedge e^c \wedge e^d \epsilon_{abcd} + \frac{i}{2} \int e^a \wedge \overline{\psi} \wedge \gamma_5 \gamma_a D\psi, \tag{4.2}$$

© Springer-Verlag GmbH Germany, part of Springer Nature 2021
G. Dall'Agata, M. Zagermann, *Supergravity*, Lecture Notes in Physics 991,
https://doi.org/10.1007/978-3-662-63980-1_4

as follows from using the duality relations on the gamma matrices derived in the Exercises to Chap. 1:

$$\frac{i}{2} \int e^a \wedge \overline{\psi} \wedge \gamma_5 \gamma_a D\psi = \frac{i}{2} \int dx^\mu \wedge dx^\nu \wedge dx^\rho \wedge dx^\sigma \left(e^a_\mu \overline{\psi}_\nu \gamma_5 \gamma_a D_\rho \psi_\sigma \right)$$

$$= -\frac{i}{2} \int d^4x \, e \, \epsilon^{\mu\nu\rho\sigma} \overline{\psi}_\nu \gamma_5 \gamma_\mu D_\rho \psi_\sigma \qquad (4.3)$$

$$= -\frac{1}{2} \int d^4x \, e \, \overline{\psi}_\mu \gamma^{\mu\nu\rho} D_\nu \psi_\rho .$$

Since we are coupling the spin 3/2 field ψ_μ to gravity, the covariant derivative in the kinetic term should a priori be the full covariant derivative, ∇, and not just the Lorentz covariant derivative, D, we have used in the above expressions. The full covariant derivative ∇ contains both the Levi–Civita connection, Γ, coupling to the vector index of the gravitino, and the spin connection, ω, coupling to the spinor index. However, even if we had used ∇, it would appear in the action only in antisymmetrized form,[1]

$$\nabla_{[\nu} \psi_{\rho]} = \partial_{[\nu} \psi_{\rho]} + \frac{1}{4} \omega^{ab}_{[\nu} \gamma_{ab} \psi_{\rho]} - \Gamma^\sigma_{[\nu\rho]} \psi_\sigma , \qquad (4.4)$$

and the last term is identically zero so that $\nabla_{[\nu} \psi_{\rho]} = D_{[\nu} \psi_{\rho]}$, and we can indeed use the Lorentz covariant derivative D in the kinetic term of the gravitino. In fact, the Levi–Civita connection in terms of Christoffel symbols will never really appear in the following.

We now discuss the invariance under supersymmetry of (4.1). We start by making one simple assumption that is motivated by our discussion in Chap. 2, namely, that the gravitino transformation rule is proportional to the (covariant) derivative of the supersymmetry parameter, so that the gravitino can be viewed as the gauge field of local supersymmetry and the theory is covariant with respect to local Lorentz transformations:

$$\delta_\epsilon \psi_\mu = M_P D_\mu \epsilon \equiv M_P \left(\partial_\mu + \frac{1}{4} \omega^{ab}_\mu \gamma_{ab} \right) \epsilon . \qquad (4.5)$$

---

[1] It is important here that Γ really denotes the torsion-free Levi–Civita connection. As we will see later, it is useful to include a torsion piece bilinear in the gravitini in the *spin connection* (but *not* in the connection Γ, which should stay torsion-free). The connections defined by Γ and ω are then no longer equivalent connections.

It is worth pointing out that (4.5) is the transformation rule that can also be inferred from the action of the supersymmetry algebra on the fields as explained in Appendix 4.A. The conjugate field satisfies

$$\delta_\epsilon \overline{\psi}_\mu = M_P \left(\partial_\mu \overline{\epsilon} - \frac{1}{4} \overline{\epsilon} \, \gamma_{ab} \, \omega_\mu^{ab} \right) \equiv M_P \, \overline{D_\mu \epsilon}. \tag{4.6}$$

Having specified only the gravitino supersymmetry transformation so far, the next thing we would like to obtain is the transformation rule of the vierbein. We could simply make an educated guess in line with our considerations leading to Eq. (2.15), but let us try to actually derive the vierbein transformation law from what we already have. From the variation of S_{EH} (3.33), we see that the only contribution with δe^a comes multiplied by the curvature R^{ab}. We therefore try to single out from δS_{RS} all possible terms that give the same type of contributions proportional to the curvature of the spin connection. Supersymmetry invariance will then determine δe^a, and we will then check the invariance of the full action.

The variation of the gravitini in the Rarita–Schwinger Lagrangian gives

$$\begin{aligned}
\delta \mathcal{L}_{RS} &= -\frac{e}{2} \overline{\psi}_\mu \gamma^{\mu\nu\rho} D_\nu \delta\psi_\rho - \frac{e}{2} \overline{\delta\psi_\mu} \gamma^{\mu\nu\rho} D_\nu \psi_\rho + \dots \\
&= -\frac{e}{2} \overline{\psi}_\mu \gamma^{\mu\nu\rho} D_\nu \delta\psi_\rho - \frac{e}{2} \overline{D_\nu \psi_\rho} \gamma^{\mu\nu\rho} \delta\psi_\mu + \dots ,
\end{aligned} \tag{4.7}$$

where we used the identity $\overline{\chi} \gamma^{\mu\nu\rho} \lambda = \overline{\lambda} \gamma^{\mu\nu\rho} \chi$ for anti-commuting Majorana spinors, and the dots refer to the variations of the vierbein, δe^a_μ, and the spin connection, ω_μ^{ab}, which we do not consider for now because they give terms that are not of the form we need. Inserting (4.5) in (4.7), we obtain

$$\delta \mathcal{L}_{RS} = -M_P \frac{e}{2} \overline{\psi}_\mu \gamma^{\mu\nu\rho} D_\nu D_\rho \epsilon - M_P \frac{e}{2} \overline{D_\nu \psi_\rho} \gamma^{\mu\nu\rho} D_\mu \epsilon + \dots \tag{4.8}$$

Integrating the last term by parts, we can replace it by

$$-\partial_\mu \left(\frac{e}{2} M_P \overline{D_\nu \psi_\rho} \gamma^{\mu\nu\rho} \epsilon \right) + \frac{e}{2} M_P \overline{D_\mu D_\nu \psi_\rho} \gamma^{\mu\nu\rho} \epsilon \tag{4.9}$$

plus terms involving derivatives of the vielbein, $D_\mu e^a_\nu$, which we also neglect in this first step, because they will not give contributions proportional to the curvature R^{ab}. The equivalence of (4.9) to the last term in (4.8) can easily be checked either by recalling that the γ-matrices are covariantly constant in the sense that

$$D_\mu \gamma^a = \partial_\mu \gamma^a + \omega_\mu^a{}_b \gamma^b + \frac{1}{4} \omega_\mu^{bc} [\gamma_{bc}, \gamma^a] = 0, \tag{4.10}$$

or that $D_\mu(\text{scalar}) = \partial_\mu(\text{scalar})$. As shown in Exercise 3.4, from the definition of the curvature, we find

$$[D_\mu, D_\nu] = \frac{1}{4} R_{\mu\nu}{}^{ab} \gamma_{ab}, \tag{4.11}$$

and therefore, using $\overline{\gamma_{ab}\psi_\rho} = -\overline{\psi}_\rho \gamma_{ab}$,

$$\overline{D_{[\mu} D_\nu \psi_{\rho]}} = -\frac{1}{8} R_{[\mu\nu}{}^{ab} \overline{\psi}_{\rho]} \gamma_{ab}. \tag{4.12}$$

Hence the variation of the Rarita–Schwinger term becomes

$$\delta \mathcal{L}_{RS} = -\frac{e}{16} M_P \overline{\psi}_\mu \gamma^{\mu\nu\rho} \gamma_{ab} \epsilon \, R_{\nu\rho}{}^{ab} - \frac{e}{16} M_P \overline{\psi}_\rho \gamma_{ab} \gamma^{\mu\nu\rho} \epsilon \, R_{\mu\nu}{}^{ab} + \dots$$

$$= -\frac{e}{16} M_P \overline{\psi}_\mu \left\{ \gamma^{\mu\nu\rho}, \gamma_{ab} \right\} \epsilon \, R_{\nu\rho}{}^{ab} + \dots$$

$$= \frac{e}{4} M_P \overline{\psi}_\mu \gamma^\mu \epsilon \, R_{\nu\rho}{}^{ab} e_a^\nu e_b^\rho + \frac{e}{2} M_P \overline{\psi}_\mu \gamma^\nu \epsilon \, R_{\nu\rho}{}^{ab} e_a^\rho e_b^\mu + \dots \tag{4.13}$$

$$= -\frac{e}{2} M_P \overline{\psi}_\mu \gamma^\nu \epsilon \left(R_\nu{}^\mu - \frac{1}{2} \delta_\nu^\mu R \right) + \dots$$

$$= -\frac{e}{2} M_P \overline{\psi}_\mu \gamma^\nu \epsilon \, G_\nu{}^\mu + \dots,$$

where the dots contain all terms that do not multiply the curvature of the spin connection, and we have used the results of Exercise 3.3 in the third line. As expected, this can be compensated by (cf. Eq. (3.33))

$$\frac{\delta \mathcal{L}_{EH}}{\delta e_\mu^a} \delta e_\mu^a = -M_P^2 \, e \, e_a^\nu \, G_\nu{}^\mu \, \delta e_\mu^a, \tag{4.14}$$

which is also proportional to the same combination of the curvature, provided we define the variation of the vielbein as

$$\delta e_\mu^a = \frac{1}{2M_P} \overline{\epsilon} \gamma^a \psi_\mu \tag{4.15}$$

(recall that $\overline{\epsilon} \gamma^a \psi_\mu = -\overline{\psi}_\mu \gamma^a \epsilon$). Note that, proceeding in this way, we did not simply guess the variation of the vielbein from our considerations in Chap. 2 but instead really *derived* it. On the other hand, we see immediately that (4.15) is indeed consistent with (2.15).

In order to complete the proof of the invariance of the action (4.1), we still need to discuss the following variations:

(i) $\dfrac{\delta \mathscr{L}_{EH}}{\delta \omega_\mu^{ab}} \delta_\epsilon \omega_\mu^{ab}$;

(ii) $\dfrac{\delta \mathscr{L}_{RS}}{\delta \omega_\mu^{ab}} \delta_\epsilon \omega_\mu^{ab}$;

(iii) $\dfrac{\delta \mathscr{L}_{RS}}{\delta e_\mu^a} \delta_\epsilon e_\mu^a$;

(iv)   Terms involving De^a from the partial integration in $\dfrac{\delta \mathscr{L}_{RS}}{\delta \psi_\mu} \delta_\epsilon \psi_\mu$.

We also need to understand and specify $\delta_\epsilon \omega^{ab}$. As we will see, the variation of the spin connection will depend on the formalism (first, second, or 1.5 order) used to prove the invariance of the action.

To do this calculation, we go back to the form expression (4.2). The variation of the action is then

$$
\delta S = \underbrace{\frac{M_P^2}{4} D\delta\omega^{ab} \wedge e^c \wedge e^d \epsilon_{abcd}}_{B1} + \underbrace{\frac{M_P^2}{2} R^{ab} \wedge \delta e^c \wedge e^d \epsilon_{abcd}}_{A1}
$$

$$
+ \underbrace{\frac{i}{2} \delta e^a \wedge \overline{\psi} \wedge \gamma_5 \gamma_a D\psi}_{B2} + \underbrace{\frac{i}{2} e^a \wedge \overline{\delta\psi} \wedge \gamma_5 \gamma_a D\psi}_{A2} \tag{4.16}
$$

$$
\underbrace{- \frac{i}{8} e^a \wedge \overline{\psi} \wedge \gamma_5 \gamma_a \gamma_{cd} \psi \wedge \delta\omega^{cd}}_{B3} + \underbrace{\frac{i}{2} e^a \wedge \overline{\psi} \wedge \gamma_5 \gamma_a D\delta\psi}_{A3}.
$$

We know from previous computations that the term $A1$, coming from $\delta \mathscr{L}_{EH}/\delta e^a$, and the terms involving $D^2\epsilon$, $D^2\psi$ coming from $\delta \mathscr{L}_{RS}/\delta\psi$ cancel. In detail, the $D^2\epsilon$-term is $A3$, where one uses the explicit expression for $\delta\psi$, and the $D^2\psi$-term can be extracted from $A2$ using the same steps that also led to (4.15). To do so, we switch the two-form $D\psi$ and the one-form $\delta\psi$, using (1.38) and the result of Exercise 1.1, so that

$$
A2 \equiv \frac{i}{2} e^a \wedge \overline{\delta\psi} \wedge \gamma_5 \gamma_a D\psi = \frac{i}{2} e^a \wedge \overline{D\psi} \wedge \gamma_5 \gamma_a \delta\psi = \frac{i}{2} M_P e^a \wedge \overline{D\psi} \wedge \gamma_5 \gamma_a D\epsilon \tag{4.17}
$$

and, integrating again by parts,

$$A2 = -M_P\, d\left(\frac{i}{2}e^a \wedge \overline{D\psi}\,\gamma_5\gamma_a\epsilon\right) + \underbrace{\frac{i}{2}\,M_P\, De^a \wedge \overline{D\psi}\,\gamma_5\gamma_a\epsilon}_{A2''}$$

$$\underbrace{-\frac{i}{2}M_P\, e^a \wedge \overline{DD\psi}\,\gamma_5\gamma_a\epsilon}_{A2'}\,. \tag{4.18}$$

The term $A2'$ then cancels $A1$ and $A3$ as before, and we are left with $\delta S = B1 + B2 + B3 + A2''$ plus boundary terms.

To proceed further, we integrate by parts the term $B1$ and get

$$B1 = \frac{M_P^2}{2}\delta\omega^{ab} \wedge De^c \wedge e^d\epsilon_{abcd} + d\left(\frac{M_P^2}{4}\delta\omega^{ab} \wedge e^c \wedge e^d\epsilon_{abcd}\right).$$

In order to write $B3$ in a very similar form, we write, using the results of the Exercise 1.1 and Eq. (1.38),

$$\overline{\psi} \wedge \gamma_5\gamma_a\gamma_{cd}\psi = \overline{\psi} \wedge \gamma_5(\gamma_{acd} + \eta_{ac}\gamma_d - \eta_{ad}\gamma_c)\psi = -i\,\overline{\psi} \wedge \gamma^e\psi\epsilon_{acde} \tag{4.19}$$

so that, after some relabeling and reordering,

$$B3 = -\frac{1}{8}\delta\omega^{ab} \wedge \overline{\psi} \wedge \gamma^c\psi \wedge e^d\epsilon_{abcd}. \tag{4.20}$$

Discarding boundary terms and inserting also (4.15) in B2, we then have

$$\delta S = \frac{M_P^2}{2}\delta\omega^{ab} \wedge \left(De^c - \frac{1}{4M_P^2}\overline{\psi} \wedge \gamma^c\psi\right) \wedge e^d\epsilon_{abcd}$$

$$+\frac{i}{4M_P}\left(\overline{\epsilon}\gamma^a\psi\right) \wedge (\overline{\psi} \wedge \gamma_5\gamma_a D\psi) + \frac{i}{2}M_P\, De^a \wedge \overline{D\psi}\,\gamma_5\gamma_a\epsilon. \tag{4.21}$$

This expression can be simplified by rewriting the second line so that the torsion piece $\left(De^a - \frac{1}{4M_P^2}\overline{\psi} \wedge \gamma^a\psi\right)$ can also be factored out. While the last term of (4.21) obviously contains a derivative of the vielbein, the other term needs a reshuffling of the gravitini in order to produce the right bilinear without derivatives. We can achieve this by using the Fierz identity

$$\psi \wedge \overline{\psi} = \frac{1}{4}\left(\overline{\psi} \wedge \gamma^a\psi\right)\gamma_a - \frac{1}{8}\left(\overline{\psi} \wedge \gamma^{ab}\psi\right)\gamma_{ab} \tag{4.22}$$

and the fact that $\gamma^c \gamma^{ab} \gamma_c = 0$ and $\gamma^c \gamma^a \gamma_c = -2\gamma^a$:

$$\frac{i}{4M_P} \left(\bar{\epsilon}\gamma^a \psi\right) \wedge \left(\bar{\psi} \wedge \gamma_5 \gamma_a D\psi\right) = -\frac{i}{8M_P}\left(\bar{\psi} \wedge \gamma_a \psi\right) \wedge \left(\overline{D\psi}\gamma_5 \gamma^a \epsilon\right). \tag{4.23}$$

Altogether, δS can then be written as

$$\delta S = \frac{M_P}{2} \left(De^a - \frac{1}{4M_P^2}\bar{\psi} \wedge \gamma^a \psi\right) \wedge \left[i\overline{D\psi}\gamma_5 \gamma_a \epsilon + M_P \delta\omega^{bc} \wedge e^d \epsilon_{abcd}\right]$$

$$= \frac{M_P}{2}\left(De^a - \frac{1}{4M_P^2}\bar{\psi} \wedge \gamma^a \psi\right) \wedge \left[-\frac{1}{6}\overline{D\psi}\gamma^{bcd}\epsilon + M_P \delta\omega^{bc} \wedge e^d\right]\epsilon_{abcd}, \tag{4.24}$$

where we have used $\gamma_5 \gamma_a = (i/6)\epsilon_{abcd}\gamma^{bcd}$. At this point the variation of the spin connection assumes a primary role, and we can try to set (4.24) to zero in various different ways.

4.1.1   Second-Order Formalism

In this case, one imposes the conventional constraint,

$$De^a = \frac{1}{4M_P^2}\bar{\psi} \wedge \gamma^a \psi, \tag{4.25}$$

which determines the spin connection, $\omega_\mu^{ab} = \hat{\omega}_\mu^{ab}(e, \psi)$, as the solution to this equation. The spin connection is thus treated from the very beginning as a dependent field, whose supersymmetry variation follows from the supersymmetry variations of e_μ^a and ψ_μ via the chain rule and the explicit functional dependence of $\hat{\omega}_\mu^{ab}(e, \psi)$. By simple inspection of (4.24), however, we see that (4.25) already implies that $\delta_\epsilon S = 0$, and we don't really need to know $\delta_\epsilon \omega_\mu^{ab}$.

Let us nevertheless use

$$T^a = \frac{1}{4M_P^2}\bar{\psi} \wedge \gamma^a \psi, \tag{4.26}$$

to solve $De^a = T^a$ for the spin connection, which yields

$$\hat{\omega}_\mu^{ab}(e, \psi) = \omega_\mu^{ab}(e) - \frac{1}{4M_P^2}\left(\bar{\psi}^{[a}\gamma_\mu \psi^{b]} - \bar{\psi}_\mu \gamma^{[a}\psi^{b]} - \bar{\psi}^{[a}\gamma^{b]}\psi_\mu\right). \tag{4.27}$$

This is used in the original approach of Ferrara, Freedman, and van Nieuwenhuizen in [1]. It is interesting to point out that the supersymmetry variation of this connection, inherited from the variations of e_μ^a and ψ_μ, does not contain derivative terms, $\partial\epsilon$, of the supersymmetry parameter:

$$\delta\hat{\omega}_\mu^{ab} = \frac{1}{M_P}\bar{\epsilon}\gamma^\rho\left(D^\sigma\psi^\tau\right)\left(2e_\rho^{[a}e_\tau^{b]}g_{\mu\sigma} - e_\tau^{[a}e_\sigma^{b]}g_{\mu\rho}\right).$$

This is the reason why $\hat{\omega}_\mu^{ab}$ is often called *supercovariant*.

4.1.2   First-Order Formalism

This is the approach followed by Deser and Zumino in their original paper [2].

Asking for the invariance of the action, $\delta S = 0$, via the vanishing of the term in square brackets in (4.24) fixes the variation of the spin connection to

$$\delta\omega_\mu^{bc} = B_\mu^{bc} - \frac{1}{2}e_\mu^c B_e^{be} + \frac{1}{2}e_\mu^b B_e^{ce}, \tag{4.28}$$

with

$$B_\mu^{bc} = \frac{i}{2M_P}\bar{\epsilon}\gamma_\mu\gamma_5 D_\rho\psi_\sigma\,\epsilon^{\rho\sigma bc}. \tag{4.29}$$

This can be extracted using the same trick that was previously used to derive the form of the spin connection in terms of the vielbein from the torsion constraint. It should be noted that (4.28) is not the same as the variation derived using second-order approach in the previous subsection. However, they become equivalent upon using the gravitino equations of motion (see Exercise 4.1).

4.1.3   1.5-Order Formalism

In the 1.5-order formalism, one uses the fact that (4.25) can be obtained as a field equation from varying the action with respect to ω_μ^{ab} (as is obvious from the terms proportional to $\delta\omega^{bc}$ in (4.24)). Thus, when we determine the supersymmetry variation of the action and require ω_μ^{ab} to be determined by

$$De^a = \frac{1}{4M_P^2}\bar{\psi}\wedge\gamma^a\psi \Leftrightarrow \frac{\delta S}{\delta\omega_\mu^{ab}} = 0,$$

we can immediately drop all terms proportional to $\delta\omega_\mu^{ab}$, as these are proportional to $\frac{\delta S}{\delta\omega_\mu^{ab}}$, which vanishes on-shell. Obviously, for the simple action we consider here, the only advantage over the second-order formalism is that we would not have to keep track of the $\delta\omega_\mu^{ab}$ terms in (4.24). Just as in the second-order formalism, the vanishing of the supersymmetry variation (4.24) is thus obtained by using (4.25), with the difference that (4.25) is now not imposed by hand but arises as a *field equation* for the independent field ω_μ^{ab}. It is in this sense that the 1.5-order formalism combines elements from the first-order formalism (the a priori independence of the field ω_μ^{ab}) and from the second-order formalism (the use of (4.25) for the cancellation of (4.24)).

It should be stressed that the 1.5-order trick of using on-shell field equations in the supersymmetry variation can only be used for auxiliary fields such as ω_μ^{ab}.

4.2 Adding a Cosmological Constant

So far we considered the construction of a supergravity action around a Minkowski background, whose non-linear completion led to Einstein gravity coupled to a gravitino field without a cosmological constant. In ordinary Einstein gravity, however, we can always add a cosmological constant Λ to obtain the action

$$S = M_P^2 \int d^4x \sqrt{-g} \left(\frac{1}{2} R - \Lambda \right) \tag{4.30}$$

and find a vacuum with

$$R_{\mu\nu} = \Lambda g_{\mu\nu} \qquad \begin{array}{l} \Lambda > 0 \leftrightarrow \text{de Sitter (dS)}, \\ \\ \Lambda < 0 \leftrightarrow \text{Anti-de Sitter (AdS)}. \end{array} \tag{4.31}$$

It is natural to ask whether these solutions and the corresponding actions can be supersymmetrized in a natural way. In this section, we will focus on pure supergravity theories (without matter multiplets). If we want to construct a supergravity action generalizing (4.30), we should be able to find a supergroup that contains the symmetry group of AdS and/or dS spacetime. We will now see that minimal supersymmetry constrains the closure of the algebra in a way that only one of the two options is consistent.[2]

---

[2] There are consistent de Sitter superalgebras with extended supersymmetries, but they do not allow for positive weight representations, and hence their realizations have the wrong sign in front of the kinetic terms of some of their fields [3,4].

(A)dS Spacetimes. These spaces are maximally symmetric spaces that can be expressed as cosets

$$dS_d = \frac{SO(1,d)}{SO(1,d-1)}, \qquad AdS_d = \frac{SO(2,d-1)}{SO(1,d-1)}. \tag{4.32}$$

Their geometrical properties can be derived by looking at these spaces as hypersurfaces embedded in a spacetime with $d + 1$ dimensions. The embedding equations can be written as

$$\mp (X_d)^2 - (X_0)^2 + \sum_{i=1}^{d-1}(X_i)^2 = \mp \ell^2, \tag{4.33}$$

in a flat $(d + 1)$-dimensional space with signature

$$ds^2 = \mp dX_d^2 - dX_0^2 + \sum_{i=1}^{d-1} dX_i^2, \tag{4.34}$$

where the upper sign describes the AdS space and the lower sign dS and where ℓ is a constant of dimension length that parameterizes the radius of curvature of the (A)dS hypersurface. The d-dimensional metrics can be obtained by choosing a proper parameterization of the $d + 1$ embedding coordinates. A typical (and useful) choice for AdS is given by global coordinates ρ, τ, ξ_i (note that time is compact $\tau \in [0, 2\pi]$)

$$\begin{aligned} X_0 &= \ell \cosh \rho \cos \tau, & X_d &= \ell \cosh \rho \sin \tau, \\ X_i &= \ell \sinh \rho \, \xi_i, & (\textstyle\sum_i \xi_i^2 &= 1), \end{aligned} \tag{4.35}$$

which give a metric covering once the entire hypersurface (4.33)

$$ds^2 = \ell^2(-\cosh^2 \rho \, d\tau^2 + d\rho^2 + \sinh^2 \rho d\Omega^2), \tag{4.36}$$

where $d\Omega^2$ is the line element on the $(d - 2)$-sphere parameterized by the constrained ξ_i coordinates. AdS spacetime as given by the hypersurface (4.33) has closed time-like curves. This can be avoided if one passes over to

(continued)

the universal covering surface. Another useful choice is given by choosing coordinates t, r, x_i,

$$X_0 = \left(1 + r^2(\ell^2 + \vec{x}^2 - t^2)\right)/(2r), \qquad X_d = \ell\, r\, t,$$

$$X_i = \ell\, r\, x_i, \quad (i = 1, \ldots, d-2),$$ (4.37)

$$X_{d-1} = \left(1 - r^2(\ell^2 - \vec{x}^2 + t^2)\right)/(2r),$$

which give a metric that only covers half of the hypersurface (4.33) but is useful in the context of the gauge/gravity correspondence,

$$ds^2 = \ell^2 \frac{dr^2}{r^2} + \ell^2 r^2 \left(-dt^2 + d\vec{x}^2\right).$$ (4.38)

From these metrics we can also derive the Ricci tensor and hence the relation between the cosmological constant in (4.31) and the radius of curvature of the (A)dS hypersurface,

$$\Lambda = -\frac{3}{\ell^2}.$$ (4.39)

Similar derivations apply to dS.

Before discussing the corresponding superalgebras, we note that the symmetry groups of both AdS and dS in d dimensions, $SO(1,d)$ and $SO(2,d-1)$, respectively, can be embedded in $SO(2, d)$. For $d = 4$, the (A)dS algebra is described by ten anti-Hermitian generators M_{AB} satisfying the commutator relations

$$[M_{AB}, M_{CD}] = -2\,\eta_{C[A}M_{B]D} + 2\,\eta_{D[A}M_{B]C},$$ (4.40)

with $\eta_{AB} = \mathrm{diag}\{-+++-\}$ for AdS and $\eta_{AB} = \mathrm{diag}\{-++++\}$ for dS space. The explicit (A)dS algebra follows by identifying $M_{5a} = \ell\, P_a$, where we split $A = \{a, 5\}$, with $a, b, \ldots = 0, 1, 2, 3$, and ℓ is the radius of curvature of (A)dS:

$$[M_{ab}, M_{cd}] = -2\,\eta_{c[a}M_{b]d} + 2\,\eta_{d[a}M_{b]c},$$

$$[P_a, M_{bc}] = 2\,\eta_{a[b}P_{c]},$$ (4.41)

$$[P_a, P_b] = \pm\frac{1}{\ell^2}\,M_{ab},$$

where the last commutator is equivalent to $[M_{5a}, M_{5b}] = -\eta_{55}M_{ab}$ and $\eta_{55} = -1$ for AdS space and $\eta_{55} = +1$ for dS; hence the upper sign is for AdS and the lower one for dS spacetime. Clearly, when $\ell \to \infty$, the (A)dS curvature goes to zero, and one gets back the Poincaré algebra.

To construct the full superalgebra, one needs to specify also the commutators with the supercharges. In particular, $[P_a, Q]$ *cannot* be zero anymore, as it used to be in the super-Poincaré case, because we would no longer close the super Jacobi identities, as

$$\underbrace{[[P_a, P_b], Q]}_{\sim M_{ab}} + \underbrace{[[P_a, Q], P_b]}_{0} - \underbrace{[[P_b, Q], P_a]}_{0} = 0. \tag{4.42}$$

We therefore need to impose new commutator relations such that the momenta do not commute with the supercharges. To respect Lorentz covariance and the graded algebra structure, the result of the commutator should be proportional to the supercharges and come with some gamma matrices. In principle there are two possibilities that respect the Majorana condition on the supercharges

$$[P_a, Q] \sim \gamma_a Q, \qquad \text{or} \qquad [P_a, Q] \sim \gamma_a \gamma_5 Q. \tag{4.43}$$

In the first case, the coefficient multiplying the right-hand side should be real, while in the second it should be imaginary. If we use a chiral notation, the sign and the ambiguities can be reabsorbed in a single dimensionful complex coefficient $\tilde{g}$. We stress this fact, because there are sometimes wrong statements in the literature about this. Once we introduce the chiral notation, the new commutators are

$$[P_a, Q_R] = -\frac{\tilde{g}}{2} \gamma_a Q_L, \qquad [P_a, Q_L] = -\frac{\tilde{g}^*}{2} \gamma_a Q_R. \tag{4.44}$$

Once we introduce these new non-trivial commutators, the super Jacobi identity can be satisfied, though only for the AdS case. This is readily seen by explicitly computing the results of the various commutators:

$$0 \stackrel{!}{=} [P_a, [P_b, Q_L]] + [Q_L, [P_a, P_b]] - [P_b, [P_a, Q_L]]$$
$$= \frac{|g|^2}{4} (\gamma_b \gamma_a - \gamma_a \gamma_b) Q_L \pm \frac{1}{\ell^2} [Q_L, M_{ab}] \tag{4.45}$$
$$= -\frac{|g|^2}{2} \gamma_{ab} Q_L \pm \frac{1}{2\ell^2} \gamma_{ab} Q_L,$$

where we used the commutation relations in (1.53) but with the anti-Hermitian generators $P_a = i \mathcal{P}_a$ and $M_{ab} = i \mathcal{M}_{ab}$. It is now clear that *only for the plus sign* we can get a solution:

$$|g|^2 = \frac{1}{\ell^2}. \tag{4.46}$$

Hence, only for AdS we can write a consistent supersymmetric completion with a single supercharge. We finally note that the closure of the super Jacobi identities requires that another commutator gets modified, namely,

$$\{Q_L, \overline{Q}_L\} = \tilde{g}^* \gamma^{ab} M_{ab}. \tag{4.47}$$

The constraint imposing that the superalgebra can be defined only for the AdS supergroup and not for dS implies a very important fact: a positive cosmological constant will *always* break supersymmetry, while a negative cosmological constant may be compatible with supersymmetry.

Although matter couplings or extended supersymmetries may allow for de Sitter vacua in a supersymmetric theory, the vacuum itself will always break supersymmetry. We will come back to supersymmetry breaking and to the vacuum selection in Chap. 7.

Once supersymmetry is broken, one could describe this phase of the theory by using non-linear realizations, as it is customary for any other symmetry whose linear action is broken. This has been the subject of intense scrutiny (see, for instance, [5–7]), and one can indeed write actions for theories with dS vacua where supersymmetry is non-linearly realized. Since an effective discussion of this topic requires some additional technical introduction to superfields in supergravity, we will not deal with it here but refer the reader to the literature on the subject, such as [8].

4.2.1 Construction of the Action

Now that we established that the anti-de Sitter group can be consistently extended to a supergroup, we would like to realize it in terms of a supersymmetric action that includes a negative cosmological constant. We will proceed in a fashion similar to what has been done in the flat case, starting from the supersymmetry transformation of the gravitino and then trying to close the action of the supersymmetry transformation on the free Lagrangian for the gravity multiplet, possibly introducing interaction terms. We therefore need to fix first the supersymmetry transformation rule of the gravitino. Since the algebra has been modified with respect to the case without cosmological constant, we expect that also the supersymmetry transformations get modified accordingly.

As detailed in Appendix 4.A, we can generically deduce the supersymmetry transformation properties of the various fields from the structure constants of the superalgebra. In fact, gravity in flat space can be thought of as a gauge theory of the Poincaré group (and by extension supergravity of the super-Poincaré group), but with some constraints needed to relate the vielbein and the spin connection degrees of freedom. From this point of view, the vielbein and the spin connection can be thought of as gauge fields for the translation generators and for the Lorentz rotations, while gravitini are the gauge fields for the supersymmetry generators. The transformation rules of the same fields under the various generators of the "gauge algebra" are then fixed by its structure constants in the usual way.

By using this trick, we can now deduce the supersymmetry transformation of the gravitino by looking at the structure constants of the AdS superalgebra coming from commutators which have a supersymmetry generator on the right-hand side. This inspection shows that a new term in the supersymmetry transformation of the gravitino should appear because of the non-zero commutator (4.44) between the translation generators and supersymmetry generators. If one interprets the spin connection term in the Lorentz covariant derivative in the original gravitino transformation (4.5) as due to the non-vanishing commutator of M_{ab} with Q, the new non-vanishing commutator (4.44) between P_a and Q should then analogously lead to an additional contribution to the gravitino transformation so as to make the transformation covariant with respect to the full AdS isometry group. In $\delta\psi_{\mu L}$, this additional contribution should be of the form given in the first equation of (4.44) contracted with the gauge field of the translation generator P_a, i.e., with the vierbein e_μ^a. We therefore should have

$$\delta\psi_{\mu L} = M_P D_\mu \epsilon_L - \frac{g}{2} M_P^2 \gamma_\mu \epsilon_R, \tag{4.48}$$

where now we have a dimensionless constant $g \in \mathbb{C}$, because of the introduction of the dimensionful M_P and M_P^2 factors. Clearly, this has to go together with the conjugate relation:

$$\delta\psi_{\mu R} = M_P D_\mu \epsilon_R - \frac{g^*}{2} M_P^2 \gamma_\mu \epsilon_L. \tag{4.49}$$

Once again we stress that g here can be any complex number, because there are sometimes wrong statements in the literature about this.

To construct the action, we start from the action (4.1) with the vierbein transformation rule

$$\delta e_\mu^a = \frac{1}{2M_P} \bar{\epsilon}_L \gamma^a \psi_{\mu R} + h.c.,$$

and (4.48) for the gravitino

$$\delta\psi_{\mu L} = M_P D_\mu \epsilon_L - \frac{g}{2} M_P^2 \gamma_\mu \epsilon_R.$$

The reason we start from the action without the cosmological constant, rather than adding explicitly the cosmological constant among the bosonic terms right from the beginning, is that it will automatically be enforced by supersymmetry in an iterative procedure at higher order in g, as will become clear momentarily.

The gravitino relation differs from the one in (4.5) by a *shift term* proportional to the constant g. Clearly this shift breaks the supersymmetry of the original action

(4.1), and we need to restore it by adding additional terms to it. In the following, we will establish again the invariance under supersymmetry of a modified action. To our knowledge this was first done in [9].

The first supersymmetry breaking effect of the shift term (proportional to g) is that of generating new terms in the variation of the Rarita–Schwinger part of the Lagrangian. To compute these terms, we use the supersymmetry variation of the conjugate gravitino, which, in form notation, reads

$$
\delta \overline{\psi}_R = M_P \overline{D \epsilon}_R + \frac{g^*}{2} M_P^2 \overline{\epsilon}_L \gamma_a e^a,
\tag{4.50}
$$

as one may easily verify. Denoting by δ_g the variations due to the $\mathcal{O}(g)$ shift term in the gravitino transformation law, the uncancelled variation of $\mathcal{L}_{RS}$ under supersymmetry is then

$$
\begin{aligned}
\delta_g \mathcal{L}_{RS} &= \frac{i}{2} e^a \wedge \delta_g \overline{\psi}_R \wedge \gamma_5 \gamma_a D \psi_L + \frac{i}{2} e^a \wedge \overline{\psi}_L \wedge \gamma_5 \gamma_a D \delta_g \psi_R + h.c. \\
&= \frac{i}{4} g^* M_P^2 \, e^a \wedge e^b \wedge \overline{\epsilon}_L \gamma_b \gamma_5 \gamma_a D \psi_L \\
&\quad - \frac{i}{4} g^* M_P^2 \, e^a \wedge \overline{\psi}_L \wedge \gamma_5 \gamma_a D \left(e^b \gamma_b \epsilon_L \right) + h.c. \\
&= \frac{i}{4} g^* M_P^2 e^a \wedge e^b \wedge \overline{\epsilon}_L \gamma_5 \gamma_{ab} D \psi_L - \frac{i}{4} g^* M_P^2 \, e^a \wedge D e^b \wedge \overline{\psi}_L \gamma_5 \gamma_a \gamma_b \epsilon_L \\
&\quad - \frac{i}{4} g^* M_P^2 \, e^a \wedge e^b \wedge \overline{\psi}_L \wedge \gamma_5 \gamma_{ab} D \epsilon_L + h.c.
\end{aligned}
\tag{4.51}
$$

Integrating by parts the first term in the last equality, we get

$$
\begin{aligned}
\delta_g \mathcal{L}_{RS} &= \frac{i}{2} g^* M_P^2 \, e^a \wedge D e^b \wedge \left(\overline{\epsilon}_L \gamma_5 \gamma_{ab} \psi_L - \frac{1}{2} \overline{\psi}_L \gamma_5 \gamma_a \gamma_b \epsilon_L \right) \\
&\quad - \frac{i}{2} g^* M_P^2 e^a \wedge e^b \wedge \overline{\psi}_L \wedge \gamma_5 \gamma_{ab} D \epsilon_L + h.c.
\end{aligned}
\tag{4.52}
$$

The first term, proportional to $D e^a$, plays a similar role as in the case without a cosmological constant and will be discussed later after Eq. (4.59). Since the remaining terms are proportional to the derivative of the supersymmetry parameter, we can try and use the supersymmetry transformation rule of the gravitini, $\delta \psi_L = M_P D \epsilon_L + \mathcal{O}(g)$, to cancel them. For this reason we add a mass-like term to the Lagrangian:

$$
\mathcal{L}_{\mathcal{M}_\psi} = \frac{i}{4} g^* M_P \, e^a \wedge e^b \wedge \overline{\psi}_L \wedge \gamma_5 \gamma_{ab} \psi_L + \frac{i}{4} g \, M_P \, e^a \wedge e^b \wedge \overline{\psi}_R \wedge \gamma_5 \gamma_{ab} \psi_R,
\tag{4.53}
$$

so that the variation of ψ in (4.53) compensates for (4.52) at order g. We point out here the reduced Planck mass factors, so that the whole coefficient has mass dimension 1, as well as the overall factor $1/4$, which is due to the double variation required to match (4.52).

While the introduction of (4.53) allows the cancellation of the $D\varepsilon$ terms in (4.52), it also gives rise to two further new variations we have to take care of. One variation comes at order g from the variation of the vierbein in (4.53). We will discuss its cancellation together with the cancellation of the De^a terms in (4.52) further below. The other new variation of (4.53) is of order g^2 and arises when the order g shift term in the gravitino transformation is used in the variation of the gravitini in (4.53). Indeed, the order g^2 variation of the gravitino mass term produces (suppressing the wedges)

$$\delta_g \mathscr{L}_{M_\psi} = -\frac{i}{8}|g|^2 M_P^3 e^a e^b \left(\overline{\psi}_L \gamma_5 \gamma_{ab} e^c \gamma_c \epsilon_R - \overline{\epsilon}_R e^c \gamma_c \gamma_5 \gamma_{ab} \psi_L\right) + h.c. \quad (4.54)$$

Putting together the gamma matrices and using the duality relation $\gamma_5 \gamma_{abc} = -i\,\epsilon_{abcd}\gamma^d$, we obtain

$$\begin{aligned}
\delta_g \mathscr{L}_{M_\psi} &= M_P^3 \frac{|g|^2}{4} e^a e^b e^c \epsilon_{abcd} \left(\overline{\psi}_L \gamma^d \epsilon_R + \overline{\psi}_R \gamma^d \epsilon_L\right) \\
&= -M_P^4 \frac{|g|^2}{2} e^a e^b e^c \epsilon_{abcd} \frac{\overline{\epsilon}_R \gamma^d \psi_L + \overline{\epsilon}_L \gamma^d \psi_R}{2M_P} \\
&= -M_P^4 \frac{|g|^2}{2} e^a e^b e^c \epsilon_{abcd} \delta e^d \\
&= -|g|^2 \frac{M_P^4}{8} \delta \left(e^a e^b e^c e^d \epsilon_{abcd}\right).
\end{aligned} \quad (4.55)$$

We further recall that $e^a e^b e^c e^d \epsilon_{abcd} = +4!\, d^4x\, e$ and then realize that we need to add a single term of order $|g|^2$ to the Lagrangian to cancel (4.55):

$$3 \int d^4x\, e\, M_P^4 |g|^2 = -M_P^2 \int d^4x\, e\, \Lambda. \quad (4.56)$$

This is a *cosmological constant term*. Notice that there is **no choice of the sign** of this cosmological constant

$$\Lambda = -3\, M_P^2\, |g|^2 = -\frac{3}{\ell^2} < 0. \quad (4.57)$$

This agrees with the discussion following from the supersymmetry algebra.

It is also extremely important to note that the variation of (4.56) does not generate terms of order g^3 and that therefore supersymmetry closes at order g^2.

The only variation left is the vierbein variation in the gravitino mass term together with the already mentioned first term in (4.52). Using

$$4(\overline{\psi}_R \wedge \gamma_{ab}\psi_R) \wedge (\overline{\varepsilon}\gamma^b\psi) = 3(\overline{\psi}_R\gamma_{ab}\varepsilon_R) \wedge (\overline{\psi} \wedge \gamma^b\psi) + (\overline{\psi}_R\varepsilon_R) \wedge (\overline{\psi} \wedge \gamma_a\psi),$$

(4.58)

which one can derive from the Fierz identities (1.48)–(1.52), one obtains for all remaining uncancelled variations

$$\delta\mathscr{L} = \frac{M_P}{2}\left(De^a - \frac{1}{4M_P^2}\overline{\psi} \wedge \gamma^a\psi\right)$$

$$\wedge\left[i\, g^* M_P\, e^b \wedge \left(\overline{\varepsilon}_L\gamma_{ba}\psi_L - \frac{1}{2}\overline{\psi}_L\gamma_b\gamma_a\varepsilon_L\right)\right.$$

$$-i\, g\, M_P\, e^b \wedge \left(\overline{\varepsilon}_R\gamma_{ba}\psi_R - \frac{1}{2}\overline{\psi}_R\gamma_b\gamma_a\varepsilon_R\right)$$

$$\left.+\epsilon_{abcd}\left(-\frac{1}{6}\overline{D\psi}\gamma^{bcd}\epsilon + M_P\delta\omega^{bc} \wedge e^d\right)\right],$$

(4.59)

where the last term proportional to ε_{abcd} is the same as in the case without cosmological constant.

This completes the proof of the invariance of the action in any of the formalisms described above. In the second-order formalism, this variation vanishes because of the torsion constraint. In the first-order formalism, we deduce from this variation the expression for $\delta\omega^{ab}$ that makes it vanish. Finally, in the 1.5 formalism, the equations of motion for the spin connection do not change, and hence once again the full Lagrangian is invariant under supersymmetry.

The final Lagrangian is therefore the following

$$\mathscr{L} = \frac{M_P^2}{2}eR - \frac{e}{2}\overline{\psi}_{\mu R}\gamma^{\mu\nu\rho}D_\nu\psi_{\rho L} - \frac{e}{2}\overline{\psi}_{\mu L}\gamma^{\mu\nu\rho}D_\nu\psi_{\rho R}$$

$$-e\, M_P\, \frac{g}{2}\overline{\psi}_{\mu R}\gamma^{\mu\nu}\psi_{\nu R} - e\, M_P\, \frac{g^*}{2}\overline{\psi}_{\mu L}\gamma^{\mu\nu}\psi_{\nu L}$$

(4.60)

$$+3\, e\, M_P^4\, |g|^2,$$

where we should remember that in the second-order formalism $\omega^{ab} = \omega^{ab}(e, \psi)$, and the final supersymmetry transformations are

$$\delta e_\mu^a = \frac{1}{2M_P}\overline{\varepsilon}_L\gamma^a\psi_{\mu R} + h.c.,$$

(4.61)

$$\delta\psi_{\mu L} = M_P D_\mu\varepsilon_L - \frac{g}{2}M_P^2\gamma_\mu\varepsilon_R.$$

(4.62)

Now that we completed the construction of a supersymmetric action for supergravity with a (negative) cosmological constant, we can make some comments.

First of all, we have seen from the construction that we have performed only minimal modifications. After shifting the supersymmetry transformation of the gravitino field, we introduced the smallest set of terms needed to cancel $\mathcal{O}(g)$ and $\mathcal{O}(g^2)$ terms in the supersymmetry variations. As already pointed out above, supersymmetry closes at order $|g|^2$. There is no need to introduce any term of order g^3 or more.

All the modifications can be summarized in three main pieces:

- A shift in the fermionic supersymmetry rules at $\mathcal{O}(g)$
- A mass-like term of $\mathcal{O}(g)$ for the fermions
- A potential term at order $\mathcal{O}(g^2)$

Although these modifications have been forced by the presence of the cosmological constant in a pure gravity theory, we will see that the pattern outlined above is actually underlying all gauged supergravity theories for models with extended supersymmetries. In fact, as we will see, in extended supergravities the appearance of non-Abelian gauge groups is tied to the presence of a non-trivial scalar potential, which may act as an effective cosmological constant. The result is that the gauging procedure introduces the same three main modifications listed above, where the mass-like term for the fermions in the general case with scalar fields becomes a Yukawa-like coupling and the cosmological constant term becomes a scalar potential (see Sect. 9.1).

It can also be seen that this scalar potential (in this case a pure cosmological constant) can be expressed as the square of the shifts of the supersymmetry variations of the fermionic fields:

$$V \, \overline{\epsilon}_L \gamma^a \epsilon_R = -3 M_P^4 |g|^2 \overline{\epsilon}_L \gamma^a \epsilon_R = -\frac{3}{2} \delta_g \overline{\psi}_{\mu R} \gamma^a \delta_g \psi_L^\mu . \tag{4.63}$$

Note the minus sign in front of the squared gravitino shifts. This identity is called the supersymmetric Ward identity [10].

4.2.2   Mass in AdS

A further crucial point in establishing susy invariance is that the gravitino has to have a mass term proportional to the square root of the cosmological constant

$$-\frac{g}{2} M_P \, \overline{\psi}_{\mu R} \gamma^{\mu\nu} \psi_{\nu R} - \frac{g^*}{2} M_P \, \overline{\psi}_{\mu L} \gamma^{\mu\nu} \psi_{\nu L} . \tag{4.64}$$

So we have a massless graviton $g_{\mu\nu}$, a "massive" gravitino ψ_μ, but we still preserve supersymmetry! This is a consequence of a general point that should always be remembered in supergravity: *Supersymmetry multiplets are degenerate in mass only in Minkowski space.* On curved spaces we can have different "masses" for different components of the same multiplet, where the quotes here refer to the fact that the whole concept of mass becomes somewhat ambiguous in curved spacetimes when the masses are comparable to the curvature scale.

Again, the superalgebra construction comes to our help in explaining this fact. Without a cosmological constant, supergravity is invariant under the super-Poincaré group. In the corresponding superalgebra, the supersymmetry generator and the momentum commute, $[Q, P] = 0$. This means that also P^2 commutes with the supersymmetry generator,

$$[P^2, Q] = 0,$$

and hence P^2, defining the mass, is a good Casimir operator, and it will take the same value on all the fields of the same supersymmetry multiplet.

This is not true for the AdS superalgebra. In this case P^2 does not commute with the supersymmetry generator, and therefore it is not the same for states in the same multiplet. Actually, P^2 is not even an SO(2,3) invariant, and therefore we cannot even classify states by that quantity.

There is, however, another good invariant for SO(2,3), which is defined as

$$\mathscr{C} = -\frac{1}{2} M_{\underline{AB}} M^{\underline{AB}}, \tag{4.65}$$

where the SO(2,3) generators are taken anti-Hermitian

$$M^\dagger_{\underline{AB}} = -M_{\underline{AB}} \tag{4.66}$$

and are represented by $M_{\underline{AB}} = \{M_{ab}, M_{5a} = \ell\, P_a\}$. Since $\mathscr{C}$ is a good invariant, we can classify states by its value. It would then be important to understand if we can associate to the values of $\mathscr{C} = -\frac{1}{2}\left(M_{ab}M^{ab} + 2\ell^2 P_a P^a\right)$ something we can call *mass* and *spin* of the field. SO(2,3) has SO(2) × SO(3) as the maximal compact subgroup. The states can thus be labeled by E and s, where E is the SO(2) $\simeq$ U(1) energy eigenvalue and s is the spin. Recall that global AdS has closed time-like curves.[3] Hence the generator of translations along the time-like circles, namely, M_{50}, can be interpreted as a dimensionless energy operator

$$H = i\ell\, P^0 = -i\, M_{50} \tag{4.67}$$

---

[3] In practice, we always consider its universal covering space, CAdS, as mentioned earlier.

and we call its eigenvalues E. The generators of three-dimensional spatial rotations are the generators M_{ij}, for $i, j = 1, 2, 3$, which commute with H. The remaining generators can be combined into three pairs of mutually conjugate operators

$$L_i^{\pm} = -i\,M_{0i} \pm M_{5i},   \tag{4.68}$$

so that the commutators become

$$[H, L_i^{\pm}] = \pm L_i^{\pm},   \tag{4.69}$$

$$[L_i^+, L_j^-] = -2H\delta_{ij} - 2M_{ij},   \tag{4.70}$$

$$[L_i^{\pm}, L_j^{\pm}] = 0.   \tag{4.71}$$

Clearly, the generators $L_i^{\pm}$ are playing the role of raising (lowering) operators for E. When applied to an eigenstate of H with eigenvalue $E = E_0$, they give states with eigenvalues $E = E_0 \pm 1$. Assuming that the spectrum of H is bounded from below, the lowest eigenvalue E_0 is realized by a state $|E_0, s_0\rangle$ satisfying $L_i^- |E_0, s_0\rangle = 0$, for any i. From this ground state, one can then construct the other states by application of products of creation operators L_i^+.

In this basis, the Casimir operator can be written as (sum over i implied)

$$\mathcal{C} = H^2 + J^2 - \frac{1}{2}L_i^+ L_i^- - \frac{1}{2}L_i^- L_i^+,   \tag{4.72}$$

where $J^2 = -\frac{1}{2}M_{ij}M^{ij}$. Using the commutator relations above and $M_{ii} = 0$, this simplifies to

$$\mathcal{C} = H(H - 3) + J^2 - L_i^+ L_i^-.   \tag{4.73}$$

On the lowest energy state, we obtain that

$$\mathcal{C}|E_0, s_0\rangle = -\frac{1}{2} M_{\underline{AB}}M^{\underline{AB}}|E_0, s_0\rangle = [E_0(E_0 - 3) + s_0(s_0 + 1)]\,|E_0, s_0\rangle.   \tag{4.74}$$

Actually, any state belonging to the corresponding unitary irreducible representation must satisfy

$$\langle E, s|\mathcal{C}|E, s\rangle = [E_0(E_0 - 3) + s_0(s_0 + 1)]\,\langle E, s|E, s\rangle,   \tag{4.75}$$

precisely because $\mathcal{C}$ is a Casimir operator of the AdS algebra. This allows us to obtain a unitarity constraint on the allowed quantum numbers. In order to get it, we use (4.75) for an excited state with energy $E = E_0 + 1$ and spin $s = s_0 - 1$ (assuming the ground state has $s_0 \geq 1$) and compare the result with the direct computation of $\mathscr{C}$ on the same state:

$$\langle E_0 + 1, s_0 - 1 | \mathscr{C} | E_0 + 1, s_0 - 1 \rangle \stackrel{(4.75)}{=} \mathscr{N}^2 \left[E_0(E_0 - 3) + s_0(s_0 + 1) \right]$$

$$\stackrel{(4.73)}{=} \langle E_0 + 1, s_0 - 1 | (E_0 + 1)(E_0 - 2) + (s_0 - 1)s_0 | E_0 + 1, s_0 - 1 \rangle$$

$$- |L_i^- | E_0 + 1, s_0 - 1 \rangle|^2,$$

$$(4.76)$$

where $\mathscr{N}^2 = \langle E_0 + 1, s_0 - 1 | E_0 + 1, s_0 - 1 \rangle$ is the norm squared of the state. This implies that

$$|L_i^- | E_0 + 1, s_0 - 1 \rangle|^2 = 2(E_0 - s_0 - 1)\, \mathscr{N}^2 \geq 0 \qquad (4.77)$$

and shows that in order to have unitary representations, one has to require

$$E_0 \geq s_0 + 1. \qquad (4.78)$$

Actually, the AdS algebra has a special unitary representation, which is the singleton,[4] for $s_0 = 0$ and $E_0 = 1/2$ (together with its partner $s_0 = 1/2$, $E_0 = 1$ state in the supersymmetric case), while all the other representations have to fulfill the above unitarity bound [12].

If we come to the definition of mass, we should stress that there is no unambiguous definition. In fact, there are two main different conventions in the literature.

Most of the older supergravity literature identifies massless representations with those living at the boundary of the unitarity bound threshold, i.e., with $E_0 = s_0 + 1$. For the ground states, this means

$$\mathscr{C} | m = 0, E_0 = s_0 + 1, s_0 \rangle$$

$$= -\frac{1}{2} M_{AB} M^{AB} | m = 0, E_0 = s_0 + 1, s_0 \rangle$$

$$= [(s_0 + 1)(s_0 - 2) + s_0(s_0 + 1)] | m = 0, E_0 = s_0 + 1, s_0 \rangle$$

$$= 2(s_0^2 - 1) | m = 0, E_0 = s_0 + 1, s_0 \rangle. \qquad (4.79)$$

---

[4] The singleton representation is a very special representation of the AdS algebra, with no Poincaré counterpart. Its four-dimensional field representations are pure gauge degrees of freedom and do not propagate in the bulk of spacetime. However, any massless field in AdS can be constructed by taking the product of two of them [11].

Setting therefore $m^2 = 0$ for these representations, one can define the mass in AdS as

$$m^2 \ell^2 = -\frac{1}{2} M_{\underline{AB}} M^{\underline{AB}} - 2(s_0^2 - 1) = E_0(E_0 - 3) - (s_0 + 1)(s_0 - 2). \quad (4.80)$$

Another natural convention that is widely used more recently, especially after the birth of the AdS/CFT correspondence, is to simply identify the mass with the result of the direct computation of the Laplace–Beltrami operator. In this case, the relations between the masses and the (E_0, s_0) quantum numbers are [13, 14]

$$m_0^2 \ell^2 = E_0(E_0 - 3), \quad (4.81)$$

$$m_{1/2} \ell = \left| E_0 - \frac{3}{2} \right|. \quad (4.82)$$

$$m_1^2 \ell^2 = (E_0 - 2)(E_0 - 1), \quad (4.83)$$

$$m_{3/2} \ell = \left| E_0 - \frac{3}{2} \right|, \quad (4.84)$$

$$m_2^2 \ell^2 = E_0(E_0 - 3). \quad (4.85)$$

In the case of scalar fields, (4.81) gives an interesting outcome if we invert the relation and write the energy E_0 as a function of the mass

$$E_0 = \frac{3}{2} \pm \sqrt{\frac{9}{4} + m^2 \ell^2}. \quad (4.86)$$

It is easy to see that in order to satisfy the unitarity bound $E_0 \geq 1$, we can have both the plus and minus signs, as long as one allows negative squared masses, $m^2 < 0$, satisfying

$$m^2 \geq -\frac{9}{4} \frac{1}{\ell^2} = -\frac{3}{4} |\Lambda|. \quad (4.87)$$

This is the *Breitenlohner–Freedman bound* [15], and it is easy to check that it corresponds to a minimum of the Casimir operator for $s = 0$. This bound is very important in establishing the stability of vacua with a negative cosmological constant. In generic field theories, scalar fields with negative mass signal instabilities of the vacuum against small fluctuations of the same fields. In a theory with many scalar fields, this is related to the fact that the scalar potential is at a local maximum or a saddle point rather than at a local minimum. However, for a gravity theory, a maximum or a saddle point of the potential with a negative value of the cosmological constant can be stable against small fluctuations, provided the negative eigenvalues of the mass matrix satisfy the Breitenlohner–Freedman bound, so that the fluctuations do not violate unitarity. The outcome of this discussion is once more that, whatever convention we use for the definition of mass, the true parameters that unambiguously label the irreducible representations are (E_0, s_0), while m is just a bookkeeping parameter related to the wave equation or to the AdS algebra.

Everything we discussed so far is valid in any matter coupled gravity theory around an anti-de Sitter vacuum. Supersymmetry further imposes additional interesting constraints. For instance, supersymmetric vacua with a negative cosmological constant always satisfy the Breitenlohner–Freedman bound, and therefore they always provide stable vacua, also for potentials unbounded from below (which is often the case in supergravity). Moreover, fields come organized in supermultiplets. In four dimensions (with minimal supersymmetry), there are three possible multiplets [12]:

- **The singleton multiplet:** this representation contains only one bosonic and one fermionic AdS representation with ground states

$$|E_0 = 1/2, s_0 = 0\rangle, \quad |E_0 = 1, s_0 = 1/2\rangle. \tag{4.88}$$

- **Regular multiplets:** here the ground states are

$$|E_0 = \Delta, s = s_0\rangle, \quad |\Delta + 1/2, s_0 + 1/2\rangle, \quad |\Delta + 1/2, s_0 - 1/2\rangle, \quad |\Delta + 1, s_0\rangle, \tag{4.89}$$

and they form **massless multiplets** if $E_0 = s_0 + 1$; otherwise they form **massive multiplets**. Notice also that the states with spin $s_0 - 1/2$ are obviously not present in the case of chiral multiplet representations.

We therefore see that, for instance, according to (4.81)–(4.85), a massive vector multiplet whose lowest energy state has $E_0 = 5/2$ and $s_0 = 1/2$ contains a massless scalar field, two fermions of masses $m\ell = 1$ and $m\ell = 2$, and a vector field of mass squared $m^2\ell^2 = 2$.

4.A Appendix: Gauging the Poincaré Algebra

When introducing general relativity as well as supergravity, we discussed the possibility of considering the vierbein and the spin connection as independent quantities. Do we have any conceptual reason behind this, in addition to the simplification of some computations? We will now see that an interesting perspective on gravity, which can help when dealing with supergravity, is that of considering gravity itself as a sort of a gauge theory where the gauge group is the Poincaré group. This analogy will work only to a certain extent, but it will be very useful for understanding many specific new features that have to be introduced when one wants to promote supersymmetry to a local symmetry of nature. In fact, supergravity *is* the gauge theory of supersymmetry, and therefore there must be a way to describe it as a theory where the gauge group is the Poincaré supergroup (or some other supergroup). For the sake of simplicity in this appendix, we set $M_P = 1$.

Consider an ordinary gauge transformation $\delta_\epsilon = \epsilon^A T_A$, where T_A are the gauge generators satisfying

$$[T_A, T_B] = f_{AB}{}^C T_C.$$

If this is a global symmetry of an action, it can be made local by introducing vector fields A_μ^A for each symmetry so that the algebra

$$[\delta(\epsilon_1^A), \delta(\epsilon_2^B)] = \delta\left(\epsilon_2^B \epsilon_1^A f_{AB}{}^C\right), \tag{4.90}$$

where $f_{AB}{}^C$ are the structure constants, has a faithful realization on them,

$$\delta_\epsilon A_\mu^A = \partial_\mu \epsilon^A + \epsilon^C A_\mu^B f_{BC}{}^A, \tag{4.91}$$

and we can introduce covariant derivatives

$$D_\mu = \partial_\mu - A_\mu^A T_A \tag{4.92}$$

acting non-trivially on fields which transform in non-trivial representations of the gauge group. The curvature, defined as

$$[D_\mu, D_\nu] = -F_{\mu\nu}^A T^A \qquad \Leftrightarrow \qquad F_{\mu\nu}^A \equiv 2\partial_{[\mu} A_{\nu]}^A + A_\mu^B A_\nu^C f_{BC}{}^A, \tag{4.93}$$

transforms covariantly:

$$\delta_\epsilon F_{\mu\nu}^A = \epsilon^C F_{\mu\nu}^B f_{BC}{}^A. \tag{4.94}$$

Let us now imagine that we want to make local the symmetries of the Poincaré group. The usual procedure is to introduce gauge fields in correspondence with the generators of the algebra. For the Poincaré algebra, this means introducing two fields, one e_μ^a with a vector gauge index and one ω_μ^{ab} with two, so as to match the gauge generators

$$A_\mu^A T_A = e_\mu^a P_a + \frac{1}{2} \omega_\mu^{ab} M_{ab}. \tag{4.95}$$

The gauge curvatures of these vectors are precisely T^a and R^{ab} as defined in (3.22) and (3.23), where the spin connection and the vielbein are so far independent fields.

This construction is perfectly legitimate. However, it clearly leads to an ordinary gauge theory and not to a gravity theory as we would like. From the T^a and R^{ab} curvatures, we could construct kinetic terms giving the propagation of independent degrees of freedom and discuss the resulting gauge theory, where the Poincaré group is realized on the vector fields as

$$\delta_P \omega^{ab} = 0, \tag{4.96}$$

$$\delta_P e^a = D\epsilon^a, \tag{4.97}$$

$$\delta_M \omega^{ab} = d\Lambda^{ab} - \omega^a{}_c \Lambda^{cb} - \Lambda^{ac} \omega_c{}^b, \tag{4.98}$$

$$\delta_M e^a = \Lambda^a{}_c e^c. \tag{4.99}$$

If, on the other hand, we want to get only the metric degrees of freedom, we have to impose a constraint between ω^{ab} and e^a. This is the *conventional constraint* or torsion constraint

$$T^a = 0. \tag{4.100}$$

This constraint however is not invariant under (4.96)–(4.97):

$$\delta_P T^a = \delta_P De^a = D\delta_P e^a = DD\epsilon^a = -R^a{}_b \epsilon^b \neq 0. \tag{4.101}$$

This means that if we impose the conventional constraint, translation symmetry is broken. Moreover, it is also clear that now the spin connection ω^{ab} cannot be treated as independent of the vielbein anymore and hence the transformation (4.96) will not be valid anymore. Since $\omega^{ab} = \omega^{ab}(e)$, the spin connection is also not invariant under translations

$$\delta_P \omega^{ab} = \int d^4 x \, \frac{\delta \omega^{ab}}{\delta e^c} \delta_P e^c \neq 0.$$

The final outcome of this discussion is that, when the conventional constraint is imposed, the Poincaré gauge algebra is deformed and translational symmetry is replaced by a new invariance under diffeomorphisms. This can be seen by considering the commutator of two translation generators on the vielbein:

$$[\delta_{P2}, \delta_{P1}] e^a = -\delta_{P[2}(D\epsilon^a_{1]}) = -\delta_{P[2} \omega^a{}_c \epsilon^c_{1]}, \tag{4.102}$$

and now $\delta_P \omega^{ab} \neq 0$. The resulting algebra then has a non-vanishing commutator

$$[P, P] \neq 0,$$

as is appropriate for general coordinate transformations, which do not commute. Actually one can check that using the constraint (4.100), the translation generators on the vielbein take the form of general coordinate transformations.

The Lie derivative of a p-form A_p along the flow of a vector field V is defined as

$$\mathcal{L}_V A_p = \lim_{t \to 0} \frac{1}{t} \left(\sigma_t^* A_p(\sigma_t(x)) - A_p(x) \right),$$

where σ_t^* is the pullback of the differential form along the flow generated by the vector field V. When applied to a scalar valued p-form, this reduces to

$$\mathcal{L}_V A_p = (\iota_V d + d \iota_V) A_p.$$

The action constructed from the curvatures and the vielbein then is invariant with respect to local Lorentz group transformations and diffeomorphisms. The infinitesimal change of a function under a diffeomorphism is given by the Lie derivative $\mathcal{L}_\epsilon$, and therefore the action is going to be invariant if

$$\mathcal{L}_\epsilon S = \int d(\iota_\epsilon \mathcal{L}) + \int \iota_\epsilon d\mathcal{L} = 0,$$

but the first term is a total derivative that can be discarded while the second is zero because $d\mathcal{L}$ has one degree more than the top form. Finally, in the construction of an action, we will not make use of a kinetic term of the form $R^{ab} \wedge \star R_{ab}$ because of the conventional constraint which makes it quartic in the derivatives. The appropriate quadratic term is the Einstein–Hilbert action above.

The method we have outlined in this section can be easily extended to generic supergravity theories by extending the Poincaré algebra to the super Poincaré algebra, by including fermionic generators and possibly other bosonic generators for the internal symmetries. The power of this approach lies in the ease of guessing the transformation laws under the various symmetries, including supersymmetry. This means that this approach can be used as a guide to derive and construct the Lagrangian and/or the equations of motion of systems respecting any symmetry group we would like to realize. Once again, we stress that one has to be careful with its application because of the constraints that will be needed to obtain a consistent gravity theory (invariant under diffeomorphisms). Imposing these constraints will break the transformation rules that do not preserve them.

4.A.1 Gauging the Super Poincaré Algebra

We end this appendix with a few remarks on the gauging of the *super* Poincaré group. We could gauge this algebra by adding new vector fields $\psi_{\mu A}$ for the fermionic generators Q^A. From the algebra we then have

$$A_\mu^A T_A = e_\mu^a P_a + \frac{1}{2} \omega_\mu^{ab} M_{ab} + \overline{\psi}_{\mu A} Q^A + \overline{\psi}_\mu^A Q_A, \qquad (4.103)$$

and we can read the supersymmetry transformations by applying

$$\delta_\epsilon A_\mu^A = \partial_\mu \epsilon^A + \epsilon^C A_\mu^B f_{BC}{}^A.$$

For instance, for $\mathcal{N} = 1$ supergravity, we would get that the spin connection is invariant,

$$\delta_\epsilon \omega^{ab} = 0, \qquad (4.104)$$

because the Lorentz generator never appears on the right-hand side of any commutator involving the supersymmetry generator. However, just like for the bosonic case, we should impose a torsional constraint in order for the vielbein and spin connection not to be independent. Doing so, we fix the form of the spin connection as $\omega^{ab} = \omega^{ab}(e, \psi)$ and check the new realization of the algebra on the fields.

One last interesting remark involves the definition of the gauge curvatures for the super-Poincaré algebra. From the structure constants of the supersymmetry algebra (1.53), one can deduce a new definition for the curvatures, including the one of the translation generators, namely, the torsion T^a. Since the translation generators P_a appear on the right-hand side of the commutator of two supercharges, the corresponding curvature definition is now

$$T^a = De^a - \frac{1}{4} \overline{\psi}^A \gamma^a \psi_A \qquad (4.105)$$

and involves a fermion bilinear. This means that imposing the conventional constraint $T^a = 0$ results in a spin connection depending on the gravitino fields. Hence, supergravity is often referred to as a theory with non-trivial torsion for the spin connection, because $T^a = 0$ implies $De^a \neq 0$.

Exercises

4.1. Prove that the supersymmetry variation of the spin connection in the first-order formalism and in the second-order formalism is equivalent upon using the gravitino equation of motion.

4.2. In the first-order formalism, compute the new piece in the variation of the spin connection due to the cosmological constant.

4.3. Find the embedding coordinates that give origin to

$$ds^2 = -dt^2 + e^{2t/\ell}\, d\vec{x}^2,$$

for de Sitter and to

$$ds^2 = -dt^2 + \ell^2 \sin^2(t/\ell)\left(d\psi^2 + \sinh^2\psi\, d\Omega^2\right),$$

for anti-de Sitter spacetime. Discuss the "cosmological" meaning of these metrics.

4.4. Check that the gauge curvatures T^a and R^{ab} coming from (4.95) and (4.93) match (3.22) and (3.23).

References

1. D.Z. Freedman, P. van Nieuwenhuizen, S. Ferrara, Progress toward a theory of supergravity. Phys. Rev. **D13**, 3214–3218 (1976)
2. S. Deser, B. Zumino, Consistent supergravity. Phys. Lett. **B62**, 335 (1976)
3. S. Ferrara, Algebraic properties of extended supergravity in de Sitter space. Phys. Lett. **B69**, 481 (1977)
4. K. Pilch, P. van Nieuwenhuizen, M.F. Sohnius, De Sitter superalgebras and supergravity. Commun. Math. Phys. **98**, 105 (1985)
5. E.A. Bergshoeff, D.Z. Freedman, R. Kallosh, A. Van Proeyen, Pure de Sitter supergravity. Phys. Rev. D **92**(8), 085040 (2015) [Erratum: Phys. Rev. D **93**(6), 069901 (2016)] [arXiv:1507.08264 [hep-th]]
6. G. Dall'Agata, E. Dudas, F. Farakos, On the origin of constrained superfields. JHEP **05**, 041 (2016) [arXiv:1603.03416 [hep-th]]
7. N. Cribiori, G. Dall'Agata, F. Farakos, From linear to non-linear SUSY and back again. JHEP **08**, 117 (2017) [arXiv:1704.07387 [hep-th]]
8. J. Wess, J. Bagger, *Supersymmetry and Supergravity* (Princeton University Press, Princeton, 1992)
9. P.K. Townsend, Cosmological constant in supergravity. Phys. Rev. **D15**, 2802–2804 (1977)
10. S. Cecotti, L. Girardello, M. Porrati, Constraints on partial superhiggs. Nucl. Phys. **B268**, 295–316 (1986)
11. M. Flato, C. Fronsdal, Lett. Math. Phys. **2**, 421 (1978); Phys. Lett. **97B**, 236 (1980); J. Geom. Phys. **6**, 294 (1988)
12. H. Nicolai, Representations of supersymmetry in Anti-de Sitter space, in *Trieste 1984, Proceedings, Supersymmetry and Supergravity '84*, 368–399 and CERN Geneva – TH. 3882 (84, REC.JUL.) 32 p.
13. S. Ferrara, C. Fronsdal, A. Zaffaroni, On N = 8 supergravity on AdS(5) and N = 4 superconformal Yang-Mills theory. Nucl. Phys. B **532**, 153–162 (1998) [arXiv:hep-th/9802203 [hep-th]]
14. O. Aharony, S.S. Gubser, J.M. Maldacena, H. Ooguri, Y. Oz, Large N field theories, string theory and gravity. Phys. Rept. **323**, 183–386 (2000) [arXiv:hep-th/9905111 [hep-th]]
15. P. Breitenlohner, D.Z. Freedman, Positive energy in anti-De Sitter backgrounds and gauged extended supergravity. Phys. Lett. B **115**, 197–201 (1982)

Part II

Matter Couplings and Phenomenology

Matter Couplings in Global Supersymmetry   **5**

In the previous chapter, we discussed $\mathcal{N} = 1$ pure supergravity, which only involves the $\mathcal{N} = 1$ supergravity multiplet. In order to include also ordinary matter fields and Yang–Mills-type gauge interactions, one has to couple the pure supergravity sector to $\mathcal{N} = 1$ chiral and vector multiplets in a way consistent with local supersymmetry. Some aspects of the coupling of free chiral multiplets to supergravity were already sketched in Chap. 2. It is the purpose of the following two chapters to extend this preliminary discussion to a more systematic and complete analysis of the general matter couplings in 4D, $\mathcal{N} = 1$ supergravity. As a preparation, we first discuss, in this chapter, general matter couplings in global supersymmetry.

5.1   Our Approach

Our way to introduce matter couplings in global and local supersymmetry is motivated by two main goals:

- Derive general matter couplings in supergravity in an as direct and physical way as possible so that it can also be readily generalized to extended supergravity and higher dimensions.
- Exhibit the differences between matter couplings in global and local supersymmetry as clearly as possible.

© Springer-Verlag GmbH Germany, part of Springer Nature 2021
G. Dall'Agata, M. Zagermann, *Supergravity*, Lecture Notes in Physics 991,
https://doi.org/10.1007/978-3-662-63980-1_5

In order to reach the first goal, we will consistently work with on-shell component fields and four-component spinors, using a combination of the Noether method and geometrical reasoning, i.e., methods that are easy to generalize also to extended supergravity and other spacetime dimensions with a minimal theoretical apparatus. In order to reach the second goal, we will introduce these methods and our conventions already at the level of global supersymmetry, i.e., in a context most readers will have some familiarity with. This also includes a discussion of non-renormalizable interactions already at the level of global supersymmetry. Most of these non-renormalizable interactions can be nicely packaged in terms of elegant geometrical structures that greatly facilitate the transition to local supersymmetry.

Our approach is to iteratively construct the most general matter couplings by first discussing the most important building blocks in isolation and then show how they can be patched together to obtain the general theory. We will do this first for global supersymmetry and then, later, in a completely analogous way also for supergravity.

5.2 Chiral Multiplets in Global Supersymmetry

In this section, we discuss the possible couplings of chiral multiplets in global supersymmetry.

5.2.1 The Renormalizable Wess–Zumino Model

We already encountered the Wess–Zumino model in Sect. 2.1. This model can be generalized in various ways. The most general *renormalizable* Wess–Zumino model describes n_C chiral multiplets (ϕ^m, χ^m) $(m = 1, \ldots, n_C)$ whose mass and interaction terms (Yukawa interactions as well as cubic and quartic scalar potentials) are all encoded in a single holomorphic function, $W(\phi^m)$, the *superpotential*. Renormalizability restricts $W(\phi^m)$ to be a polynomial of at most cubic order. In the following, we use barred indices $\overline{m}, \overline{m}, \ldots$ to denote the complex conjugates of the scalar fields, ϕ^m, and the right-handed projections of the Majorana fermions, χ^m, and write:

$$\phi^{\overline{n}} \equiv (\phi^n)^*$$

$$\partial_m \equiv \frac{\partial}{\partial \phi^m}, \qquad \partial_{\overline{m}} \equiv \frac{\partial}{\partial \phi^{\overline{m}}} \tag{5.1}$$

$$\chi_R^{\overline{m}} \equiv P_R \chi^m, \qquad \chi_L^m \equiv P_L \chi^m. \tag{5.2}$$

The Lagrangian then is[1,2]

$$\mathcal{L} = -\delta_{m\bar{n}}\Big[(\partial_\mu \phi^m)(\partial^\mu \phi^{\bar{n}}) + \overline{\chi}_L^m \not{\partial} \chi_R^{\bar{n}} + \overline{\chi}_R^{\bar{n}} \not{\partial} \chi_L^m\Big]$$

$$- (\partial_m \partial_n W)\overline{\chi}_L^m \chi_L^n - (\partial_{\bar{m}} \partial_{\bar{n}} W^*)\overline{\chi}_R^{\bar{m}} \chi_R^{\bar{n}} \tag{5.3}$$

$$- V_F,$$

with supersymmetry transformations given by

$$\delta\phi^m = \overline{\epsilon}_L \chi_L^m, \tag{5.4}$$

$$\delta\phi^{\bar{m}} = \overline{\epsilon}_R \chi_R^{\bar{m}}, \tag{5.5}$$

$$\delta\chi_L^m = \frac{1}{2}\not{\partial}\phi^m \epsilon_R - \frac{1}{2}\delta^{m\bar{n}}(\partial_{\bar{n}} W^*)\epsilon_L, \tag{5.6}$$

$$\delta\chi_R^{\bar{m}} = \frac{1}{2}\not{\partial}\phi^{\bar{m}} \epsilon_L - \frac{1}{2}\delta^{\bar{m}n}(\partial_n W)\epsilon_R. \tag{5.7}$$

The scalar potential in (5.3) is

$$V_F = \delta^{m\bar{n}}(\partial_m W)(\partial_{\bar{n}} W^*), \tag{5.8}$$

which is manifestly non-negative. This potential is called *F-term potential*, as it arises, in an off-shell formulation, from the auxiliary fields of the chiral multiplets, which are often denoted as F^m and which are on-shell proportional to $\delta^{m\bar{n}}(\partial_{\bar{n}} W^*)$. Note that these quantities also appear as an additional contribution (also called "fermionic shifts") in the supersymmetry variation of the chiral fermions in (5.6) and (5.7). The fact that the scalar potential is a sum of squares of these fermionic shifts is a generic feature of supersymmetric theories that we already saw in (4.63) and that we are going to encounter again in the following.

If one does not insist on renormalizability, e.g., in the context of low-energy effective field theory descriptions, one could admit the following generalizations:

- An arbitrary holomorphic superpotential $W(\phi^m)$ (i.e., one that is not necessarily a *cubic* polynomial)

---

[1] In terms of superfields, Φ^n, this corresponds to $\int d^4\theta\, \delta_{mn}\Phi^m(\Phi^n)^\dagger + \big(\int d^2\theta\ W(\Phi) + \text{h.c.}\big)$ after integrating out auxiliary fields.

[2] In the following we will often denote the Kronecker symbol as $\delta_{m\bar{n}}$ to make manifest covariance of the equations, even though one should interpret it as δ_{mn}.

- Non-minimal kinetic terms,

$$\delta_{m\bar{n}}(\partial_\mu\phi^m)(\partial^\mu\phi^{\bar{n}}) \rightarrow g_{m\bar{n}}(\phi,\phi^*)(\partial_\mu\phi^m)(\partial^\mu\phi^{\bar{n}}) \tag{5.9}$$

- Higher derivative terms (i.e., terms with more than two spacetime derivatives)

For standard applications in particle physics with small external momenta and small field gradients, higher derivative terms are usually irrelevant and will not be discussed further in this book. Since we already discussed the rôle of the superpotential (all equations above are equally valid for an arbitrary holomorphic $W(\phi^m)$), we now turn to the consequences of non-minimal kinetic terms. We first recall a few basic geometrical features of supersymmetric non-linear sigma models. The reader familiar with this may jump ahead to Sect. 5.3.

5.2.2   Non-linear Sigma Models I: The Holonomy Group

Let us first consider n_s *real* scalar fields $\varphi^i(x)$ $(i = 1, \ldots, n_s)$. The most general Lagrangian for this field content with at most two spacetime derivatives contains a scalar potential and a non-minimal kinetic term:

$$\mathcal{L} = -\frac{1}{2}g_{ij}(\varphi)(\partial_\mu\varphi^i)(\partial^\mu\varphi^j) - V(\varphi), \tag{5.10}$$

where the coefficients g_{ij} may depend on the scalars φ^k, as indicated. Under general field redefinitions $\varphi^i \rightarrow \widetilde{\varphi}^i(\varphi^j)$, the kinetic term transforms into the analogous expression $-\frac{1}{2}\widetilde{g}_{ij}(\widetilde{\varphi})(\partial_\mu\widetilde{\varphi}^i)(\partial^\mu\widetilde{\varphi}^j)$, with

$$\widetilde{g}_{ij}(\widetilde{\varphi}) = \frac{\partial\varphi^k}{\partial\widetilde{\varphi}^i}g_{kl}(\varphi(\widetilde{\varphi}))\frac{\partial\varphi^l}{\partial\widetilde{\varphi}^j} \tag{5.11}$$

so that $g_{ij}(\varphi)$ can be interpreted as a metric on the scalar manifold (or "target space"), $\mathcal{M}_{\text{scalar}}$, which is the space parameterized by all possible values of the scalar fields $(\varphi^1, \ldots, \varphi^{n_s})$. Unitarity of this field theory requires $g_{ij}(\varphi)$ to be positive definite (i.e., Riemannian). A very simple example for a scalar manifold is given by the Euclidean two-dimensional plane, $\mathcal{M}_{\text{scalar}} \cong \mathbb{R}^2$, which can be equivalently parameterized by, for instance, two real Cartesian field variables (φ^1, φ^2) with metric $g_{ij} = \delta_{ij}$, or by fields $R(x)$ and $\theta(x)$ corresponding to polar coordinates with $g_{RR} = 1$, $g_{\theta\theta} = R^2(x)$, $g_{R\theta} = 0$, or by a complex field $z = \frac{1}{\sqrt{2}}(\varphi^1 + i\varphi^2)$ with $g_{z\bar{z}} = 1$, $g_{zz} = g_{\bar{z}\bar{z}} = 0$. Obviously, the field dependence or independence of the metric $g_{ij}(\varphi)$ strongly depends on the chosen field variables. For a generic, possibly curved, scalar manifold, it is usually not possible to choose global Cartesian field variables with $g_{ij}(\varphi) = \delta_{ij}$. The coordinate independent flatness criterion for $\mathcal{M}_{\text{scalar}}$ is the vanishing of its Riemann tensor, $R_{ij}{}^k{}_l$, which is defined in terms of

Christoffel symbols, Γ_{ij}^k, in the usual way,

$$\Gamma_{ij}^k = \frac{1}{2} g^{kl} \left(\partial_i g_{jl} + \partial_j g_{il} - \partial_l g_{ij} \right), \tag{5.12}$$

$$R_{ij}{}^k{}_l = \partial_i \Gamma_{jl}^k + \Gamma_{im}^k \Gamma_{jl}^m - (i \leftrightarrow j), \tag{5.13}$$

where $\partial_i \equiv \partial / \partial \varphi^i$.

5.2.2.1 The Holonomy Group of the Scalar Manifold

The Christoffel symbols, Γ_{ij}^k, define a covariant derivative, $\mathscr{D}_i$, on $\mathscr{M}_{\text{scalar}}$, e.g.,

$$\mathscr{D}_i V^j \equiv \partial_i V^j + \Gamma_{ik}^j V^k \tag{5.14}$$

for a tangent vector, V^j, on $\mathscr{M}_{\text{scalar}}$. For a curve $\gamma : [0, 1] \rightarrow \mathscr{M}_{\text{scalar}}$ on the scalar manifold with tangent vectors $\left(\frac{d\gamma}{dt}(t) \right)$, one can then define the notion of parallel transport in the usual way, i.e., by saying that V^i is parallel transported along γ, if

$$\left(\frac{d\gamma^i}{dt} \right) \mathscr{D}_i V^j = 0 \qquad \forall t \in [0, 1]. \tag{5.15}$$

If the curve is closed, $\gamma(0) = \gamma(1)$, the vectors $V^j(\gamma(0))$ and $V^j(\gamma(1))$ live in the same tangent space and can hence be compared via the linear transformation that connects them,

$$V^j(\gamma(1)) = M^j{}_i(\gamma) V^i(\gamma(0)). \tag{5.16}$$

The matrix $M^j{}_i(\gamma)$ depends in general on the chosen path γ and is called the *holonomy* along γ. Since the metric is covariantly constant, $\mathscr{D}_i g_{jk} = 0$, the parallel transport by the Γ connection is length preserving and hence $M^j{}_i(\gamma) \in SO(n_s)$.

The *holonomy group*, $\text{Hol}(\mathscr{M}_{\text{scalar}})$, of the scalar manifold $\mathscr{M}_{\text{scalar}}$ is then defined to be the group formed by the matrices $M^j{}_i(\gamma)$ of all possible closed curves[3] γ.

Obviously, $\text{Hol}(\mathscr{M}_{\text{scalar}}) \subseteq SO(n_s)$, and a generic Riemannian manifold usually also sweeps out the full orthogonal group. Manifolds with additional structure, however, often have restricted holonomy groups, $\text{Hol}(\mathscr{M}_{\text{scalar}}) \subset SO(n_s)$. A simple example is given by a flat manifold, which has a trivial holonomy group that just consists of the identity transformation. As we will see in a moment, the scalar manifolds that appear in supersymmetric field theories also have a

---

[3] More precisely, one defines the holonomy group, $\text{Hol}(p)$, at a point $p \in \mathscr{M}_{\text{scalar}}$ to be the group generated by all $M^j{}_i(\gamma)$ of curves that begin and end at p. This pointwise holonomy group is the same for all points on $\mathscr{M}_{\text{scalar}}$ if (as we will always assume) $\mathscr{M}_{\text{scalar}}$ is connected, so that one can elevate it to the holonomy group, $\text{Hol}(\mathscr{M}_{\text{scalar}})$, of the entire manifold.

restricted holonomy group. In fact, the peculiar mathematical structures of many
scalar manifolds in supersymmetry can be fully characterized by their restricted
holonomy groups. The holonomy groups themselves can often be derived by very
simple arguments that follow essentially from the supermultiplet structure. Thus, the
holonomy group of a scalar manifold is a very important and convenient concept
from which many features of supersymmetric non-linear sigma models can be
derived quite easily.

5.2.3 Non-linear Sigma Models II: Fermions and Supersymmetry

Let us consider an (oversimplified) toy model consisting of n_s real scalars $\varphi^i(x)$
and n_s Majorana "fermions", $\psi^i(x)$, $(i, j, \ldots = 1, \ldots, n_s)$. We have put the word
"fermions" in quotes here because we are temporarily neglecting any complication
that may arise from chirality properties. Also, we are neglecting the fact that the
above field content contains twice as many fermions as allowed by an honest chiral
supermultiplet, because the scalars are real here. These are our temporary over-
simplifications, which we will correct soon. Assume supersymmetry transformation
rules of the schematic form

$$\delta\varphi^i = \bar{\epsilon}\psi^i, \tag{5.17}$$

$$\delta\psi^i = \slashed{\partial}\varphi^i\epsilon. \tag{5.18}$$

These equations motivate the interpretation of the fermions ψ^i_α as tangent vectors
on the scalar manifold (while they are at the same time spinors in spacetime, as
indicated by the (usually suppressed) spinor index α).[4]
  Now consider the Lagrangian

$$\mathscr{L} = -\frac{1}{2}g_{ij}(\varphi)(\partial_\mu\varphi^i)(\partial^\mu\varphi^j) - \frac{1}{2}g_{ij}(\varphi)\overline{\psi}^i\slashed{\partial}\psi^j. \tag{5.19}$$

Varying the bosonic term, one obtains

$$\delta\mathscr{L}_{\text{bos}} = \underbrace{g_{ij}(\delta\varphi^i)\Box\varphi^j}_{A} + \underbrace{\delta\varphi^k(\partial_\mu\varphi^i)(\partial^\mu\varphi^j)g_{kl}\Gamma^l_{ij}}_{B}, \tag{5.20}$$

where the term B comes from varying the field-dependent metric, $\delta g_{ij} = g_{ij,k}\delta\varphi^k$,
as well as from partial integrations that act on g_{ij}.

---

[4] More properly speaking, they are sections in the tangent bundle of $\mathscr{M}_{\text{scalar}}$ and sections in the
spacetime spinor bundle.

Using $\overline{\delta\psi}^i \slashed{\partial}\psi^j = -\overline{\partial_\mu\psi}^j \gamma^\mu \delta\psi^i$ and a partial integration, one obtains for the variation of the fermionic term:

$$\delta\mathscr{L}_{\text{fer}} = \underbrace{-g_{ij}\overline{\psi}^i \epsilon \Box\varphi^j}_{C} \underbrace{-\frac{1}{2}g_{ij,k}(\partial_\mu\varphi^k)(\partial_\nu\varphi^i)\overline{\psi}^j(\gamma^{\mu\nu} + \eta^{\mu\nu})\epsilon}_{D} + \mathcal{O}(\psi^3)\text{-terms},$$

(5.21)

where the term D is due to partial integration acting on the field-dependent metric g_{ij}, and the $\mathcal{O}(\psi^3)$-terms come from the variation of g_{ij}.

We see that, just as in the case with minimal kinetic terms, the two terms A and C cancel. However, the field dependence of g_{ij} introduces the uncancelled terms B, D, and the $\mathcal{O}(\psi^3)$-terms. In order to cancel B and D, one has to add a new term to the Lagrangian:

$$\mathscr{L}_{\text{new}} = -\frac{1}{2}g_{ij}\overline{\psi}^i(\slashed{\partial}\varphi^l)\Gamma^j_{lm}\psi^m.$$

(5.22)

One can indeed verify that varying the fermions in $\mathscr{L}_{\text{new}}$ precisely cancels B and D, whereas the variation of the scalar field-dependent terms gives rise to additional $\mathcal{O}(\psi^3)$-terms. The cancellation of the $\mathcal{O}(\psi^3)$-terms would require adding quartic terms in ψ^i to the Lagrangian and terms of the form $\psi\psi\epsilon$ to the transformation of ψ^i, which, however, we do not want to discuss here. Instead, we would like to focus on the geometric meaning of the new term (5.22). This term can be interpreted as a covariantization of the derivative of the fermions in the original Lagrangian (5.19),

$$-\frac{1}{2}g_{ij}\overline{\psi}^i \slashed{\partial}\psi^j \rightarrow -\frac{1}{2}g_{ij}\overline{\psi}^i \slashed{\mathscr{D}}\psi^j,$$

(5.23)

where

$$\mathscr{D}_\mu\psi^j \equiv \partial_\mu\psi^j + (\partial_\mu\varphi^l)\Gamma^j_{lm}\psi^m.$$

(5.24)

More precisely, this derivative is a covariant derivative with respect to *arbitrary coordinate transformations* on $\mathscr{M}_{\text{scalar}}$, provided the fermions transform as tangent vectors on $\mathscr{M}_{\text{scalar}}$,

$$\varphi^i \rightarrow \widetilde{\varphi}^i(\varphi^j), \quad \psi^i \rightarrow \widetilde{\psi}^i = \frac{\partial\widetilde{\varphi}^i}{\partial\varphi^j}\psi^j \quad \Longrightarrow \quad \mathscr{D}_\mu\psi^i \rightarrow \mathscr{D}_\mu\widetilde{\psi}^i = \frac{\partial\widetilde{\varphi}^i}{\partial\varphi^j}\mathscr{D}_\mu\psi^j.$$

(5.25)

The kinetic term (5.23) then transforms into $-\frac{1}{2}\widetilde{g}_{ij}\overline{\widetilde{\psi}}^i \slashed{\mathscr{D}}\widetilde{\psi}^j$ with $\widetilde{g}_{ij}$ as in (5.11), supporting our earlier interpretation of the ψ^i as components of a tangent vector on $\mathscr{M}_{\text{scalar}}$.

It should be stressed that the above covariantization with respect to coordinate transformations on $\mathcal{M}_{\text{scalar}}$ has nothing to do with a covariantization with respect to arbitrary spacetime coordinate transformations (we still work in flat Minkowski spacetime) or Yang–Mills-type gauge transformations.

5.2.4 4D Supersymmetry and Kähler Manifolds

While the above toy model was sufficient for the purpose of understanding the source of the covariantization of the fermionic derivatives with respect to general coordinate transformations on $\mathcal{M}_{\text{scalar}}$, it was still an oversimplified toy model. When dealing with four-dimensional $\mathcal{N} = 1$ supersymmetry, the scalar fields naturally combine into *complex* scalars, and the number and chirality of the fermions are important.

Indeed, the $\mathcal{N} = 1$ supersymmetry transformation laws for free chiral multiplets are (cf. Eqs. (5.4)–(5.7))

$$\delta\phi^m = \bar{\epsilon}_L \chi_L^m, \qquad \delta\phi^{\overline{m}} = \bar{\epsilon}_R \chi_R^{\overline{m}},$$

$$\delta\chi_L^m = \frac{1}{2}\displaystyle{\not}\partial\phi^m \epsilon_R, \qquad \delta\chi_R^{\overline{m}} = \frac{1}{2}\displaystyle{\not}\partial\phi^{\overline{m}} \epsilon_L, \tag{5.26}$$

which show that the scalars ϕ^m are superpartners of the left-handed components, χ_L^m, whereas the complex conjugate scalars, $\phi^{\overline{m}}$, transform into the right-handed components, $\chi_R^{\overline{m}}$. As chirality is a *spacetime* property, the internal geometry of $\mathcal{M}_{\text{scalar}}$ cannot interfere with it and should respect the above natural splitting of the fields. This has profound consequences for the geometry of $\mathcal{M}_{\text{scalar}}$. For one thing it implies that $\mathcal{M}_{\text{scalar}}$ must be a *complex* manifold, i.e., it looks locally like $\mathbb{C}^{n_C}$ and can be covered by mutually biholomorphic coordinate systems. Furthermore, the holonomy group $\text{Hol}(\mathcal{M}_{\text{scalar}})$ should respect the natural splitting expressed in the supersymmetry transformation laws (5.26), i.e., it should not mix ϕ^m with $\phi^{\overline{m}}$. Thus, if we combine $\delta\phi^m$ and $\delta\phi^{\overline{m}}$ into a $2n_C$-dimensional column vector, an element, M, of the holonomy group of $\mathcal{M}_{\text{scalar}}$ has to be block diagonal:

$$\begin{pmatrix} \delta\phi^m \\ \delta\phi^{\overline{m}} \end{pmatrix} \to M \begin{pmatrix} \delta\phi^m \\ \delta\phi^{\overline{m}} \end{pmatrix} = \begin{pmatrix} A^m{}_n\, \delta\phi^n \\ A^*{}^{\overline{m}}{}_{\overline{n}}\, \delta\phi^{\overline{n}} \end{pmatrix}, \tag{5.27}$$

or,

$$M = \begin{pmatrix} A & 0 \\ 0 & A^* \end{pmatrix}, \tag{5.28}$$

where, a priori, $A \in GL(n_C, \mathbb{C})$.

Switching to the $2n_C$-dimensional real basis $(\mathrm{Re}(\delta\phi^n), \mathrm{Im}(\delta\phi^m))$, $GL(n_C, \mathbb{C})$ is naturally embedded into $GL(2n_C, \mathbb{R})$ as the subgroup given by the $(2n_C \times 2n_C)$-matrices

$$M' = \begin{pmatrix} \mathrm{Re}(A) & -\mathrm{Im}(A) \\ \mathrm{Im}(A) & \mathrm{Re}(A) \end{pmatrix}. \tag{5.29}$$

Recalling that holonomies must also be orthogonal to preserve the metric, we have the additional requirement $M' \in SO(2n_C) \subset GL(2n_C, \mathbb{R})$ so that, putting everything together,

$$\mathrm{Hol}(\mathscr{M}_{\text{scalar}}) \subset GL(n_C, \mathbb{C}) \cap SO(2n_C) \subset U(n_C), \tag{5.30}$$

where the last inclusion follows because $SO(2n_C)$ is compact and the maximal compact subgroup of $GL(n_C, \mathbb{C})$ is $U(n_C)$.[5]

Complex manifolds of complex dimension n_C whose holonomy group is contained in $U(n_C)$ are called *Kähler manifolds*. We thus learn that the scalar manifold of chiral multiplets in global $\mathscr{N} = 1$ supersymmetry must be a Kähler manifold.

Kähler manifolds can also be defined in a different way by means of differential conditions on certain globally defined invariant tensors, and this definition is very useful to make explicit some of the properties of such manifolds. We therefore review the steps necessary to arrive at this alternative, though equivalent, definition.

The first step is to consider a complex manifold. A complex manifold of complex dimension n_C is a differentiable manifold of real dimension $2n_C$ that can be covered by local complex coordinate systems, ϕ^m $(m = 1, \ldots, n_C)$, such that the transition functions on overlapping coordinate patches are biholomorphic. On a complex manifold, one has a natural and well-defined tensor field, $\mathscr{J}$, of type $(1,1)$ given by

$$\mathscr{J} = i\, d\phi^m \otimes \frac{\partial}{\partial \phi^m} - i\, d\phi^{\bar{m}} \otimes \frac{\partial}{\partial \phi^{\bar{m}}}, \tag{5.31}$$

where $d\phi^m$ and $\partial/\partial\phi^m$ and their complex conjugates are the coordinate bases of the complexified cotangent and tangent spaces. Being a $(1,1)$ tensor field, $\mathscr{J}$ maps at each point of the manifold tangent vectors to tangent vectors via $\mathscr{J} \cdot V := \iota_V \mathscr{J}$ and acts on $\partial_m \equiv \partial/\partial\phi^m$ and $\partial_{\bar{m}} \equiv \partial/\partial\phi^{\bar{m}}$ as

$$\mathscr{J} \cdot \partial_m = i\partial_m, \qquad \mathscr{J} \cdot \partial_{\bar{m}} = -i\partial_{\bar{m}}. \tag{5.32}$$

---

[5] The intersection of $GL(n_C, \mathbb{C})$ and $SO(n_C)$ inside $GL(2n_C, \mathbb{R})$ depends on how $SO(2n_C)$ is embedded inside $GL(2n_C, \mathbb{R})$ relative to the embedding (5.29) of $GL(n_C, \mathbb{C})$. The intersection is maximal if the metric to be preserved by $SO(2n_C)$ in the basis $(\mathrm{Re}(\delta\phi^n), \mathrm{Im}(\delta\phi^m))$ is the unit matrix $\mathbb{1}_{2n_C}$. In that case, the orthogonality condition reads $M'^T M' = \mathbb{1}_{2n_C}$, which is equivalent to $A^\dagger A = \mathbb{1}_{n_C}$, which means that in that case the intersection of $GL(n_C, \mathbb{C})$ and $SO(2n_C)$ sweeps out the full $U(n_C)$.

$\mathscr{J}$ thus squares at each point to minus the identity operator on the corresponding tangent space,

$$\mathscr{J}^2 = -\mathrm{id} \tag{5.33}$$

and is called an *almost complex structure* on the complex manifold.

The name "almost" here refers to the fact that the converse of the above statement is in general not true: A differentiable manifold of real dimension $2n_C$ with real coordinates φ^i ($i = 1, \ldots, 2n_C$) and an everywhere well-defined (1,1) tensor field

$$\mathscr{J} = d\varphi^i \otimes \frac{\partial}{\partial \varphi^j}\, \mathscr{J}_i{}^j, \tag{5.34}$$

satisfying

$$\mathscr{J}_i{}^k\, \mathscr{J}_k{}^j = -\delta_i^j, \tag{5.35}$$

does not necessarily admit compatible complex coordinate systems with biholomorphic transition functions and is consequently called only an *almost complex manifold*. To give rise to a *complex manifold*, $\mathscr{J}$ has to satisfy an additional differential identity, which we don't need here, however, because we always assume a complex manifold from the start.

If a complex manifold carries a metric, g, that is compatible with $\mathscr{J}$ in the sense that, in terms of real coordinates φ^i,

$$\mathscr{J}_i{}^k\, \mathscr{J}_j{}^l\, g_{kl} = g_{ij} \tag{5.36}$$

or, equivalently,

$$\mathscr{J}_i{}^k g_{kj} = -\mathscr{J}_j{}^k g_{ik}, \tag{5.37}$$

one calls the complex manifold a *Hermitian manifold*. In terms of the complex coordinates, $\phi^m, \phi^{\bar{m}}$, the components of a Hermitian metric satisfy

$$g_{mn} = 0, \tag{5.38}$$

$$g_{\bar{m}\bar{n}} = 0, \tag{5.39}$$

$$g_{m\bar{n}} = g_{\bar{n}m} = (g_{\bar{m}n})^* = (g_{n\bar{m}})^*, \tag{5.40}$$

as one easily verifies by inserting the complex basis vectors ∂_m, $\partial_{\bar{m}}$ into the coordinate independent version, $g(\mathscr{J}\cdot\, ,\, \mathscr{J}\cdot\,) = g(\cdot, \cdot)$, of (5.36). Thus, the line element can be written as

$$ds^2 = d\varphi^i \otimes d\varphi^j\, g_{ij} = \left(d\phi^m \otimes d\phi^{\bar{n}} + d\phi^{\bar{n}} \otimes d\phi^m \right) g_{m\bar{n}}. \tag{5.41}$$

From these ingredients, we can in turn define the *fundamental two-form* or *Kähler form* on a Hermitian manifold,

$$J = \frac{1}{2} d\varphi^i \wedge d\varphi^j \, J_{ij} \equiv \frac{1}{2} d\varphi^i \wedge d\varphi^j \, \mathscr{J}_i{}^k g_{kj}, \qquad (5.42)$$

which can also be expressed in complex coordinates as

$$J = i \, d\phi^m \wedge d\phi^{\bar{n}} \, g_{m\bar{n}}(\phi^p, \phi^{\bar{q}}). \qquad (5.43)$$

A Kähler manifold is then defined as a complex manifold with Hermitian metric and closed fundamental form, namely,

$$dJ = 0, \qquad (5.44)$$

which means that J is also a *symplectic form* on $\mathscr{M}_{\text{scalar}}$.[6] We will later make use of this symplectic structure on the scalar manifold by exploiting some analogies with the Hamiltonian formulation of classical mechanics on phase space.

Condition (5.44) implies that locally there must exist a *real* function, $K(\phi^n, \phi^{\bar{m}})$, called the *Kähler potential*, such that

$$g_{m\bar{n}} = \partial_m \partial_{\bar{n}} K, \qquad (5.45)$$

$$J = i \, \partial_m \partial_{\bar{n}} K \, d\phi^m \wedge d\phi^{\bar{n}}. \qquad (5.46)$$

Note that the metric and the Kähler form do not change for any transformation that maps the Kähler potential to a new one by the addition of the real part of a holomorphic function:

$$K(\phi^n, \phi^{\bar{n}}) \rightarrow K(\phi^n, \phi^{\bar{n}}) + h(\phi^n) + h^*(\phi^{\bar{n}}). \qquad (5.47)$$

These are the *Kähler transformations*, which are also used to match different local expressions of the Kähler potential.

The restricted form of the metric further implies that the non-vanishing components of the Levi–Civita connection are

$$\Gamma^l_{mn} = g^{l\bar{k}} \partial_m g_{n\bar{k}}, \qquad \Gamma^{\bar{l}}_{\overline{mn}} = g^{\bar{l}k} \partial_{\overline{m}} g_{\bar{n}k}, \qquad (5.48)$$

where $g^{l\bar{k}}$ denotes the inverse of the matrix $g_{l\bar{k}}$. This further implies simple expressions for the Riemann tensor,

$$R_{\bar{n}m}{}^l{}_k = \partial_{\bar{n}} \Gamma^l_{mk}, \qquad \text{etc.} \qquad (5.49)$$

---

[6] Kähler manifolds can in fact equivalently be defined as manifolds that are complex and real symplectic at the same time.

In particular, the restricted form of the Riemann tensor (i.e., the fact that when indices are lowered one never has pairs of indices of the same type) shows that the Lie algebra of the holonomy group, generated by the curvature tensor, is indeed contained in $u(n_C)$.

From the above considerations, it follows that the geometry of the scalar manifold of $\mathcal{N} = 1$ chiral multiplets can be parameterized by a suitable Kähler potential K. The most general globally $\mathcal{N} = 1$ supersymmetric Wess–Zumino model of n_C chiral multiplets can therefore be expressed entirely in terms of the two functions $K(\phi^m, \phi^{\overline{m}})$ and $W(\phi^m)$:

$$
\begin{aligned}
\mathcal{L} = &-g_{m\overline{n}}\Big[(\partial_\mu \phi^m)(\partial^\mu \phi^{\overline{n}}) + \overline{\chi}_L^m \mathcal{D}\!\!\!/\, \chi_R^{\overline{n}} + \overline{\chi}_R^{\overline{n}} \mathcal{D}\!\!\!/\, \chi_L^m \Big] \\
&- (\mathcal{D}_m \partial_n W) \overline{\chi}_L^m \chi_L^n - (\mathcal{D}_{\overline{m}} \partial_{\overline{n}} W^*) \overline{\chi}_R^{\overline{m}} \chi_R^{\overline{n}} \\
&- g^{m\overline{n}} (\partial_m W)(\partial_{\overline{n}} W^*) + \mathcal{O}(\chi^4),
\end{aligned}
\tag{5.50}
$$

where

$$
\mathcal{D}_\mu \chi_R^{\overline{n}} \equiv \partial_\mu \chi_R^{\overline{n}} + (\partial_\mu \phi^{\overline{m}}) \Gamma^{\overline{n}}_{\overline{m}\,\overline{l}} \chi_R^{\overline{l}},
\tag{5.51}
$$

$$
\mathcal{D}_\mu \chi_L^n \equiv \partial_\mu \chi_L^n + (\partial_\mu \phi^m) \Gamma^n_{ml} \chi_L^l,
\tag{5.52}
$$

$$
\mathcal{D}_m \partial_n W \equiv \partial_m \partial_n W - \Gamma^l_{mn} \partial_l W,
\tag{5.53}
$$

$$
\mathcal{D}_{\overline{m}} \partial_{\overline{n}} W^* \equiv (\mathcal{D}_m \partial_n W)^*.
\tag{5.54}
$$

These new derivatives are covariant with respect to arbitrary *holomorphic* coordinate transformations,

$$
\phi^m \to \widetilde{\phi}^m(\phi^n), \qquad \chi_L^m \to \widetilde{\chi}_L^m = \frac{\partial \widetilde{\phi}^m}{\partial \phi^n} \chi_L^n
\tag{5.55}
$$

(and analogously for the conjugate fields) on the scalar manifold, so that the entire Lagrangian is form invariant under such field redefinitions.

The above Lagrangian is further invariant under the following supersymmetry transformations

$$
\delta \phi^m = \overline{\epsilon}_L \chi_L^m,
\tag{5.56}
$$

$$
\delta \phi^{\overline{m}} = \overline{\epsilon}_R \chi_R^{\overline{m}},
\tag{5.57}
$$

$$
\delta \chi_L^m = \frac{1}{2} \partial\!\!\!/\, \phi^m \epsilon_R - \frac{1}{2} g^{m\overline{n}} (\partial_{\overline{n}} W^*) \epsilon_L + \mathcal{O}(\chi \chi \epsilon),
\tag{5.58}
$$

$$
\delta \chi_R^{\overline{m}} = \frac{1}{2} \partial\!\!\!/\, \phi^{\overline{m}} \epsilon_L - \frac{1}{2} g^{\overline{m}n} (\partial_n W) \epsilon_R + \mathcal{O}(\chi \chi \epsilon).
\tag{5.59}
$$

Note that in *global* supersymmetry, K and W are completely independent functions and that the metric $g_{m\bar{n}}$, and hence the Lagrangian $\mathscr{L}$, are invariant under the Kähler transformations introduced in (5.47). As we will see in the next chapter, the independence of Kähler and superpotential will be lost when one couples matter to supergravity.

5.3 Globally Supersymmetric Gauge Theories

In this section, we recall the form of globally supersymmetric theories that involve vector multiplets (either alone or in connection with chiral multiplets).

5.3.1 Super Maxwell Theory

An on-shell $\mathscr{N} = 1$ vector multiplet contains one vector field, $A_\mu(x)$, and one Majorana fermion, $\lambda(x)$, ("gaugino"). Pure super Maxwell theory describes the coupling of n_V such vector multiplets (A_μ^I, λ^I) $(I, J, \ldots = 1, \ldots, n_V)$ with Abelian field strengths $F_{\mu\nu}^I \equiv \partial_\mu A_\nu^I - \partial_\nu A_\mu^I$. The Lagrangian is

$$\mathscr{L} = -\frac{1}{4}\delta_{IJ} F_{\mu\nu}^I F^{\mu\nu J} - \frac{1}{2}\delta_{IJ} \bar{\lambda}^I \displaystyle{\not}\partial \lambda^J. \tag{5.60}$$

The corresponding action is invariant under the supersymmetry transformations

$$\delta A_\mu^I = -\frac{1}{2}\bar{\epsilon}\gamma_\mu\lambda^I, \tag{5.61}$$

$$\delta\lambda^I = \frac{1}{4}\gamma^{\mu\nu} F_{\mu\nu}^I \epsilon, \tag{5.62}$$

as will be verified in the exercises.

5.3.2 Super Yang-Mills Theory

The generalization to pure $\mathscr{N} = 1$ super Yang-Mills theory is obtained by making the replacements

$$F_{\mu\nu}^I \rightarrow \mathscr{F}_{\mu\nu}^I \equiv 2\partial_{[\mu}A_{\nu]}^I + f_{JK}{}^I A_\mu^J A_\nu^K, \tag{5.63}$$

$$\partial_\mu\lambda^I \rightarrow \widehat{\partial}_\mu\lambda^I \equiv \partial_\mu\lambda^I + A_\mu^J f_{JK}{}^I \lambda^K, \tag{5.64}$$

where $f_{JK}{}^I$ are the structure constants of the gauge algebra. Here and in the following, we will use a hat on a derivative to indicate that it is a gauge covariant derivative in the conventional sense. The symbol $\mathscr{D}$ will instead be reserved for derivatives that are covariant with respect to scalar field redefinitions as explained in the previous subsection. The above covariant derivative of the gaugini reflects the fact that the gaugini have to transform in the adjoint representation of the gauge group, just as their superpartners, the vector fields. We mention here that we have implicitly set all gauge couplings equal to 1. For general values of the gauge couplings, g_I, the δ_{IJ} in the kinetic terms has to be replaced by $\delta_{IJ}g_I^{-2}$ (no sum), where g_I can be different for each gauge group factor.

5.3.3 Coupling Super Maxwell/Yang–Mills Theories to Chiral Multiplets

We now discuss the general couplings of n_V vector multiplets $\left(A_\mu^I, \lambda^I\right)$ ($I, J, \ldots = 1, \ldots, n_V$) and n_C chiral multiplets ($m, n, \ldots = 1, \ldots n_C$). The coupling of these two sectors can occur in two ways:

1. Non-minimal kinetic terms for the vector fields and gaugini:

$$
\delta_{IJ}\left[-\frac{1}{4}\mathscr{F}_{\mu\nu}^I \mathscr{F}^{\mu\nu J} - \frac{1}{2}\bar{\lambda}^I \widehat{\partial}\lambda^J \right] \rightarrow N_{IJ}(\phi^m, \phi^{\overline{m}})\left[-\frac{1}{4}\mathscr{F}_{\mu\nu}^I \mathscr{F}^{\mu\nu J} - \frac{1}{2}\bar{\lambda}^I \widehat{\partial}\lambda^J \right],
\tag{5.65}
$$

where $N_{IJ}(\phi^m, \phi^{\overline{m}})$ denotes a scalar field-dependent kinetic matrix
2. Gauge couplings for the scalars and fermions in the chiral multiplets:

$$
\begin{aligned}
\partial_\mu \phi^m &\rightarrow \widehat{\partial}_\mu \phi^m = \partial_\mu \phi^m + A_\mu^I(\ldots), \\
\mathscr{D}_\mu \chi^m &\rightarrow \widehat{\mathscr{D}}_\mu \chi^m = \mathscr{D}_\mu \chi^m + A_\mu^I(\ldots),
\end{aligned}
\tag{5.66}
$$

where $\widehat{\partial}_\mu$ and $\widehat{\mathscr{D}}_\mu$ denote suitable gauge covariant derivatives that contain minimal couplings to the vector fields A_μ^I as specified by some yet to be determined couplings $(\ldots)$

We will now discuss these two types of couplings in turn.

5.3.4 Non-minimal Kinetic Terms for Vector Multiplets: The Gauge Kinetic Function

A modification of the kinetic terms of the vector multiplets with non-trivial functions $N_{IJ}(\phi^m, \phi^{\overline{m}})$ can be made consistent with supersymmetry only when the matrix elements $N_{IJ}(\phi^m, \phi^{\overline{m}})$ are the real parts (or imaginary, depending on conventions) of *holomorphic* functions, $f_{IJ}(\phi^m)$,

$$N_{IJ}\left(\phi^m, \phi^{\overline{m}}\right) = \mathrm{Re}\left(f_{IJ}(\phi^m)\right). \tag{5.67}$$

The set of $f_{IJ}(\phi^m)$ is usually referred to as the *gauge kinetic function* (or *gauge kinetic matrix*).

Moreover, supersymmetry also requires a term of the form

$$\frac{1}{4}(\mathrm{Im}\, f_{IJ})\left[\frac{1}{2}\epsilon^{\mu\nu\rho\sigma}\mathscr{F}^I_{\mu\nu}\mathscr{F}^J_{\rho\sigma} - i\,\widehat{\partial}_\mu(\overline{\lambda}^I\gamma_5\gamma^\mu\lambda^J)\right]. \tag{5.68}$$

The first term in (5.68) can be viewed as a generalized field-dependent θ-angle term. Note that, for constant f_{IJ}, the entire expression (5.68) is a total spacetime derivative that would not change the classical field equations.

A non-trivial gauge kinetic function $f_{IJ}(\phi^m)$ also requires additional fermionic terms, e.g., terms of the form $\overline{\lambda}\lambda$, $\overline{\chi}\lambda\mathscr{F}_{\mu\nu}$, but it leaves the scalar potential unchanged: $V = V_F$, with $V_F = g^{m\overline{n}}(\partial_m W)(\partial_{\overline{n}}\overline{W})$, as before. Instead of giving the full Lagrangian here, we will first also include genuine gauge interactions of the form (5.66) and then display the most general globally supersymmetric Lagrangian with vector and chiral multiplets and at most two spacetime derivatives in total. In order to do this, however, we first have to understand the rôle of symmetries in non-linear σ-models, in particular in those that are based on Kähler manifolds.

5.3.5 Non-linear σ-Models III: Global and Local Symmetries

Let us again consider a toy model consisting of n_s real scalar fields φ^i ($i, j, \cdots = 1, \ldots, n_s$) with a Lagrangian of the form

$$\mathscr{L} = -\frac{1}{2}g_{ij}(\varphi)(\partial_\mu\varphi^i)(\partial^\mu\varphi^j). \tag{5.69}$$

An internal symmetry of this Lagrangian is a map $\mathscr{I} : \mathscr{M}_{\text{scalar}} \to \mathscr{M}_{\text{scalar}}$, $\varphi^i \to \widetilde{\varphi}^i(\varphi^j)$, that preserves the metric on $\mathscr{M}_{\text{scalar}}$:

$$\widetilde{g}_{lk} \equiv g_{ij}\frac{\partial\varphi^i}{\partial\widetilde{\varphi}^l}\frac{\partial\varphi^j}{\partial\widetilde{\varphi}^k} = g_{lk}, \tag{5.70}$$

or, in other words, an *isometry* of $\mathscr{M}_{\text{scalar}}$. The simplest example is the case $\mathscr{M}_{\text{scalar}} = \mathbb{R}^{n_s}$ with $g_{ij} = \delta_{ij}$. The isometries in this case consist of orthogonal field rotations,

$$\varphi^i \to M^i{}_j\varphi^j, \qquad M^i{}_j \in SO(n_s) \tag{5.71}$$

and rigid translations ("shift symmetries"),

$$\varphi^i(x) \to \varphi^i(x) + a^i, \qquad a^i \in \mathbb{R}^{n_s}, \tag{5.72}$$

which together generate the isometry group Iso $(\mathcal{M}_{\text{scalar}}) \cong SO(n_s) \ltimes \mathbb{R}^{n_s}$, with $\ltimes$ denoting a semidirect product.

At the infinitesimal level, isometries are generated by the corresponding Killing vectors, $\xi_I^i(\varphi)$ $(I = 1, \dots, \dim(\text{Iso}(\mathcal{M}_{\text{scalar}})))$, on $\mathcal{M}_{\text{scalar}}$,

$$\varphi^i \to \varphi^i + \alpha^I \xi_I^i(\varphi). \tag{5.73}$$

Here, α^I are infinitesimal real parameters, and the ξ_I^i satisfy the Killing equation

$$\mathcal{L}_{\xi_I} g = 0 \Leftrightarrow g_{ij,k}\xi_I^k + g_{kj}\xi_{I,i}^k + g_{ik}\xi_{I,j}^k = 0, \tag{5.74}$$

$$\Leftrightarrow \mathcal{D}_i \xi_{Ij} + \mathcal{D}_j \xi_{Ii} = 0, \tag{5.75}$$

where a comma denotes a partial derivative with respect to a scalar field, and $\mathcal{L}_{\xi_I}$ is the Lie derivative along ξ_I. The Killing equation is the infinitesimal version of (5.70) and hence ensures the invariance of the Lagrangian (5.69) under (5.73) if $\alpha^I = const$.

We recall that Killing vectors are the generators of a Lie algebra

$$[\xi_I, \xi_J] = f_{IJ}{}^K \xi_K, \tag{5.76}$$

where $[\xi_I, \xi_J]^i = \xi_I^j \partial_j \xi_J^i - \xi_J^j \partial_j \xi_I^i$. In the above example with $\mathcal{M}_{\text{scalar}} \cong \mathbb{R}^{n_s}$, the isometry group is the Euclidean group in n_s dimensions, where the orthogonal rotations are generated by

$$\delta\varphi^i = \alpha^{\tilde{I}} \underbrace{T_{\tilde{I}}^i{}_j \varphi^j}_{=\xi_{\tilde{I}}^i(\varphi)}, \qquad T_{\tilde{I}}^i{}_j \in \mathfrak{so}(n_s), \tag{5.77}$$

whereas infinitesimal shift symmetries are generated by constant Killing vectors

$$\delta\varphi^i = \alpha^{\hat{I}} \underbrace{d_{\hat{I}}^i}_{=\xi_{\hat{I}}^i}, \qquad d_{\hat{I}}^i = const. \tag{5.78}$$

Here, we have split the index $I = (\tilde{I}, \hat{I})$, corresponding to the rotations labeled by $\tilde{I} = 1, \dots, \dim(\mathfrak{so}(n_s))$ and the translations with $\hat{I} = 1, \dots, n_s$.

The above symmetries of the non-linear sigma model are *global* internal symmetries in the sense that the infinitesimal symmetry parameters α^I have to be *constant* (i.e., spacetime *in*dependent).

If we allow for spacetime-*dependent* parameters, $\alpha^I = \alpha^I(x)$, the Lagrangian (5.69) is no longer invariant and transforms into terms proportional to $\partial_\mu \alpha^I$. In order to make (5.69) invariant also under such *local* (or, "*gauge*") symmetries, we have to turn the partial derivatives of the scalar fields into gauge covariant derivatives $\widehat{\partial}_\mu$:

$$\partial_\mu \varphi^i \to \widehat{\partial}_\mu \varphi^i \equiv \partial_\mu \varphi^i - A_\mu^I \xi_I^i(\varphi). \tag{5.79}$$

This gauge covariant derivative transforms under the gauge transformations

$$\delta_{\text{gauge}} \varphi^i = \alpha^I(x) \xi_I^i(\varphi), \tag{5.80}$$

$$\delta_{\text{gauge}} A_\mu^I = \partial_\mu \alpha^I(x) + f_{JK}{}^I A_\mu^J \alpha^K(x), \tag{5.81}$$

as

$$\delta_{\text{gauge}} \left(\widehat{\partial}_\mu \varphi^i \right) = \alpha^I(x) \left(\partial_j \xi_I^i \right) \widehat{\partial}_\mu \varphi^j, \tag{5.82}$$

which then again leaves the Lagrangian (5.69) invariant due to the Killing equation (5.74).

In the above example with $\mathcal{M}_{\text{scalar}} \cong \mathbb{R}^{n_s}$, the gauge covariant derivatives become

$$\widehat{\partial}_\mu \varphi^i = \partial_\mu \varphi^i - A_\mu^{\widetilde{I}} T_{\widetilde{I}}{}^i{}_j \varphi^j - A_\mu^{\widehat{I}} d_{\widehat{I}}^i. \tag{5.83}$$

We emphasize again that the gauge covariantization (5.79) should not be confused with the covariantization with respect to general field reparameterizations $\varphi^i \to \widetilde{\varphi}^i(\varphi^j)$ that we encountered earlier, e.g., for the fermionic derivative

$$\mathscr{D}_\mu \psi^i \equiv \partial_\mu \psi^i + (\partial_\mu \varphi^j) \Gamma_{jk}^i \psi^k \tag{5.84}$$

in a supersymmetric non-linear sigma model. This derivative is covariant with respect to such general field reparameterizations (under which the fermions transform as in Eq. (5.25)), but not yet with respect to the *gauge* transformations (5.80), (5.81). Moreover, the connection in (5.79) corresponds to an elementary vector field, while the one in (5.84) is a composite connection built from scalar fields and their derivatives. In order to see that (5.84) is not gauge covariant, one recalls that the (gauge covariantized) supersymmetry transformation law of the fermions ψ^i would be[7]

$$\delta_{\text{susy}} \psi^i = \widehat{\partial} \varphi^i \epsilon. \tag{5.85}$$

---

[7] We remind the reader that this is still a toy model where we are neglecting any chirality properties of the fermions. The complete model will be discussed below.

Consistency with (5.82) then demands that the fermions also transform with the derivative of the Killing vectors under gauge transformations,

$$\delta_{\text{gauge}} \psi^i = \alpha^I(x)(\partial_j \xi_I^i) \psi^j. \tag{5.86}$$

Because of this, the proper gauge covariantization of (5.84) is

$$\begin{aligned}
\mathscr{D}_\mu \psi^i \to \widehat{\mathscr{D}}_\mu \psi^i &\equiv \partial_\mu \psi^i + (\widehat{\partial_\mu \varphi^j}) \Gamma_{jk}^i \psi^k - A_\mu^I (\partial_j \xi_I^i) \psi^j \\
&= \mathscr{D}_\mu \psi^i - A_\mu^I (\mathscr{D}_j \xi_I^i) \psi^j,
\end{aligned} \tag{5.87}$$

as will be verified in the exercises.

5.3.6   Killing Prepotentials, D-Terms, and the General Globally Supersymmetric Lagrangian

In the above toy model, we have neglected that the scalar manifold in a proper supersymmetric non-linear sigma model in 4D has an additional structure, namely, it must be a Kähler manifold of complex dimension n_C, where n_C is the number of chiral multiplets. This means, in particular, that the metric on the scalar manifold derives locally from a Kähler potential, $g_{m\bar{n}} = \partial_m \partial_{\bar{n}} K$, where K itself is defined only modulo Kähler transformations $K \to K + h(\phi) + h^*(\phi^*)$.

The isometries on a Kähler manifold should therefore have the following properties [1]:

1. Iso ($\mathscr{M}_{\text{scalar}}$) should respect the complex structure and not mix ϕ^m and $\phi^{\overline{m}}$:

$$\delta_{\text{gauge}} \phi^m = \alpha^I \xi_I^m(\phi^n), \tag{5.88}$$

$$\delta_{\text{gauge}} \phi^{\overline{m}} = \alpha^I \xi_I^{\overline{m}}(\phi^{\overline{n}}), \tag{5.89}$$

where $\xi_I^m(\phi^n)$ is a *holomorphic* Killing vector and $\xi_I^{\overline{m}} \equiv (\xi_I^m)^*$.

2. Iso ($\mathscr{M}_{\text{scalar}}$) should leave the Kähler potential invariant up to Kähler transformations:

$$\delta_I K \equiv \xi_I^m \partial_m K + \xi_I^{\overline{m}} \partial_{\overline{m}} K = r_I(\phi^m) + r_I^*(\phi^{\overline{m}}), \tag{5.90}$$

where the $r_I(\phi^m)$ are some holomorphic functions corresponding to Kähler transformations (5.47) with $h(\phi^m) = r_I(\phi^m)$ that measure the non-invariance of the Kähler potential under gauge transformations. The condition (5.90) is equivalent to the fact that the Killing vectors preserve the metric $g_{m\bar{n}}$.

Since the holomorphic Killing vectors preserve the metric on $\mathcal{M}_{\text{scalar}}$ as well as its complex structure, they also preserve the Kähler form J:

$$\mathcal{L}_{\xi_I} J = \iota_{\xi_I} dJ + d\iota_{\xi_I} J = 0, \tag{5.91}$$

where we expressed the Lie derivative, $\mathcal{L}_{\xi_I}$, on a differential form in terms of the inner product, ι_{ξ_I}, and the exterior derivative.

For a Kähler manifold, one has $dJ = 0$ so that *locally* there must exist functions $\mathcal{P}_I$ with

$$\iota_{\xi_I} J = d\mathcal{P}_I. \tag{5.92}$$

The functions $\mathcal{P}_I$ are called *Killing prepotentials*. They are real, because the full Killing vectors $\xi_I = \xi_I^m \partial_m + \xi_I^{\overline{m}} \partial_{\overline{m}}$ and the Kähler form J are real. Viewing J as a symplectic form and $\mathcal{M}_{\text{scalar}}$ as the phase space of a fictitious mechanical system, the Killing prepotentials simply correspond to the *moment maps*[8] whose gradients generate the flows along the integral curves of the ξ_I [2]. In terms of its holomorphic and anti-holomorphic components, (5.92) becomes

$$i\xi_I^m g_{m\overline{n}} = \partial_{\overline{n}} \mathcal{P}_I,$$
$$-i\xi_I^{\overline{n}} g_{m\overline{n}} = \partial_m \mathcal{P}_I. \tag{5.93}$$

Recall now that $g_{m\overline{n}} = \partial_m \partial_{\overline{n}} K$, and $\partial_{\overline{n}} \xi_I^m = 0$. This implies that $i\xi_I^m g_{m\overline{n}} = \partial_{\overline{n}} \left(i\xi_I^m \partial_m K \right)$ so that

$$\begin{aligned}
\mathcal{P}_I &= i\xi_I^m \partial_m K - i\, r_I(\phi) \\
&= -i\xi_I^{\overline{n}} \partial_{\overline{n}} K + i\, r_I^*(\phi^*) \\
&= \frac{i}{2} \left[\xi_I^m \partial_m K - \xi_I^{\overline{n}} \partial_{\overline{n}} K - \left(r_I(\phi) - r_I^*(\phi^*) \right) \right]
\end{aligned} \tag{5.94}$$

where the equalities are due to $\mathcal{P}_I = \mathcal{P}_I^*$, and $r_I(\phi)$ denote some as yet arbitrary holomorphic functions. This expression is compatible with the request (5.90) that the Kähler potential be invariant up to Kähler transformations as follows from

$$\delta_\xi K = \xi_I^m \partial_m K + \xi_I^{\overline{m}} \partial_{\overline{m}} K = -i\mathcal{P}_I + r_I + i\mathcal{P}_I + r_I^* = r_I + r_I^*. \tag{5.95}$$

The integration functions r_I are then completely determined up to imaginary constants, $\Delta r_I = -i\,\eta_I$, called *Fayet–Iliopoulos terms*. If Eq. (5.95) were the only constraints on the Killing prepotentials, the Fayet–Iliopoulos constants would

---

[8] In classical mechanics, if the moment maps Poisson commute with the Hamiltonian, they are conserved along the time evolution generated by the Hamiltonian.

be completely undetermined for each gauge group generator. There is, however, another identity the $\mathscr{P}_I$ have to satisfy, the *equivariance condition*,

$$\{\mathscr{P}_I, \mathscr{P}_J\} = f_{IJ}{}^K \mathscr{P}_K, \tag{5.96}$$

where on the left-hand side are the Poisson brackets of the prepotentials, defined as the inner product of the corresponding Killing vectors with the fundamental two-form,

$$\{\mathscr{P}_I, \mathscr{P}_J\} \equiv -J(\xi_I, \xi_J) = -\iota_{\xi_J}\iota_{\xi_I}J = \iota_{\xi_I}\iota_{\xi_J}J, \tag{5.97}$$

which is also equivalent to

$$\{\mathscr{P}_I, \mathscr{P}_J\} = -i\xi_I^m g_{m\bar{n}}\bar{\xi}_J^{\bar{n}} + i\xi_J^m g_{m\bar{n}}\bar{\xi}_I^{\bar{n}} = \xi_I^m \partial_m \mathscr{P}_J + \bar{\xi}_I^{\bar{n}}\partial_{\bar{n}}\mathscr{P}_J$$

$$= \delta_I \mathscr{P}_J = -\delta_J \mathscr{P}_I. \tag{5.98}$$

The equivariance condition is proven in the next subsection. The prepotentials enter explicitly in the supersymmetric Lagrangian via the *D-term*[9]

$$D^I \equiv (\mathrm{Re}\,f)^{-1\,IJ}\mathscr{P}_J \qquad (\text{``D-term''}), \tag{5.99}$$

which, among other places, appear in the supersymmetry transformations of the gaugini as a shift term and in the scalar potential, where they contribute the so-called D-term potential,

$$V_D = \frac{1}{2}(\mathrm{Re}\,f_{IJ})D^I D^J = \frac{1}{2}(\mathrm{Re}\,f)^{-1}\mathscr{P}_I\mathscr{P}_J. \tag{5.100}$$

5.3.6.1 Proof of the Equivariance Condition

As we saw above, the equivariance condition (5.96) is rather important in fixing the form of the prepotentials and hence of the D-terms. Its origin can be understood as a consequence of gauge invariance, and for the sake of completeness, we sketch here the proof, which may be omitted in a first reading.

Taking the exterior derivative of the right-hand side of (5.97) and recalling that on scalar valued differential forms (from now on $\iota_I \equiv \iota_{\xi_I}$) the Lie derivative acts as $\mathcal{L}_I = \iota_I d + d\iota_I$, one obtains

$$d\{\mathscr{P}_I, \mathscr{P}_J\} = d\iota_I\iota_J J = (\mathcal{L}_I\iota_J - \iota_I d\iota_J) J, \tag{5.101}$$

where the last term vanishes identically because $\iota_I J = d\mathscr{P}_I$.

---

[9] Similar to the F-terms, the above D-terms are on-shell expressions for a certain auxiliary field, often called $D(x)$, of an off-shell vector supermultiplet.

Moreover, we know that isometries must preserve the Kähler form, i.e., $\mathcal{L}_I J = 0$, and therefore we can rewrite the previous expression as

$$d\{\mathscr{P}_I, \mathscr{P}_J\} = [\mathcal{L}_I, \iota_J] J. \tag{5.102}$$

This can be further manipulated using the algebraic identity $[\mathcal{L}_I, \iota_J] = \iota_{[I,J]}$ and the fact that Killing vectors generate a Lie algebra, $[\xi_I, \xi_J] = f_{IJ}{}^K \xi_K$. Using these two identities, we see that

$$d\{\mathscr{P}_I, \mathscr{P}_J\} = \iota_{[I,J]} J = f_{IJ}{}^K \iota_K J = f_{IJ}{}^K d\mathscr{P}_K, \tag{5.103}$$

and hence obtain the equivariance condition up to additive constants,

$$\{\mathscr{P}_I, \mathscr{P}_J\} = f_{IJ}{}^K \mathscr{P}_K + c_{IJ}, \tag{5.104}$$

where the constants $c_{IJ} = -c_{JI}$ should also satisfy

$$f_{[IJ}{}^L c_{K]L} = 0, \tag{5.105}$$

because Poisson brackets fulfill the Jacobi identity.

The last step is to show that such constants should either disappear or be reabsorbed in the definition of the prepotentials, in case this is possible (see below). To understand this, we recall that the invariance of the gauge kinetic term, $R_{IJ} \mathscr{F}^I_{\mu\nu} \mathscr{F}^{\mu\nu\,J}$, where $R_{IJ} \equiv \mathrm{Re}\,(f_{IJ})$, requires the kinetic matrix to transform according to

$$\delta_K R_{IJ} = \mathcal{L}_K R_{IJ} \overset{!}{=} 2 f_{K(I}{}^L R_{J)L} \tag{5.106}$$

in order to compensate the transformations of the field strengths, which also transform with the structure constants. The inverse of the gauge kinetic matrix, $R^{IJ} \equiv (R_{IJ})^{-1}$, enters in the definition of the D-term potential, $V_D \sim \mathscr{P}_I R^{IJ} \mathscr{P}_J$, so that the transformations $\delta_I R^{JK} = -2 f_{IL}^{(K} R^{J)L}$ and $\delta_I \mathscr{P}_J = \mathcal{L}_I \mathscr{P}_J = \{\mathscr{P}_I, \mathscr{P}_J\} = f_{IJ}{}^K \mathscr{P}_K + c_{IJ}$ imply that gauge invariance requires

$$c_{IJ} R^{JK} \mathscr{P}_K \overset{!}{=} 0. \tag{5.107}$$

This cannot be satisfied for arbitrary scalar field values when $c_{IJ} \neq 0$ and therefore imposes that either the c_{IJ} have to be absorbed in the definition of the prepotentials (for semi-simple gauge groups, see below), or they lead to inconsistent gaugings.

Let us first consider the case of semi-simple gauge groups. For semi-simple gauge groups, the condition (5.105) implies that the constants c_{IJ} must also satisfy

$$c_{IJ} = f_{IJ}{}^K c_K, \tag{5.108}$$

for some constants c_K, which is related to the fact that semi-simple Lie algebras have a trivial second cohomology group. If one now shifts $\mathscr{P}_K$ by c_K (which is allowed because a priori $\mathscr{P}_K$ is only defined up to additive constants by (5.92)), one can eliminate the coefficients c_{IJ} from the Poisson bracket (5.104) and reproduce the required equivariance condition (5.96) without c_{IJ}. Note that this is equivalent to adding a very specific Fayet–Iliopoulos term to $\mathscr{P}_K$ even though the gauge group is non-Abelian. The important point here is that the Fayet–Iliopoulos constant is *fixed* by the equivariance condition.

Let us now consider Abelian gaugings. For Abelian gaugings, there are two possible cases. The first case corresponds to situations where no c_{IJ} appears in the Poisson bracket (5.104) right from the beginning. In that case, the gauging is consistent for arbitrary Fayet–Iliopoulos constants, because no c_{IJ} is generated by such shifts in $\mathscr{P}_K$ due to the vanishing structure constants. The second case corresponds to situations where a non-trivial c_{IJ} does appear in (5.104). Due to the vanishing structure constants for Abelian gaugings, it is not possible to absorb this c_{IJ} by adding Fayet–Iliopoulos constants to the $\mathscr{P}_K$. In this case, the gauging is thus inconsistent, because the scalar potential is not gauge invariant, no matter how one chooses the Fayet–Iliopoulos constant.

We will now give two examples that illustrate the above points. The first example covers the case where a non-Abelian gauging requires a specific Fayet–Iliopoulos constant to eliminate a non-trivial c_{IJ}. The second example covers the case where an Abelian gauging is inconsistent due to a non-vanishing c_{IJ}.

5.3.6.2 Two Examples of Gaugings with Fayet–Iliopoulos Constants

As an example of the existence of non-trivial integration constants c_{IJ} for non-Abelian gauge groups and their absorption by the prepotential, we follow [3], where the gauge group $G = SU(2)$ results from the gauging of the isometries of the scalar manifold $\mathscr{M}_{\text{scalar}} = \mathbb{CP}^1$ with Kähler potential

$$ K = -3\nu \log\left[-\frac{1}{3}(1 + \phi\phi^*)^{-\frac{1}{3}} \right], \tag{5.109}$$

where ν is an arbitrary positive real coefficient for the time being. In details, the isometry group $SU(2)$ acts non-linearly on the complex scalar field ϕ

$$ \delta_1\phi = \frac{i}{2}(\phi^2 - 1), \qquad \delta_2\phi = \frac{1}{2}(\phi^2 + 1), \qquad \delta_3\phi = -i\phi, \tag{5.110}$$

so that $[\xi_I, \xi_J] = \epsilon_{IJK}\xi_K$, and we can obtain the Killing prepotentials from (5.94).

Let us start by checking the transformation of the Kähler potential under such isometries, recalling that it can vary by a Kähler transformation $\delta_I K = r_I + r_I^*$. We see that

$$ \delta_1 K = i\frac{\nu}{2}(\phi - \phi^*), \qquad \delta_2 K = \frac{\nu}{2}(\phi + \phi^*), \qquad \delta_3 K = 0, \tag{5.111}$$

which fixes

$$r_1 = i\frac{v}{2}\phi + i\,\eta_1, \qquad r_2 = \frac{v}{2}\phi + i\,\eta_2, \qquad r_3 = i\,\eta_3, \tag{5.112}$$

where η_I are real constants. The prepotentials then follow by applying (5.94):

$$\mathscr{P}_I = \left\{ \frac{v}{2}\frac{\phi + \phi^*}{1 + \phi\phi^*} + \eta_1, \ -i\frac{v}{2}\frac{\phi - \phi^*}{1 + \phi\phi^*} + \eta_2, \ v\frac{\phi\phi^*}{(1 + \phi\phi^*)} + \eta_3 \right\}. \tag{5.113}$$

In these expressions we still have three undetermined real constants η_I. The equivariance condition forces these constants to vanish with the exception of η_3, which has to be $\eta_3 = -\frac{v}{2}$, so that

$$\{\mathscr{P}_1, \mathscr{P}_2\} = \delta_1 \mathscr{P}_2 = \mathscr{P}_3. \tag{5.114}$$

We therefore see that in this example, the existence of a non-trivial constant shift of the prepotential (usually called Fayet–Iliopoulos term) is necessary in order to satisfy the equivariance condition, even though the gauge group is non-Abelian.

We now provide a different example [4], where the non-trivial constants c_{IJ} can *not* be absorbed in the prepotentials. This example starts from the complex plane as scalar manifold, with Kähler potential $K = \phi\phi^*$. As gauge group, we choose the Abelian group given by the product of two shift symmetries:

$$\delta_1\phi = \lambda, \qquad \delta_2\phi = i\kappa. \tag{5.115}$$

The Kähler potential is not invariant under such isometries, and hence one has non-trivial integration functions $r_I(\phi)$, which moreover allow for arbitrary constant Fayet–Iliopoulos terms:

$$\delta_1 K = \ \lambda(\phi + \phi^*) \qquad \Rightarrow \qquad r_1 = \lambda\phi + i\eta_1,$$
$$\delta_2 K = -i\kappa(\phi - \phi^*) \qquad \Rightarrow \qquad r_2 = -i\kappa\phi + i\eta_2. \tag{5.116}$$

Once again, we can construct the respective Killing prepotentials using (5.94)

$$\mathscr{P}_1 = i\,\delta_1\phi\,\partial_\phi K - ir_1 = -i\lambda(\phi - \phi^*) + \eta_1, \tag{5.117}$$

$$\mathscr{P}_2 = i\,\delta_2\phi\,\partial_\phi K - ir_2 = -\kappa(\phi + \phi^*) + \eta_2, \tag{5.118}$$

and compute the equivariance condition

$$\{\mathscr{P}_1, \mathscr{P}_2\} = \delta_1 \mathscr{P}_2 = -2\lambda\kappa = c_{12} \neq 0. \tag{5.119}$$

As one can see, the constant coefficient c_{12} has to be there if one wants to gauge both isometries, and, since we have vanishing structure constants, it cannot be absorbed into the definition of the prepotentials $\mathscr{P}_I$, i.e., the c_{IJ} are independent of any Fayet–Iliopoulos shifts η_I. How do we deal with such a case then? As we discussed above, one can see that requiring full gauge invariance of the action and in particular of the gauge kinetic term and the potential forbids such extensions, and therefore one has to understand the previous example as an *inconsistent* gauge group choice for a supersymmetric theory.

5.3.6.3 Lagrangian and Susy Rules

In order to summarize the results of this section, we now present the most general $\mathscr{N} = 1$ globally supersymmetric Lagrangian of n_C chiral and n_V vector multiplets with at most two spacetime derivatives[10]

$$
\begin{aligned}
\mathscr{L}_{\text{global}} = & -g_{m\bar{n}}\Big[(\widehat{\partial}_\mu\phi^m)(\widehat{\partial}^\mu\phi^{\bar{n}}) + \overline{\chi}_L^m\,\widehat{\mathscr{D}}\chi_R^{\bar{n}} + \overline{\chi}_R^{\bar{n}}\,\widehat{\mathscr{D}}\chi_L^m\Big] \\
& + (\operatorname{Re}f_{IJ})\Big[-\frac{1}{4}\mathscr{F}_{\mu\nu}^I\mathscr{F}^{\mu\nu\,J} - \frac{1}{2}\overline{\lambda}^I\,\widehat{\mathscr{D}}\lambda^J\Big] \\
& + \frac{1}{8}(\operatorname{Im}f_{IJ})\Big[\mathscr{F}_{\mu\nu}^I\mathscr{F}_{\rho\sigma}^J\epsilon^{\mu\nu\rho\sigma} - 2i\widehat{\partial}_\mu(\overline{\lambda}^I\gamma_5\gamma^\mu\lambda^J)\Big] \qquad (5.120) \\
& + \Big\{-\frac{1}{4}f_{IJ,m}\mathscr{F}_{\mu\nu}^I\overline{\chi}_L^m\gamma^{\mu\nu}\lambda_L^J + \frac{i}{2}D^I f_{IJ,m}\overline{\chi}_L^m\lambda^J \\
& \quad + \frac{1}{4}(\partial_m W)g^{m\bar{n}}f_{IJ,\bar{n}}^*\overline{\lambda}_R^I\lambda_R^J \\
& \quad - (\mathscr{D}_m\partial_n W)\overline{\chi}_L^m\chi_L^n - 2\xi_I^{\bar{n}}g_{m\bar{n}}\overline{\lambda}^I\chi_L^m + \text{h.c.}\Big\} \\
& - V(\phi^m, \phi^{\bar{n}}) + \mathscr{L}_{4F},
\end{aligned}
$$

where $\mathscr{L}_{4F}$ denotes four Fermion terms, and

$$
\mathscr{D}_m\partial_n W \equiv \partial_m\partial_n W - \Gamma_{mn}^l\partial_l W, \tag{5.121}
$$

$$
V(\phi^m, \phi^{\bar{n}}) = V_F + V_D = g^{m\bar{n}}(\partial_m W)(\partial_{\bar{n}}W^*) + \frac{1}{2}(\operatorname{Re}f_{IJ})D^I D^J \geq 0. \tag{5.122}
$$

---

[10] In certain theories with non-gauge invariant gauge kinetic functions, generalized *Chern–Simons terms* may also occur, which are terms of the form $A^I \wedge A^J \wedge dA^K$ and $A^I \wedge A^J \wedge A^K \wedge A^L$ [5].

The supersymmetry transformations are (neglecting three Fermion terms):

$$\delta\phi^m = \bar{\epsilon}_L \chi^m_L,$$

$$\delta\chi^m_L = \frac{1}{2}\,\widehat{\partial\!\!\!/}\phi^m \epsilon_R - \frac{1}{2}g^{m\bar{n}}(\partial_{\bar{n}} W^*)\epsilon_L,$$

$$\delta A^I_\mu = -\frac{1}{2}\bar{\epsilon}\gamma_\mu \lambda^I, \qquad\qquad\qquad (5.123)$$

$$\delta\lambda^I = \frac{1}{4}\gamma^{\mu\nu}\mathscr{F}^I_{\mu\nu}\epsilon + \frac{i}{2}\gamma_5 D^I \epsilon.$$

5.3.6.4 A Familiar Special Case: Canonical Kähler Potential and Minimal Gauge Kinetic Function

In the special case

$$K = \delta_{m\bar{n}}\phi^m \phi^{\bar{n}} \qquad \text{and} \qquad f_{IJ} = \delta_{IJ}, \qquad\qquad (5.124)$$

the Lagrangian (5.120) simplifies to

$$\mathscr{L}_{\text{global}} = -\delta_{m\bar{n}}\Big[(\widehat{\partial}_\mu\phi^m)(\widehat{\partial}^\mu\phi^{\bar{n}}) + \overline{\chi}^m_L\,\widehat{\partial\!\!\!/}\chi^{\bar{n}}_R + \overline{\chi}^{\bar{n}}_R\,\widehat{\partial\!\!\!/}\chi^m_L\Big]$$

$$+\delta_{IJ}\Big[-\frac{1}{4}\mathscr{F}^I_{\mu\nu}\mathscr{F}^{\mu\nu\,J} - \frac{1}{2}\bar{\lambda}^I\,\widehat{\partial\!\!\!/}\lambda^J\Big] \qquad\qquad (5.125)$$

$$-\Big\{(\partial_m\partial_n W)\overline{\chi}^m_L\chi^n_L + 2\xi^{\bar{n}}_I\delta_{m\bar{n}}\bar{\lambda}^I\chi^m_L + \text{h.c.}\Big\} - (V_F + V_D),$$

where

$$\widehat{\partial}_\mu\chi^m_L \equiv \partial_\mu\chi^m_L - A^I_\mu(\partial_n\xi^m_I)\chi^n_L \qquad\qquad (5.126)$$

is the gauge covariant derivative of the chiral fermions (remembering that the Christoffel symbols on $\mathscr{M}_{\text{scalar}}$ now vanish).

Assuming that the gauge action on the scalar fields is linear, i.e.,

$$\delta_I\phi^m = -i\,T_I{}^m{}_n\phi^n,$$

$$\delta_I\phi^{\bar{m}} = +i\,T_I{}^{\bar{m}}{}_{\bar{n}}\phi^{\bar{n}}, \qquad\qquad (5.127)$$

where $T_I{}^m{}_n$ is Hermitian (i.e., $T_I{}^m{}_l\delta_{m\bar{n}} = \delta_{l\bar{k}}T_I{}^{\bar{k}}{}_{\bar{n}}$), and $T_I{}^{\bar{m}}{}_{\bar{n}} \equiv (T_I{}^m{}_n)^*$, one reads off the Killing vectors

$$\xi^m_I = -i\,T_I{}^m{}_n\phi^n, \qquad \xi^{\bar{m}}_I = +i\,T_I{}^{\bar{m}}{}_{\bar{n}}\phi^{\bar{n}}. \qquad\qquad (5.128)$$

Due to the Hermiticity of the $T_I{}^m{}_n$, the Kähler potential is gauge invariant. According to (5.90), the holomorphic functions r_I can then at most be an imaginary constant, $-i\eta_I$ (the Fayet–Iliopoulos terms) provided the corresponding gauge group factor is Abelian. The Killing prepotentials are then

$$\mathcal{P}_I = i[\xi_I^m \partial_m K - r_I] = [\phi^{\bar{n}} T_{I\,\bar{n}m} \phi^m - \eta_I], \tag{5.129}$$

where $T_{I\,\bar{n}m} \equiv T_I{}^l{}_m \delta_{l\bar{n}}$. The gauge covariant derivative (5.126) then reduces to the familiar expression

$$\widehat{\partial}_\mu \chi_L^m = \partial_\mu \chi_L^m + i\, A_\mu^I \, T_I{}^m{}_n \, \chi_L^n, \tag{5.130}$$

and the D-term potential becomes

$$V_D = \frac{1}{2} \sum_I \left[\phi^{\bar{n}} T_{I\,\bar{n}m} \phi^m - \eta_I \right]^2, \tag{5.131}$$

with η_I non-zero at most for Abelian factors.

5.4   Supersymmetry Breaking

Before concluding this chapter, we collect here a few remarks on supersymmetry breaking in globally supersymmetric field theories.

This is a deep and extremely well-studied subject, and entire books could be written only on this topic. Here we discuss only some very basic facts that will be strikingly different in supergravity.

Global supersymmetry can be broken explicitly or spontaneously. In the first case, one loses the restrictions, but also the control on the theory, which comes from supersymmetry. Some control is retained, however, if the supersymmetry breaking terms are "soft," by which one means that they do not generate quadratic divergences in quantum loop corrections and hence preserve some of the good UV behavior of theories with unbroken supersymmetry.

An attractive way to generate such soft supersymmetry breaking terms is by considering a supersymmetric Lagrangian that exhibits *spontaneous* supersymmetry breaking at a certain energy scale, M_S. The effective field theory below that energy scale will then contain in general soft supersymmetry breaking terms.

Spontaneous supersymmetry breaking means that one considers the theory in a vacuum state $|0\rangle$ that is not left invariant by all the supersymmetry transformations, $Q|0\rangle \neq 0$. The supersymmetry transformation of a field operator $\Phi(x)$ is given by $i\,[\bar{\epsilon}Q, \Phi(x)] \equiv \delta_{\text{susy}} \Phi(x)$, so that there must be at least one field $\Phi(x)$ with $\langle \delta_{\text{susy}} \Phi \rangle \neq 0$. If we assume a maximally symmetric spacetime, only scalar fields could possibly have a (constant) non-vanishing vacuum expectation value (vev), whereas the vev of all fermions and tensor fields must vanish. As bosonic fields transform into fermionic fields, we thus automatically have $\langle \delta_{\text{susy}} \Phi_{\text{bos}} \rangle =$

$\langle$Fermions$\rangle$ $=$ 0, and all supersymmetry breaking can only come from a non-vanishing $\langle \delta_{\text{susy}} \Phi_{\text{fer}} \rangle$.

Inspection of the supersymmetry transformations of the chiral fermions and the gaugini in (5.123) then shows that spontaneous supersymmetry breaking in global supersymmetry could occur via a non-vanishing F-term vev $\langle F \rangle \cong \langle \partial W \rangle \neq 0$ and/or a non-vanishing D-term vev $\langle D \rangle \neq 0$. Hence a useful order parameter of supersymmetry breaking is the scalar potential $V = |F|^2 + D^2 \geq 0$:

$$V(\langle \phi^m \rangle, \langle \phi^{\overline{m}} \rangle) > 0 \qquad \Leftrightarrow \qquad \text{susy breaking,}$$
$$V(\langle \phi^m \rangle, \langle \phi^{\overline{m}} \rangle) = 0 \qquad \Leftrightarrow \qquad \text{susy preserved.} \tag{5.132}$$

It should be noted that the F- and D-term contributions are completely independent in global supersymmetry. Supersymmetry is broken whenever at least one of the two is non-vanishing, and a positive cosmological constant is always generated.

The scale of supersymmetry breaking at tree level is defined by $M_S^2 \equiv \sqrt{\langle V \rangle}$, which becomes $M_S^2 \sim F$ for pure F-term SUSY breaking or $M_S^2 \sim D$ for pure D-term SUSY breaking.

In the case of spontaneous supersymmetry breaking, we should also note that several aspects of the resulting effective theory are dictated by the supertrace of the squared masses

$$Str \mathcal{M}^{2n} = \sum_{J} (-1)^{2J} (2J + 1) \mathcal{M}_J^{2n}, \tag{5.133}$$

where $\mathcal{M}_J$ is the mass of a field of spin J, and we are summing over all fields and spins, with opposite signs for bosons and fermions. For instance, for a renormalizable model, the supertrace of the squared masses computed at tree level in the non-supersymmetric vacuum by the action is proportional to the charge matrix of the chiral fields, $\mathcal{Q}$, and the expectation value of the D-terms [6]:

$$Str \mathcal{M}^2 = 2 \, tr \mathcal{Q} \langle D \rangle. \tag{5.134}$$

This imposes phenomenological constraints on the supersymmetry breaking patterns and the resulting spectrum of particles. Another example is the one-loop effective potential, which can be computed for an explicit cutoff Λ:

$$V_1 = V_0 + \frac{1}{64\pi^2} Str \mathcal{M}^0 \Lambda^4 + \frac{1}{32\pi^2} Str \mathcal{M}^2 \Lambda^2 + \frac{1}{64\pi^2} Str \mathcal{M}^4 \log \frac{\mathcal{M}^2}{\Lambda^2} + \dots \tag{5.135}$$

where the dots stand for contributions with negative powers of Λ. For spontaneous supersymmetry breaking, the quartic divergence is absent because of the equal number of bosonic and fermionic degrees of freedom. The same is generally true for the second most divergent term, which is the quadratically divergent contribution, proportional to Str $\mathcal{M}^2$. In the Standard Model, Str $\mathcal{M}^2$ depends on the Higgs field and induces a quadratically divergent contribution to the Higgs squared mass, the well-known source of the gauge hierarchy problem. In unbroken supersymmetry, Str $\mathcal{M}^2$ is identically vanishing, and, following (5.134), its vanishing persists if we have broken supersymmetry without anomalous U(1) factors. However, without anomalous U(1) factors, Str $\mathcal{M}^2 = 0$, and therefore, at least some of the sleptons and squarks should be lighter than some of the SM fermions (a more detailed discussion can be found in [7]). Hence direct supersymmetry breaking within the Standard Model sector is problematic.

The usual way out is to consider explicit supersymmetry breaking terms coming from the relation of the observable SM sector with some hidden supersymmetry breaking sector. The sparticle spectrum then depends on the messenger of supersymmetry breaking and on whether the corrections arise from direct couplings or radiative corrections. As we will discuss in Chap. 7, the non-renormalizable interactions that arise from the coupling to supergravity provide a compelling example for the first possibility.

Exercises

5.1. Prove that the Lagrangian (5.60) transforms under supersymmetry into

$$\delta \mathcal{L} = J^\rho \partial_\rho \epsilon \quad (= 0 \text{ in global supersymmetry})$$

with the supercurrent

$$J^\rho \equiv -\frac{1}{4} \delta_{IJ} F^I_{\mu\nu} \overline{\lambda}^J \gamma^\rho \gamma^{\mu\nu}.$$

5.2. Compute non-vanishing components of the Levi–Civita connection and of the Riemann tensor in terms of the metric for a Kähler manifold.

5.3. Show that the derivative $\widehat{\mathcal{D}}_\mu$ in (5.87) is simultaneously covariant with respect to field reparameterizations (5.25) and with respect to gauge transformations (5.80), (5.81), and (5.86).

5.4. Consider a model with Kähler potential

$$K = |\phi|^2 + |A|^2 + |B|^2$$

and superpotential

$$W = \phi AB.$$

The three scalar fields are charged under a U(1) symmetry, which is gauged and generates a Fayet–Iliopoulos term η. Their charges are $Q_\phi = 0$, $Q_A = 1$, and $Q_B = -1$. The kinetic term for the vector field is canonical.

(a) Compute the scalar potential.
(b) Find the supersymmetric and non-supersymmetric vacua.
(c) Compute the boson and fermion masses at the vacua.
(d) Find the range where the non-supersymmetric vacuum can be a minimum (at least in the field directions A and B).
(e) Discuss the type of supersymmetry breaking.

References

1. J. Bagger, E. Witten, The gauge invariant supersymmetric nonlinear sigma model. Phys. Lett. B **118**, 103–106 (1982)
2. H. Goldstein, *Classical Mechanics*, 2nd edn. (Addison-Wesley, Boston, 1980). ISBN 978-0-201-02918-5
3. R. Kallosh, L. Kofman, A.D. Linde, A. Van Proeyen, Superconformal symmetry, supergravity and cosmology. Class. Quant. Grav. **17**, 4269–4338 (2000) [arXiv:hep-th/0006179 [hep-th]]
4. C.M. Hull, A. Karlhede, U. Lindstrom, M. Rocek, Nonlinear σ models and their gauging in and out of superspace. Nucl. Phys. B **266**, 1–44 (1986)
5. J. De Rydt, J. Rosseel, T.T. Schmidt, A. Van Proeyen, M. Zagermann, Symplectic structure of N = 1 supergravity with anomalies and Chern-Simons terms. Class. Quant. Grav. **24**, 5201–5220 (2007). 0705.4216
6. S. Ferrara, L. Girardello, F. Palumbo, A general mass formula in broken supersymmetry. Phys. Rev. **D20**, 403 (1979)
7. S. Weinberg, *The Quantum Theory of Fields*, vol. 3. Supersymmetry (Cambridge University Press, Cambridge, 2000), 419 p.

Matter Couplings in Supergravity

6

In this chapter, we study the coupling of 4D, $\mathcal{N} = 1$ chiral and vector multiplets to supergravity, using the geometric language developed in the previous chapter. Our main emphasis will be on the difference between the matter couplings in global and local supersymmetry.

6.1   New Supergravity Couplings

As we have seen in the previous section, the general $\mathcal{N} = 1$ globally supersymmetric theory of chiral and vector multiplets is completely specified by the following data:

- The numbers, n_C and n_V, of, respectively, chiral and vector multiplets
- The Kähler potential $K(\phi^m, \phi^{\overline{m}})$ that determines the geometry of $\mathcal{M}_{\text{scalar}}$
- The holomorphic superpotential $W(\phi^m)$ that encodes the self-interactions of the chiral multiplets
- The holomorphic gauge kinetic function $f_{IJ}(\phi^m)$ related to the kinetic terms of the vector multiplets
- The action of the gauge group on $\mathcal{M}_{\text{scalar}}$, as specified by the holomorphic Killing vectors $\xi_I^m(\phi^n)$ and the corresponding Killing prepotentials $\mathcal{P}_I(\phi^m, \phi^{\overline{m}})$
- The real Fayet–Iliopoulos terms, η_I, which might be non-zero for *Abelian* gauge group factors

When we couple such a theory to supergravity, making it locally supersymmetric, there will be additional couplings of the matter multiplets to the supergravity multiplet but also new and modified couplings among the fields of the matter multiplets themselves [1–3]. All these additional or modified couplings are still completely specified by the data that already specified a theory in global supersymmetry and that were mentioned above. As we will now explain, they will appear in the Lagrangian with inverse powers of M_P.

© Springer-Verlag GmbH Germany, part of Springer Nature 2021

G. Dall'Agata, M. Zagermann, *Supergravity*, Lecture Notes in Physics 991,
https://doi.org/10.1007/978-3-662-63980-1_6

6.1.1   Coupling Chiral Multiplets to Supergravity

Let us start with the modifications that are necessary in order to make a globally supersymmetric Wess–Zumino model also invariant under local supersymmetry. The globally supersymmetric Lagrangian is (cf. (5.50))

$$
\mathscr{L}_{WZ} = -g_{m\bar{n}} \left[(\partial_\mu \phi^m)(\partial^\mu \phi^{\bar{n}}) + \overline{\chi}_L^m \overline{\mathscr{D}} \chi_R^{\bar{n}} + \overline{\chi}_R^{\bar{n}} \overline{\mathscr{D}} \chi_L^m \right]
$$

$$
- (\mathscr{D}_m \partial_n W) \overline{\chi}_L^m \chi_L^n - (\mathscr{D}_{\bar{m}} \partial_{\bar{n}} W^*) \overline{\chi}_R^{\bar{m}} \chi_R^{\bar{n}} \tag{6.1}
$$

$$
- g^{m\bar{n}} (\partial_m W)(\partial_{\bar{n}} W^*) + \mathscr{O}(\chi^4).
$$

If we now allow for a spacetime-dependent supersymmetry parameter $\epsilon = \epsilon(x)$, the derivative in the kinetic terms of the fermions χ^m will produce new terms when it acts on $\epsilon(x)$ coming from the supersymmetry transformations of the chiral fermions

$$
\delta \chi_L^m = \frac{1}{2} \slashed{\partial} \phi^m \epsilon_R - \frac{1}{2} g^{m\bar{n}} (\partial_{\bar{n}} W^*) \epsilon_L + \mathscr{O}(\chi\chi\epsilon), \tag{6.2}
$$

$$
\delta \chi_R^{\bar{m}} = \frac{1}{2} \slashed{\partial} \phi^{\bar{m}} \epsilon_L - \frac{1}{2} g^{\bar{m}n} (\partial_n W) \epsilon_R + \mathscr{O}(\chi\chi\epsilon). \tag{6.3}
$$

The result is an uncancelled variation of the form

$$
\delta \mathscr{L}_{WZ} = \overline{J}_R^\mu \, \partial_\mu \epsilon_R + \overline{J}_L^\mu \, \partial_\mu \epsilon_L, \tag{6.4}
$$

where the supercurrents are

$$
\overline{J}_L^\mu = -g_{m\bar{n}} \overline{\chi}_L^m \gamma^\mu \slashed{\partial} \phi^{\bar{n}} + \overline{\chi}_R^{\bar{n}} \gamma^\mu \partial_{\bar{n}} W^*, \qquad J_R^\mu = (J_L^\mu)^c. \tag{6.5}
$$

As we have already shown in Chap. 2 for the special case of a free Wess–Zumino model, the cancellation of these terms is achieved by adding the Noether couplings to the gravitino

$$
\mathscr{L}_{\text{Noether}} = -\frac{1}{M_P} \left[\overline{J}_R^\mu \psi_{\mu R} + \overline{J}_L^\mu \psi_{\mu L} \right]. \tag{6.6}
$$

Using $\delta \psi_\mu = M_P \, \partial_\mu \epsilon$, one then finds that everything cancels modulo terms that come from the variation of the supercurrents themselves:

$$
\delta(\mathscr{L}_{WZ} + \mathscr{L}_{\text{Noether}}) = -\frac{1}{M_P} \left[(\delta \overline{J}_R^\mu) \psi_{\mu R} + \text{h.c.} \right] \tag{6.7}
$$

These terms are of the form

$$\delta(\mathscr{L}_{WZ} + \mathscr{L}_{\text{Noether}}) = -\delta g^{\mu\nu} T_{\mu\nu} + Z_1 + Z_2. \qquad (6.8)$$

Just as discussed in Chap. 2, $T_{\mu\nu}$ is the energy momentum tensor of $\mathscr{L}_{WZ}$, and the new field $g_{\mu\nu}$ is identified with the spacetime metric, signaling the necessity for a coupling to gravity. The minimal coupling to a dynamical metric is achieved by covariantizing everything with respect to general spacetime coordinate and local Lorentz transformations and by adding the pure supergravity Lagrangian. The metric variation of this covariantized Lagrangian then precisely cancels the first term in (6.8), and the theory would be supersymmetric if there weren't also the two additional terms Z_1 and Z_2 in Eq. (6.8) that we have neglected so far. As we will now show, these two terms are actually quite important, as they lead to additional M_P^{-2}-suppressed interactions between the fields of the chiral multiplets themselves that have some far-reaching consequences.

In order to make this more precise, let us first state what Z_1 and Z_2 are:

$$Z_1 = -\frac{e}{2M_P} g_{m\bar{n}} \overline{\psi}_\mu \gamma^{\mu\nu\rho} \gamma_5 \epsilon \, (\partial_\nu \phi^m)(\partial_\rho \phi^{\bar{n}}), \qquad (6.9)$$

$$Z_2 = \frac{e}{M_P} \left[\overline{\psi}_{\mu L} \gamma^{\mu\nu} \epsilon_L \, (\partial_\nu W^*) + \overline{\psi}_{\mu R} \gamma^{\mu\nu} \epsilon_R \, (\partial_\nu W) \right]. \qquad (6.10)$$

The first term Z_1 comes from the variations of the form $\delta\chi_L^m \sim \frac{1}{2}\slashed{\partial}\phi^m \epsilon_R$ in J_L^μ and its conjugate, which give rise to terms with three antisymmetrized gamma matrices as well as terms with one gamma matrix. The latter are part of the energy momentum tensor terms in (6.8) (because $\delta g_{\mu\nu}$ involves only one gamma matrix), whereas the terms with three antisymmetrized gamma matrices are precisely given by Z_1. The first term of Z_2 is due to the variations $\delta\chi_L^m \sim -\frac{1}{2}g^{m\bar{n}}(\partial_{\bar{n}}W^*)\epsilon_L$ in the first term in (6.5) and due to the variation $\delta\chi_R^{\overline{m}} \sim \frac{1}{2}\slashed{\partial}\phi^{\overline{m}}\epsilon_L$ in the second term in (6.5). The second term in Z_2 arises from the analogous variations of J_R^μ.

We will now see that the cancellation of Z_1 and Z_2 requires the introduction of new terms with important consequences.

6.1.2 The Kähler Covariant Derivative

In order to cancel Z_1, we first rewrite it by using the relation between the metric of the scalar manifold and the Kähler potential:

$$Z_1 \sim \overline{\psi}_\mu \gamma^{\mu\nu\rho} \gamma_5 \epsilon \, (\partial_\nu \phi^m) \left(\partial_\rho \phi^{\bar{n}} \right) \partial_m \partial_{\bar{n}} K$$

$$= \overline{\psi}_\mu \gamma^{\mu\nu\rho} \gamma_5 \epsilon \, \frac{1}{2} \left(\partial_\rho \phi^{\bar{n}} \partial_\nu \partial_{\bar{n}} K - \partial_\rho \phi^m \partial_\nu \partial_m K \right). \qquad (6.11)$$

Using the last expression, we then integrate by parts the spacetime derivative that acts on the Kähler potential. This produces in particular terms where the derivative acts on ϵ and terms where it acts on $\overline{\psi}_\mu$. The former term is

$$\frac{e}{2M_P} \overline{\psi}_\mu \gamma^{\mu\nu\rho} \gamma_5 (D_\nu \epsilon) \frac{1}{2} \left(\partial_\rho \phi^{\bar{n}} \partial_{\bar{n}} K - \partial_\rho \phi^m \partial_m K \right). \tag{6.12}$$

We now repeat our old trick and simply add the negative of this term (times a factor $1/2$) to the Lagrangian, but with $D_\nu \epsilon$ replaced by ψ_ν,

$$\mathcal{L}_{\text{Kähler cov}} = -\frac{e}{2} \overline{\psi}_\mu \gamma^{\mu\nu\rho} \left(\frac{i}{2M_P^2} Q_\nu(\phi) \gamma_5 \right) \psi_\rho, \tag{6.13}$$

where Q_ν is a composite vector field,

$$Q_\nu(\phi) \equiv \frac{i}{2} \left[(\partial_{\bar{n}} K) \partial_\nu \phi^{\bar{n}} - (\partial_m K) \partial_\nu \phi^m \right]. \tag{6.14}$$

Varying the two gravitini in this expression would then precisely cancel (6.12).

The cancellation of the remaining term in Z_1, where the derivative acts on $\overline{\psi}_\mu$, will be discussed later (see footnote 1 in this chapter).

The new interaction term $\mathcal{L}_{\text{Kähler cov}}$, however, now poses another problem: as one easily verifies, it is not invariant under Kähler transformations $K \rightarrow K + h + h^*$. $\mathcal{L}_{\text{Kähler cov}}$ would thus seem to single out a particular Kähler potential, even though a specific Kähler potential is not an intrinsic geometrical object on a Kähler manifold. In general, the Kähler potential is in fact only locally defined and requires Kähler transformations on the overlaps of local coordinate patches. So if the Lagrangian was not Kähler invariant, the physics would in general also be different for different coordinate patches of the scalar manifold.

To understand the resolution of this problem, we observe that the term $\mathcal{L}_{\text{Kähler cov}}$ can be absorbed into the Rarita–Schwinger action by modifying the covariant derivative with a new term,

$$\mathcal{L}_{\text{RS}} + \mathcal{L}_{\text{Kähler cov}} = -\frac{e}{2} \overline{\psi}_\mu \gamma^{\mu\nu\rho} \mathscr{D}_\nu(\omega, Q) \psi_\rho, \tag{6.15}$$

where

$$\mathscr{D}_{[\nu}(\omega, Q) \psi_{\rho]} \equiv D_{[\nu}(\omega) \psi_{\rho]} + \frac{i}{2M_P^2} Q_{[\nu} \gamma_5 \psi_{\rho]}, \tag{6.16}$$

with D_ν being the Lorentz covariant derivative. To understand the significance of this modification, one notes that Q_μ transforms under Kähler transformations like a U(1) connection:

$$Q_\mu \rightarrow Q_\mu + \partial_\mu \text{Im}(h). \tag{6.17}$$

More precisely, Q_μ is a composite U(1) connection, i.e., it is not an elementary vector field, but rather a function of the scalar fields and their derivatives.

We now see that we can render the Lagrangian invariant if we require that Kähler transformations, $K \rightarrow K + h + h^*$, be accompanied by chiral rotations of the gravitino:

$$\psi_\mu \rightarrow \exp\left[-\frac{i}{2M_P^2}\mathrm{Im}(h(\phi))\gamma_5\right]\psi_\mu. \tag{6.18}$$

Indeed, the derivative (6.16) then transforms covariantly,

$$\mathscr{D}_{[\mu}\psi_{\rho]} \rightarrow \exp\left[-\frac{i}{2M_P^2}\mathrm{Im}(h(\phi))\gamma_5\right]\mathscr{D}_{[\mu}\psi_{\rho]}, \tag{6.19}$$

and the combination (6.15) is Kähler (and obviously also locally Lorentz) invariant. These geometric arguments thus suggest that, in supergravity, Kähler transformations on the scalar manifold also act on the gravitino as a chiral U(1) symmetry, with Q_μ being the corresponding (composite) U(1) connection. If this is to make sense, this non-trivial action of Kähler transformations on the gravitini should also be compatible with supersymmetry. As we will now show, this requirement will lead to further interesting differences with respect to global supersymmetry and provides further consistency checks.

First we note that if the gravitino transforms under Kähler transformations, the consistency with the supersymmetry transformation law $\delta\psi_\mu \sim M_P D_\mu\epsilon$ also requires that ϵ transforms under Kähler transformations,

$$\epsilon \rightarrow \exp\left[-\frac{i}{2M_P^2}\mathrm{Im}(h(\phi))\gamma_5\right]\epsilon \tag{6.20}$$

and that its derivative (as it appears in $\delta\psi_\mu$) should also be covariantized,[1]

$$\mathscr{D}_\mu(\omega, Q)\epsilon \equiv D_\mu(\omega)\epsilon + \frac{i}{2M_P^2}Q_\mu\gamma_5\epsilon. \tag{6.21}$$

This in turn implies, because of $\delta\chi_L^m = \frac{1}{2}\partial\!\!\!/\phi^m\epsilon_R + \ldots$, that also the chiral fermions transform under Kähler transformations,

$$\chi^m \rightarrow \exp\left[+\frac{i}{2M_P^2}\mathrm{Im}(h(\phi))\gamma_5\right]\chi^m, \tag{6.22}$$

---

[1] This is indeed confirmed by computing the gravitino variations of $\mathscr{L}_{RS}$ with the new Kähler covariant transformation law, $\delta\psi_\mu \sim M_P\mathscr{D}_\mu\epsilon$, which leads to a new term that precisely cancels the remaining uncancelled part of Z_1 (i.e., the part of Z_1 with a derivative acting on $\overline{\psi}_\mu$).

and that their derivatives have to be Lorentz, $\mathcal{M}_{\text{scalar}}$ reparameterization, and Kähler covariant, e.g.,[2]

$$\mathcal{D}_\mu \chi_L^m \equiv D_\mu \chi_L^m + (\partial_\mu \phi^n) \Gamma_{nl}^m \chi_L^l - \frac{i}{2M_P^2} Q_\mu \chi_L^m. \tag{6.23}$$

Note that there is a different sign in (6.22) (and hence also in (6.23)) compared to the corresponding terms of the gravitino or the supersymmetry transformation parameter (cf. (6.18) and (6.20) as well as (6.16) and (6.21)). This sign difference arises because one has to move the γ_5 matrix in (6.20) through one gamma matrix in the supersymmetry transformation $\delta \chi_L^m = \frac{1}{2} \partial\!\!\!/ \phi^m \epsilon_R + \dots$.

Although we will discuss gauge multiplets later, we already mention here that $\delta \lambda^I \sim \frac{1}{4} \gamma^{\mu\nu} \mathcal{F}_{\mu\nu}^I \epsilon + \dots$ implies that also the gaugini transform non-trivially under Kähler transformations (with the same sign as ψ_μ and ϵ)

$$\delta \lambda^I \to \exp\left[-\frac{i}{2M_P^2} \text{Im}(h(\phi)) \gamma_5 \right] \lambda^I \tag{6.24}$$

and that likewise all their derivatives have to be properly covariantized with respect to Kähler transformations (again with the same sign as for ψ_μ and ϵ).

To conclude, all fermion fields and not just the gravitino are charged with respect to a composite chiral U(1) symmetry that is related to Kähler transformations and that is not present in the global case. It should be emphasized that in the limit of global supersymmetry, $M_P \to \infty$, these chiral rotations become trivial, as is signaled by the inverse powers of M_P. This is consistent with the rigid supersymmetry results of the previous chapter, where this chiral composite U(1) is not encountered.

Interestingly, the above non-trivial transformations of the fermions under Kähler transformations also imply that the superpotential and its derivatives have to transform as we will show in Sect. 6.1.3. The result is that

$$W \to \exp\left[-\frac{1}{M_P^2} h(\phi^m) \right] W(\phi^m) \tag{6.25}$$

and its derivatives have to be Kähler covariantized as follows

$$\partial_n W \to e^{\frac{K}{2M_P^2}} \mathcal{D}_n W \equiv e^{\frac{K}{2M_P^2}} \left[\partial_n + \frac{(\partial_n K)}{M_P^2} \right] W. \tag{6.26}$$

---

[2] For the sake of simplicity, we do not introduce a new symbol for the Kähler covariantized derivative and still call it $\mathcal{D}_\mu$.

To summarize: The cancellation of Z_1 by adding $\mathcal{L}_{\text{Kähler cov}}$ gives rise to the interpretation that the fermions and the superpotential should transform non-trivially under Kähler transformations. In order to ensure this, all derivatives of the fermions and the superpotential have to be Kähler covariantized, and the superpotential terms have to be dressed with an exponential of the Kähler potential. One can show that all these modifications are indeed also necessary for the cancellation of various other variations we have not discussed here in detail. In general, we define the Kähler covariant derivatives in field space as

$$D_m \Phi = \left(\partial_m + \frac{p}{M_P^2} \partial_m K \right) \Phi,$$

$$\overline{D}_{\overline{m}} \Phi = \left(\partial_{\overline{m}} - \frac{p}{M_P^2} \partial_{\overline{m}} K \right) \Phi,$$

$$(6.27)$$

where p is the Kähler "charge" of the field Φ.

6.1.3 Additional Bare Superpotential Terms

In global supersymmetry, all superpotential terms always appear with at least one derivative with respect to the scalar fields (see, e.g., the Lagrangian (5.120)). As we saw in the previous subsection, the coupling to supergravity (in particular the cancellation of the term Z_1) requires a Kähler covariantization of these derivatives of W, which then introduces "bare" W-terms inside these Kähler covariant derivatives, i.e., W-terms that are not differentiated with respect to any scalar field. In this subsection, we show that there are additional "bare" superpotential terms in the Lagrangian and the supersymmetry transformation laws. Their necessity follows from the cancellation of the term Z_2 to which we now turn.

In order to cancel the Z_2-term

$$Z_2 \equiv \frac{e}{M_P} \left[\overline{\psi}_{\mu L} \gamma^{\mu \nu} \epsilon_L (\partial_\nu W^*) + \overline{\psi}_{\mu R} \gamma^{\mu \nu} \epsilon_R (\partial_\nu W) \right] \tag{6.28}$$

we proceed as we did for Z_1 and first perform an integration by parts. This will then give again terms with a derivative acting on the supersymmetry parameter ϵ and terms where the derivative acts on the gravitini $\overline{\psi}_\mu$. To cancel the former, we then again add to the Lagrangian a term where the derivatives of ϵ are replaced by gravitini, or, more precisely,

$$\frac{e}{2M_P^2} \left[W^* \overline{\psi}_{\mu L} \gamma^{\mu \nu} \psi_{\nu L} + W \overline{\psi}_{\mu R} \gamma^{\mu \nu} \psi_{\nu R} \right]. \tag{6.29}$$

This term is an obvious mass-like term for the gravitino, and therefore, following the rules we have learned in the case of pure supergravity in the presence of a cosmological constant, we have to further modify the variation of the gravitino field by adding a new term of the form

$$\delta_{\text{new}} \psi_{\mu L} \sim \frac{1}{2M_P} W \gamma_\mu \epsilon_R. \tag{6.30}$$

This new variation applied to the Rarita–Schwinger action also gives the term required to cancel the second piece coming from the partial integration of Z_2, namely, the term with the derivative acting on the gravitino.

Before proceeding further, let us come back to the Kähler covariantization of the superpotential terms and prove (6.25). Subjecting (6.30) to Kähler transformations tells us that the left-hand side transforms as

$$\exp\left[-\frac{1}{4M_P^2} (h(\phi) - h^*(\phi)) \right], \tag{6.31}$$

while the epsilon parameter on the right-hand side transforms with the opposite sign due to the opposite chirality:

$$\exp\left[+\frac{1}{4M_P^2} (h(\phi) - h^*(\phi)) \right]. \tag{6.32}$$

At this point it is obvious that in order for the Kähler transformation to be compatible with supersymmetry, we need to transform also the superpotential, as we already mentioned earlier. The superpotential, on the other hand, is a *holomorphic* function by construction and hence can transform only with a holomorphic factor,

$$W \to \exp\left[-\frac{\alpha}{M_P^2} (h(\phi)) \right] W, \tag{6.33}$$

where α is a real constant. In order to get the same rotation on the left- and on the right-hand side of (6.30), we still need something that transforms under Kähler transformations with the exponential of $h + h^*$, like the exponential of the Kähler potential itself, $e^{\beta K/M_P^2}$. The right coefficients follow then by equating the two sides:

$$-\frac{1}{4M_P^2} (h(\phi) - h^*(\phi)) = +\frac{1}{4M_P^2} (h(\phi) - h^*(\phi)) - \frac{\alpha}{M_P^2} (h(\phi))$$

$$+ \frac{\beta}{M_P^2} (h(\phi) + h^*(\phi)). \tag{6.34}$$

This fixes $\alpha = 1$ and $\beta = 1/2$ and tells us that we have to replace the superpotential with the combination

$$e^{K/(2M_P^2)}W \tag{6.35}$$

and that indeed W transforms under Kähler transformations as in (6.25).

Coming back to the check of supersymmetry invariance, we now see that the new transformation law for the gravitino (6.30) applied to the new bilinear term (6.29) gives a new variation of the form $|W|^2\overline{\psi}\gamma\epsilon$. Not too surprisingly, this can then finally be cancelled by adding a new contribution $\sim -e|W|^2$ to the scalar potential and varying the vierbein determinant e. This is the generalization of the procedure derived in Sect. 4.2 for the case of a constant superpotential, i.e., for pure supergravity with a cosmological constant.

Although it may be hard to believe, it turns out that, after proper Kähler covariantizations, the above modifications are sufficient to ensure also the cancellations of all the other variations we have not considered explicitly here.

The end result is the Lagrangian

$$
\begin{aligned}
e^{-1}\mathscr{L} = {} & \frac{M_P^2}{2} R(e, \omega(e)) - \frac{1}{2}\overline{\psi}_\mu \gamma^{\mu\nu\rho}\mathscr{D}_\nu(\omega(e), Q)\psi_\rho \\[4pt]
& - g_{m\bar{n}}\left[(\partial_\mu \phi^m)(\partial^\mu \phi^{\bar{n}}) + \overline{\chi}_L^m \not{\mathscr{D}}\chi_R^{\bar{n}} + \overline{\chi}_R^{\bar{n}}\not{\mathscr{D}}\chi_L^m\right] \\[4pt]
& - \left\{e^{K/2M_P^2}(\mathscr{D}_m\mathscr{D}_n W)\overline{\chi}_L^m \chi_L^n + \text{h.c.}\right\} \\[4pt]
& + \frac{1}{M_P}\left\{g_{m\bar{n}}\overline{\psi}_{\mu L}\gamma^\nu\gamma^\mu \chi_L^m(\partial_\nu \phi^{\bar{n}}) + \overline{\psi}_{\mu R}\gamma^\mu \chi_L^m e^{K/2M_P^2}\mathscr{D}_m W + \text{h.c.}\right\} \\[4pt]
& + \frac{1}{2M_P^2}\left\{e^{K/2M_P^2}W\overline{\psi}_{\mu R}\gamma^{\mu\nu}\psi_{\nu R} + \text{h.c}\right\} - V(\phi^m, \phi^{\bar{n}}),
\end{aligned}
\tag{6.36}
$$

with the scalar potential given by the sum of two contributions

$$V = e^{K/M_P^2}\left[g^{m\bar{n}}(\mathscr{D}_m W)(\mathscr{D}_{\bar{n}}W^*) - \frac{3|W|^2}{M_P^2}\right], \tag{6.37}$$

where the first term is the Kähler covariantization of the F-terms from global supersymmetry and the second is a genuine contribution from gravitational couplings, in the sense that it is a variation of the vierbein determinant that leads to a cancellation of the $|W|^2$ terms mentioned after (6.35). In the Lagrangian (6.36), the first line is the Kähler covariantization of the pure supergravity action. The second and third line correspond to the Kähler and spacetime covariant Wess–Zumino action (without the potential). Note that now

$$\mathscr{D}_m \mathscr{D}_n W = \left(\partial_m + \frac{\partial_m K}{M_P^2}\right)\left[\left(\partial_n + \frac{\partial_n K}{M_P^2}\right) W\right] - \Gamma_{mn}{}^p \left(\partial_p + \frac{\partial_p K}{M_P^2}\right) W.$$

(6.38)

The fourth line is the Kähler and spacetime covariant Noether coupling of the supercurrents to the gravitino, $\mathscr{L}_{\text{Noether}} = -\frac{1}{M_P}\left[\bar{J}_R^\mu \psi_{\mu R} + \bar{J}_L^\mu \psi_{\mu L}\right]$ (with the fermions moved into a different order). The fifth line, finally, contains the W-dependent extra terms as well as the (Kähler covariantized) scalar potential of the Wess–Zumino model.

The supersymmetry transformation rules, up to three fermion terms, are

$$\delta e_\mu^a = \frac{1}{2M_P}\bar{\epsilon}\gamma^a \psi_\mu,$$

$$\delta \psi_{\mu L} = M_P \mathscr{D}_\mu(\omega(e), Q_\nu)\epsilon_L + \frac{1}{2M_P}e^{K/2M_P^2} W \gamma_\mu \epsilon_R,$$

$$\delta \phi^m = \bar{\epsilon}_L \chi_L^m,$$

$$\delta \chi_L^m = \frac{1}{2}\slashed{\partial}\phi^m \epsilon_R - \frac{1}{2}g^{m\bar{n}}e^{K/2M_P^2}(\mathscr{D}_{\bar{n}}W^*)\epsilon_L.$$

(6.39)

Obviously, in the $M_P \to \infty$ limit, these equations reduce to the globally supersymmetric theory discussed earlier in Eqs. (5.50) and (5.56)–(5.59). One also notices that truncating out the chiral multiplets and keeping a constant superpotential $e^{\frac{K}{2M_P^2}}W = -g\,M_P^3$ gives back the pure supergravity Lagrangian with cosmological constant, Eq. (4.60).

Note further that in supergravity the Kähler potential and the superpotential are no longer independent, as one can shift terms back and forth via Kähler transformations. In fact, as long as W is not equal to zero, one can even make the superpotential equal to M_P^3 by performing a Kähler transformation with $h(\phi) = M_P^2 \log(W/M_P^3)$ (cf. Eq. (6.25)). More generally, instead of using the two functions K and W, one can express the entire Lagrangian in terms of the function

$$\mathscr{G} = K + M_P^2 \log \frac{|W|^2}{M_P^6},$$

(6.40)

which is manifestly Kähler invariant. For instance, the part of the scalar potential coming from the superpotential becomes

$$V = e^{\mathscr{G}/M_P^2}\left(M_P^2 g^{m\bar{n}}\mathscr{G}_m \mathscr{G}_{\bar{n}} - 3M_P^4\right).$$

(6.41)

Note, however, that by doing so, one cannot recover the $W = 0$ case, which has to be discussed separately. Hence the usefulness of leaving explicit both K and W in our approach.

6.1.4 Inclusion of Vector Multiplets

The inclusion of vector multiplets requires the following changes:

1. All terms that were already present in global supersymmetry (cf. Eq. (5.120)) are also present in supergravity, but they all have to be made spacetime and Kähler covariant.
2. A new Noether coupling of the vector multiplet supercurrent

$$\bar{J}^\mu_{VM} \equiv e\bar{\lambda}^J \left[-\frac{1}{4} (\mathrm{Re}\, f_{IJ}) \mathcal{F}^I_{\nu\rho} \gamma^\mu \gamma^{\nu\rho} - \frac{i}{2} \mathcal{P}_J \gamma^\mu \gamma_5 \right] \tag{6.42}$$

to the gravitino has to be introduced:

$$\mathcal{L}'_{\text{Noether}} = -\frac{1}{M_P} \bar{J}^\mu_{VM} \psi_\mu \tag{6.43}$$

in order to cancel terms of the form

$$\delta\mathcal{L} = \bar{J}^\mu_{VM} \partial_\mu \epsilon \tag{6.44}$$

that arise due to the derivative in $-\frac{1}{2}e\,(\mathrm{Re}\, f_{IJ}) \bar{\lambda}^I \widehat{\partial}\lambda^J$ when it acts on the ϵ in $\delta\lambda$.

3. The composite Kähler connection Q_μ receives an additional contribution proportional to $A^I_\mu \mathcal{P}_I$ for each of the gauged isometries:

$$Q_\mu = Q_\mu(\phi^m, \phi^{\bar{n}}, A^I_\mu) = \frac{i}{2} \left((\partial_{\bar{n}} K) \partial_\mu \phi^{\bar{n}} - (\partial_m K) \partial_\mu \phi^m \right) + A^I_\mu \mathcal{P}_I \tag{6.45}$$

$$= \frac{i}{2} \left((\partial_{\bar{n}} K) \widehat{\partial}_\mu \phi^{\bar{n}} - (\partial_m K) \widehat{\partial}_\mu \phi^m \right)$$

$$+ A^I_\mu \mathrm{Im}(r_I), \tag{6.46}$$

where the last equality follows from the form (5.94) of the prepotentials (we will see more on this in Sect. 6.1.5). This additional term is needed, e.g., in order to cancel a variation proportional to $\mathcal{F}^J_{\mu\nu} \mathcal{P}_J \bar{\epsilon} \gamma^{\mu\nu\rho} \gamma_5 \psi_\rho$ that occurs in the variation $-\frac{1}{M_P} \delta(\bar{J}^\mu_{VM}) \psi_\mu$ and is not of the form $-\delta g^{\mu\nu} T_{\mu\nu}$.

It should be noted that this additional contribution to Q_μ has another important consequence. Namely, if one shifts the Killing prepotential $\mathcal{P}_I$ of an Abelian factor by a Fayet–Iliopoulos constant, $\mathcal{P}_I \to \mathcal{P}_I + \eta_I$, one introduces new chiral gauge interactions for all fermions, including, e.g., the gravitino,

$$\mathcal{D}_\mu \psi_\nu \to \mathcal{D}_\mu \psi_\nu + \frac{i}{2M_P^2} A_\mu^I \eta_I \gamma_5 \psi_\nu, \tag{6.47}$$

which can easily lead to quantum anomalies [4]. Thus, the introduction of Fayet–Iliopoulos constants in $\mathcal{N} = 1$ supergravity requires some care. We will actually see later on that in supergravity the Fayet–Iliopoulos terms are related to the non-invariance of the superpotential under gauge transformations.

Ignoring four fermion terms, the end result of all these modifications is the following general matter-coupled Lagrangian[3]

$$
\begin{aligned}
e^{-1}\mathcal{L} = {}& \frac{M_P^2}{2} R(e, \omega(e)) - \frac{1}{2}\overline{\psi}_\mu \gamma^{\mu\nu\rho} \mathcal{D}_\nu(\omega(e), Q)\psi_\rho \\
& - g_{m\bar{n}}\left[(\widehat{\partial}_\mu \phi^m)(\widehat{\partial}^\mu \phi^{\bar{n}}) + \overline{\chi}_L^m \, \widehat{\mathcal{D}} \chi_R^{\bar{n}} + \overline{\chi}_R^{\bar{n}} \, \widehat{\mathcal{D}} \chi_L^m \right] \\
& + (\mathrm{Re}\, f_{IJ})\left[-\frac{1}{4}\mathcal{F}_{\mu\nu}^I \mathcal{F}^{\mu\nu J} - \frac{1}{2}\overline{\lambda}^I \, \widehat{\mathcal{D}} \lambda^J \right] \\
& + \frac{1}{8}(\mathrm{Im}\, f_{IJ})\left[\mathcal{F}_{\mu\nu}^I \mathcal{F}_{\rho\sigma}^J \epsilon^{\mu\nu\rho\sigma} - 2i\, \widehat{\mathcal{D}}_\mu (e\overline{\lambda}^I \gamma_5 \gamma^\mu \lambda^J) \right] \\
& + \left\{ -\frac{1}{4} f_{IJ,m}\mathcal{F}_{\mu\nu}^I \overline{\chi}_L^m \gamma^{\mu\nu} \lambda_L^J + \frac{i}{2}D^I f_{IJ,m}\overline{\chi}_L^m \lambda^J \right. \\
& + \frac{1}{4}e^{K/2M_P^2}(\mathcal{D}_m W)g^{m\bar{n}} f_{IJ,\bar{n}}^* \overline{\lambda}_R^I \lambda_R^J \\
& \left. - e^{K/2M_P^2}(\mathcal{D}_m \mathcal{D}_n W)\overline{\chi}_L^m \chi_L^n - 2\xi_I^{\bar{n}} g_{m\bar{n}}\overline{\lambda}^I \chi_L^m + \text{h.c.} \right\} \\
& + \frac{1}{4M_P}(\mathrm{Re}\, f_{IJ})\overline{\psi}_\mu \gamma^{\nu\rho}\gamma^\mu \lambda^J \mathcal{F}_{\nu\rho}^I \\
& + \left\{ \frac{1}{M_P} g_{m\bar{n}}\overline{\psi}_{\mu L}\gamma^\nu \gamma^\mu \chi_L^m (\widehat{\partial}_\nu \phi^{\bar{n}}) + \text{h.c.} \right\} \\
& + \frac{1}{M_P}\left\{ \overline{\psi}_{\mu R}\gamma^\mu \left[\frac{i}{2}\lambda_L^I \mathcal{P}_I + \chi_L^m e^{K/2M_P^2}\mathcal{D}_m W \right] + \text{h.c.} \right\} \\
& + \frac{1}{2M_P^2}\left\{ e^{K/2M_P^2} W\overline{\psi}_{\mu R}\gamma^{\mu\nu}\psi_{\nu R} + \text{h.c} \right\} - V(\phi^m, \phi^{\bar{n}}), \tag{6.48}
\end{aligned}
$$

---

[3] As mentioned earlier, when the gauge kinetic function is not gauge invariant, generalized Chern–Simons terms of the form $A^I \wedge A^J \wedge dA^K$ and $A^I \wedge A^J \wedge A^K \wedge A^L$ may be possible. Their form, however, is the same as in global supersymmetry [5].

with the scalar potential

$$V = e^{K/M_P^2} \left[g^{m\bar{n}} (\mathcal{D}_m W)(\mathcal{D}_{\bar{n}} \overline{W}) - \frac{3|W|^2}{M_P^2} \right] + \frac{1}{2} (\mathrm{Re}\, f_{IJ}) D^I D^J. \tag{6.49}$$

The supersymmetry transformation rules, up to three fermion terms, are

$$\delta e_\mu^a = \frac{1}{2M_P} \bar{\epsilon} \gamma^a \psi_\mu,$$

$$\delta \psi_{\mu L} = M_P \mathcal{D}_\mu(\omega(e), Q) \epsilon_L + \frac{1}{2M_P} e^{K/2M_P^2} W \gamma_\mu \epsilon_R,$$

$$\delta \phi^m = \bar{\epsilon}_L \chi_L^m,$$

$$\delta \chi_L^m = \frac{1}{2} \widehat{\partial} \phi^m \epsilon_R - \frac{1}{2} g^{m\bar{n}} e^{K/(2M_P^2)} (\mathcal{D}_{\bar{n}} W^*) \epsilon_L,$$

$$\delta A_\mu^I = -\frac{1}{2} \bar{\epsilon} \gamma_\mu \lambda^I,$$

$$\delta \lambda^I = \frac{1}{4} \gamma^{\mu\nu} \mathcal{F}_{\mu\nu}^I \epsilon + \frac{i}{2} \gamma_5 D^I \epsilon. \tag{6.50}$$

It is again easy to see that, in the global limit, $M_P \to \infty$, the above equations reduce to the globally supersymmetric theory discussed in Eqs. (5.120) and (5.123).

For completeness, we display the full (i.e., local Lorentz, scalar coordinate, Kähler, and gauge covariant) derivative of λ^I and χ^m:

$$\widehat{\mathcal{D}}_\mu \chi_L^m = D_\mu \chi_L^m + (\widehat{\partial}_\mu \phi^n) \Gamma_{nl}^m \chi_L^l - A_\mu^I (\partial_n \xi_I^m) \chi_L^n - \frac{i}{2M_P^2} Q_\mu \chi_L^m,$$

$$\widehat{\mathcal{D}}_\mu \lambda^I = D_\mu \lambda^I + A_\mu^J f_{JK}^I \lambda^K + \frac{i}{2M_P^2} Q_\mu \gamma_5 \lambda^I, \tag{6.51}$$

where, as usual, D_μ denotes the Lorentz covariant derivative. The full covariant derivatives of ψ_μ and ϵ are just as for λ^I, except for the gauge covariantization term $A_\mu^J f_{JK}^I \lambda^K$, which is absent for these fermions (hence, we can omit the hat on their derivatives).

More details on the action and the four Fermi terms can be found in [5, 6].

6.1.5 More on D-Terms

Although the D-terms and the D-term potential take the same form as in global supersymmetry, local supersymmetry does have some interesting implications also for the D-terms. To understand this, we recall that the general matter-coupled supergravity Lagrangian is invariant under Kähler transformations that act at the same time on the Kähler potential, the superpotential, and the fermions. As in global supersymmetry, a gauge transformation therefore does not necessarily have to leave the Kähler potential invariant but may in general transform it with a Kähler transformation,

$$\delta_{\text{gauge}} K \equiv \xi_I^m \partial_m K + \xi_I^{\overline{m}} \partial_{\overline{m}} K = r_I + r_I^*. \tag{6.52}$$

However, in supergravity theories this requires a non-trivial action also on the superpotential (if $W \neq 0$)

$$\delta_{\text{gauge}} W \equiv \xi_I^m \partial_m W = -\frac{r_I}{M_P^2} W, \tag{6.53}$$

so that the combination $\mathscr{G}$ in (6.40) remains invariant:

$$\delta_{\text{gauge}} \mathscr{G} = \xi_I^m \partial_m \mathscr{G} + \xi_I^{\overline{m}} \partial_{\overline{m}} \mathscr{G} = 0 \tag{6.54}$$

For all the points in field space where $W \neq 0$, we can then also rewrite r_I as

$$r_I = -M_P^2 \, \xi_I^m \frac{\partial_m W}{W}, \tag{6.55}$$

so that the Killing prepotentials can be also expressed in terms of the gauge-invariant quantity $\mathscr{G}$:

$$\mathscr{P}_I = i \xi_I^m \partial_m K - i r_I = i \xi_I^m \left[\partial_m K + \frac{M_P^2 \partial_m W}{W} \right] = i \xi_I^m \frac{M_P^2 \mathscr{D}_m W}{W} = i \xi_I^m \partial_m \mathscr{G}. \tag{6.56}$$

The D-terms are thus

$$D^I = i \, (\text{Re } f)^{-1IJ} \, \xi_J^m \, \partial_m \mathscr{G}, \tag{6.57}$$

and the total scalar potential with F-terms and D-terms can be written in a very compact and suggestive form

$$V = e^{\mathscr{G}/M_P^2} \left[h^{m\overline{n}} \mathscr{G}_m \mathscr{G}_{\overline{n}} - 3 \right] M_P^2, \tag{6.58}$$

where

$$h^{m\overline{n}} \equiv g^{m\overline{n}} + \frac{e^{-\mathscr{G}/M_P^2}}{2} \, (\text{Re } f)^{-1IJ} \, \xi_I^m \xi_J^{\overline{n}}, \tag{6.59}$$

so that the new metric contains the Kähler metric giving the F-term potential (6.41) and the additional term coming from the D-term potential.

We can now comment on some of the differences between global and local supersymmetry, which may lead to relevant physical differences.

First of all, we see that in supergravity, D-terms and F-terms are not independent of one another; rather the D-terms are (for $W \neq 0$) a particular combination of the F-terms (6.57).

The next important difference concerns the Fayet–Iliopoulos constants. Just as in global supersymmetry, the gauge transformation (6.52) of the Kähler potential fixes r_I only up to an additive imaginary constant $i \eta_I$ and hence $\mathscr{P}_I$ up to an additive real constant η_I. The equivariance condition (5.96), together with gauge invariance of the other terms in the action, again restricts the possible values for these constants except for U(1) factors. The difference to global supersymmetry now is that the superpotential W also transforms under gauge transformations as in (6.53) so that a shift of r_I by an additive constant $i \eta_I$ implies that W transforms with an additional phase factor under the corresponding U(1) transformation. In other words, changing η_I changes the U(1) charge of W. Note that such U(1) transformations due to FI constants may even occur when the Kähler potential is invariant under this U(1) factor, because, according to (6.52), this only implies $r_I(\phi) = i \eta_I$.

Another important effect of a FI constant is that it leads to a chiral U(1) transformation of the fermions, as follows from their non-trivial transformations under Kähler transformations described in Sect. 6.1.2. As explained around (6.47), this may then easily lead to anomalous gauge couplings and requires some care.

6.1.6 The Gradient Flow Relations

In supergravity theories, supersymmetry imposes some interesting relations between various quantities. Particularly interesting are the relations between the potential and the supersymmetry transformations of the fermions as well as the relations between the supersymmetry transformations of the fermions themselves, known as *gradient flow relations* [7].

When the metric and the scalar fields are the only non-trivial fields, the supersymmetry rules can be written as

$$\delta \psi_{\mu L} = M_P \mathscr{D}_\mu \epsilon_L + \gamma_\mu S \epsilon_R, \tag{6.60}$$

$$\delta \chi_L^m = \frac{1}{2} \widehat{\partial} \phi^m \epsilon_R + \mathscr{N}^m \epsilon_L, \tag{6.61}$$

$$\delta \lambda_L^I = \frac{1}{4} \gamma^{\mu\nu} \mathscr{F}_{\mu\nu}^I \epsilon_L + N^I \epsilon_L. \tag{6.62}$$

Here, S is the fermionic shift in the gravitino transformation,

$$S \equiv \frac{1}{2 M_P} e^{\frac{K}{2 M_P^2}} W, \tag{6.63}$$

whereas the shift of the chiralini is the Kähler covariantized F-term,

$$\mathcal{N}^m \equiv -\frac{1}{2} g^{m\bar{n}} e^{\frac{K}{2M_P^2}} \mathcal{D}_{\bar{n}} W^*, \qquad \mathcal{N}^{\bar{m}} \equiv (\mathcal{N}^m)^*, \qquad (6.64)$$

and the shift of the gaugini is given by the D-terms,

$$N^I \equiv \frac{i}{2} D^I. \qquad (6.65)$$

These fermionic shifts also enter various bilinear terms in the fermions in the Lagrangian as well as the scalar potential, which is simply given by the squares of the shifts in (6.60)–(6.62), with appropriate signs and prefactors:

$$V = -12SS^* + 4g_{m\bar{n}} \mathcal{N}^m \mathcal{N}^{\bar{n}} + 2\mathrm{Re}(f_{IJ}) N^I N^{*J}. \qquad (6.66)$$

Writing the potential in this way makes a number of interesting properties of the scalar potential obvious, as we will explain in our discussion on spontaneous supersymmetry breaking in Sect. 7.1.2. Here, we would like to list instead a number of useful differential relations between the shifts. These are often called gradient flow relations because for supersymmetric configurations, the behavior of the scalar fields is fixed in terms of gradients of the gravitino shift and the other fermion shifts and fermion masses are further determined by additional gradients in field space.

Let us start by noting that the gravitino shift S is covariantly holomorphic[4]

$$\mathcal{D}_{\bar{m}} S = 0. \qquad (6.67)$$

The shift of matter fields is on the other hand equal to the Kähler covariant derivative of the shift of the gravitino with respect to the scalar fields,

$$\mathcal{D}_m S = -\frac{2}{M_P} g_{m\bar{n}} \mathcal{N}^{\bar{n}}. \qquad (6.68)$$

Furthermore, full covariant derivatives of the shift of the chiralini give the mass matrices for the gravitini

$$\mathcal{D}_m \mathcal{N}^n = -\delta^n_m \frac{S}{M_P}, \qquad (6.69)$$

or for the chiralini

$$\mathcal{D}_m \mathcal{N}^{\bar{n}} g_{l\bar{n}} = \frac{1}{2} \mathcal{M}_{ml}, \qquad (6.70)$$

---

[4] Recall the definition (6.27) for the covariant derivatives, which here are taken in fields space. In the current context $p(S) = -p(S^*) = \frac{1}{2}$.

where

$$\mathcal{M}_{mn} = e^{K/2M_P^2}(\mathcal{D}_m \mathcal{D}_n W). \tag{6.71}$$

Finally, the full covariant derivative of the shift of the gaugini is proportional to the gauged isometries:

$$\mathcal{D}_m \left(N^{I*} \mathrm{Re}\, f_{IJ} \right) = \frac{1}{2} g_{m\bar{n}} \xi_J^{\bar{n}}. \tag{6.72}$$

The origin of these relations is obviously related to the request of invariance of the action under supersymmetry variations. Indeed all these relations are needed to compensate for the supersymmetry variation of the scalar fields in the potential and of the mass terms. For instance, schematically, we can see that

$$\frac{\partial V}{\partial \phi^m} \sim S \mathcal{N}_m + \mathcal{M}_{mn} \mathcal{N}^n + g_{m\bar{n}} \xi_I^{\bar{n}} N^I, \tag{6.73}$$

and therefore the gradient flow relations mentioned above are necessary to close supersymmetry, because such a relation is needed for the cancellation of $\delta\mathcal{L}_{\text{fermion mass}}$ against δV.

6.1.7 Final Remarks

The discussion on the matter couplings we outlined in this chapter clearly depends heavily on what kind of matter we allow in these couplings. More general Lagrangians could be obtained by introducing additional matter multiplets, such as tensor multiplets. The tensor fields in such tensor multiplets, however, can in general be dualized to either massless scalar or massive vector fields so that the theory will be eventually of the standard form we described above. On the other hand, these dualities are often non-perturbative or may require complicated field redefinitions. It may therefore be interesting to study these models directly with tensor fields, also because such tensor fields naturally arise from string theory compactifications We are not going to do so in the following, though we would like to show here explicitly the duality relations.

The best way to exhibit the duality relations in four dimensions between tensor and scalar fields (when massless) or between tensor and vector fields (when massive) is by means of Lagrange multipliers. Consider the action

$$S = \frac{1}{2} \int V_1 \wedge \star V_1 - \int B_2 \wedge dV_1, \tag{6.74}$$

where V_1 is a generic one-form and B_2 is the two-form built from an antisymmetric tensor field $B_{\mu\nu}$. From the variation of the action, we get two equations of motion

$$\frac{\delta S}{\delta B_2} = 0, \qquad \Leftrightarrow \qquad dV_1 = 0, \tag{6.75}$$

$$\frac{\delta S}{\delta V_1} = 0, \qquad \Leftrightarrow \qquad \star V_1 = dB_2 \equiv H_3. \tag{6.76}$$

Recall that in four dimensions $\star^2 F_p = -(-1)^{p(4-p)} F_p$. If one solves the first equation, then $V_1 = d\phi$ locally, and one can use the second equation to replace the two-form B in the action to go back to the action for a free scalar field

$$S = \frac{1}{2} \int d\phi \wedge \star d\phi = -\frac{1}{2} \int d^4x \, e \, \partial_\mu \phi \partial^\mu \phi. \tag{6.77}$$

On the other hand, the solution to the second equation (6.76) gives a replacement rule for the one-form, V_1, in terms of B_2 and leads to the standard kinetic term for B_2:

$$S = \frac{1}{2} \int H_3 \wedge \star H_3. \tag{6.78}$$

Thus, we see that in four spacetime dimensions a massless scalar field is equivalent to a massless antisymmetric tensor field. This implies that in a supersymmetric theory one can replace one of the scalar fields in a massless chiral multiplet by a two-form field, producing an off-shell linear representation of supersymmetry called tensor multiplet, which contains a scalar field, a fermion, and a tensor field. In theories with extended or with on-shell representations of supersymmetry, the number of scalar fields that can be dualized simultaneously may vary, and one can then obtain tensor multiplets, double-tensor multiplets, and also vector-tensor multiplets. Although we do not have a direct interest in these multiplets here, it is important to mention that tensor fields do play an important role in extended supergravities when the gauging procedure is performed, as we explain in Chap. 9.

For the massive case, we can also produce a duality relation between a two-form tensor and a vector field. Once again, we can introduce the duality relation by means of Lagrange multipliers:

$$S = -\frac{1}{2} \int \left[H_3 \wedge \star H_3 - 2H_3 \wedge \star dB_2 + m^2 B_2 \wedge \star B_2 \right]. \tag{6.79}$$

The equations of motion for H_3 and B_2 are respectively

$$\frac{\delta S}{\delta H_3} = 0, \qquad \Leftrightarrow \qquad H_3 - d B_2 = 0, \tag{6.80}$$

$$\frac{\delta S}{\delta B_2} = 0, \qquad \Leftrightarrow \qquad d(\star H_3) + m^2 \star B_2 = 0. \tag{6.81}$$

The first one can be solved in the usual way giving back the Lagrangian for a massive tensor

$$S = -\frac{1}{2} \int [d B_2 \wedge \star d B_2 - m^2 B_2 \wedge \star B_2]. \tag{6.82}$$

The second one is instead solved by

$$B_2 = \frac{1}{m^2} \star d \star H_3, \tag{6.83}$$

which, inserted back in the action, gives the action for a massive one form if we identify $A_1 = \star H_3$:

$$S = \int \left[\frac{1}{m^2} d A_1 \wedge \star d A_1 - A_1 \wedge \star A_1 \right]. \tag{6.84}$$

Generically, we do not introduce massive fields directly in our Lagrangian, but rather vector fields become massive by means of gauge symmetry breaking, where some of the scalar fields that are charged under a gauge symmetry are eaten by the vector fields. Whenever this happens, we cannot dualize anymore the (massless) scalar fields involved in the Higgs mechanism as described before. However, we can see that, just like massless scalars are dual to massless tensor fields, the massive vectors fields resulting from the spontaneous symmetry breaking are dual to massive tensor fields. Once we are in the broken phase, the massive vector can be described by a Stückelberg coupling everywhere in field space, and therefore we can always find a field redefinition such that any of the scalar fields involved in the gauging is coupled electrically to a vector field as

$$A_\mu \equiv \partial_\mu \phi + e \tilde{A}_\mu, \tag{6.85}$$

which may be seen as the definition of the massive vector field A_μ, so that (6.84) becomes

$$S = \int \left[\frac{e^2}{m^2} d \tilde{A}_1 \wedge \star d \tilde{A}_1 - \left(d\phi + e \tilde{A}_1 \right) \wedge \star \left(d\phi + e \tilde{A}_1 \right) \right]. \tag{6.86}$$

The dual procedure by which the tensor field gets a mass is similar, but with vector fields coupled to tensors

$$B_{\mu\nu} \equiv 2\,\partial_{[\mu}A_{\nu]} + m\,\tilde{B}_{\mu\nu}, \tag{6.87}$$

so that (6.82) becomes

$$S = -\frac{1}{2}\int\left[d\tilde{B}_2 \wedge \star d\tilde{B}_2 - \left(dA_1 + m\,\tilde{B}_2\right)\wedge\star\left(dA_1 + m\,\tilde{B}_2\right)\right]. \tag{6.88}$$

Other interesting couplings arise by considering higher derivative corrections to the Lagrangian, which may become important when one discusses supergravity as an effective theory.

For what concerns us, we stop here, because we have introduced all the main ingredients needed for extracting the most relevant phenomenological differences between supergravity and global supersymmetry.

6.2 Kähler–Hodge Manifolds

Gravity interactions have an impact also on the mathematical properties of the scalar manifold, which remains a Kähler manifold, but now of a restricted type. Since this is a more technical aspect, first-time readers may want to skip this section.

As seen in Sect. 6.1.2, local supersymmetry and invariance under Kähler transformations, $K \to h + h^*$, require that the fermions and the superpotential transform non-trivially under Kähler transformations:

$$\psi_\mu, \epsilon, \lambda^I \to \exp\left[-i\frac{\mathrm{Im}\,h}{2M_P^2}\gamma_5\right]\psi_\mu, \epsilon, \lambda^I, \tag{6.89}$$

$$\chi^m \to \exp\left[+i\frac{\mathrm{Im}\,h}{2M_P^2}\gamma_5\right]\chi^m, \tag{6.90}$$

$$W \to \exp\left[-\frac{h}{M_P^2}\right]W. \tag{6.91}$$

In general, the Kähler potential is not a globally defined function on the Kähler manifold but may be subject to Kähler transformations when one switches between two overlapping coordinate patches U_A and U_B:

$$K_A = K_B + h_{AB} + h_{AB}^*, \tag{6.92}$$

where K_A and K_B are the local Kähler potentials on U_A and U_B, respectively, and h_{AB} is a holomorphic function on the overlap $U_A \cap U_B$.

Because of (6.91), this means that W likewise needs to be patched together on the manifold by means of Kähler transformations

$$W_A = \exp\left[-\frac{h_{AB}}{M_P^2}\right] W_B \qquad (6.93)$$

and hence cannot in general be viewed as a simple function on $\mathcal{M}_{\text{scalar}}$, but rather as a section of a holomorphic line bundle, $\mathcal{L}$, over $\mathcal{M}_{\text{scalar}}$.[5] In order for such a line bundle to be well defined, the local transition functions must fit together in a globally consistent way. The important point for us is that the local transition functions are not arbitrary holomorphic functions, but instead they are the exponentials of the functions h_{AB} that also describe how the local Kähler potentials K_A, K_B are patched together. This relates the global consistency condition of the line bundle gluings to a global restriction on the possible Kähler geometries that could occur in supergravity.

The same is true for fermions, from which we get the strongest restrictions. Unlike the superpotential, the fermions transform with the exponential of $i\,\text{Im}h$ which means that they can be viewed as sections in a principal U(1) bundle that is associated with the line bundle $\mathcal{L}$. The connection on this U(1)-bundle is essentially (6.14) but now viewed as a connection on the scalar manifold,

$$Q \equiv \frac{i}{2}\left[(\partial_{\bar{n}}K)d\phi^{\bar{n}} - (\partial_m K)d\phi^m\right], \qquad (6.94)$$

whose curvature is related to the Kähler form

$$dQ = J. \qquad (6.95)$$

Similar to the Dirac magnetic monopole (see also the next subsection for an explicit example), the flux of the field strength of this connection through any topologically non-trivial two sphere in $\mathcal{M}_{\text{scalar}}$ must then be quantized, and this in turn implies that the Kähler form J has to obey a similar non-trivial quantization condition. In the supergravity literature, one therefore sometimes uses the special term "Kähler–Hodge manifolds" to denote Kähler manifolds that admit a line bundle, $\mathcal{L}$, whose principal bundle has a curvature equal to the Kähler form. To summarize: In 4D, $\mathcal{N} = 1$ supergravity, the scalar manifold of the chiral multiplets cannot just be an arbitrary Kähler manifold but must be a Kähler–Hodge manifold, i.e., a Kähler

---

[5] A line bundle is simply a vector bundle where the fiber is a one-dimensional vector space, $\mathbb{R}$ or $\mathbb{C}$. A holomorphic line bundle is a line bundle with fiber $\mathbb{C}$ over a complex base manifold (here our Kähler manifold $\mathcal{M}_{\text{scalar}}$) where the transition function between two local trivializations can be chosen to be holomorphic. According to (6.93), this is the case for the superpotential W, which is hence a section of a holomorphic line bundle, $\mathcal{L}$, over $\mathcal{M}_{\text{scalar}}$.

manifold that admits a holomorphic line bundle whose curvature is related to the Kähler form. In global supersymmetry, by contrast, the superpotential and the fermions do not transform under Kähler transformations. In the terminology of this subsection, this corresponds to a trivial line bundle with flat curvature that is unrelated to the Kähler form, which hence remains unconstrained. Thus, any Kähler manifold of the right dimension can serve as a target space for the chiral multiplet scalars in global supersymmetry.

Just like the $\mathcal{N} = 1$ case outlined here, also for $\mathcal{N} > 1$ supersymmetry, the transition from global to local supersymmetry imposes different constraints on the scalar manifold. Once again this leads to the fact that $\mathcal{N} > 1$ supergravity theories cannot be described in terms of $\mathcal{N} = 1$ models. For instance, $\mathcal{N} = 2$ supergravity allows quaternionic scalar manifolds (their definition will be given later), which are not Kähler. Hence they cannot be used as the scalar σ-model of an $\mathcal{N} = 1$ theory. On the other hand, the rigid $\mathcal{N} = 2$ counterpart is given by hyper-Kähler manifolds, which are also Kähler. Hence in rigid $\mathcal{N} = 2$ supersymmetry, one can always view the theory as a special classes of $\mathcal{N} = 1$ models.

6.2.1   An Example: Quantization of Newton's Constant

Before discussing other properties of the supergravity action and its differences with respect to global supersymmetry, we now present an explicit example of how global conditions constrain the scalar manifold in supergravity, illustrating some of the abstract discussion of the previous subsection. One curious non-trivial effect of these global conditions is that Newton's constant can become quantized in these models [8] in a sense to be described below.

As an example of a non-trivial manifold admitting two cycles, consider

$$\mathcal{M}_{\text{scalar}} = \mathbb{CP}^1 \simeq S^2.$$

This manifold can be locally parameterized by one complex scalar field ϕ (which we now take of canonical dimension zero after a suitable rescaling with M_P). More precisely, we need to cover the manifold with two coordinate patches. We start with the stereographic projection of the sphere onto the complex plane from one of the poles as one coordinate system (Fig. 6.1). A good Kähler potential for this manifold is then $K/M_P^2 = N \log(1 + \phi\phi^*)$, resulting in the metric

$$g_{\phi\bar{\phi}} = \frac{N}{(1 + \phi\phi^*)^2}. \tag{6.96}$$

This gives an effective theory whose bosonic sector reads

$$M_P^2 \int d^4x \, e \left[\frac{1}{2}R - g_{\phi\bar{\phi}} \partial_\mu \phi \partial^\mu \phi^* \right]. \tag{6.97}$$

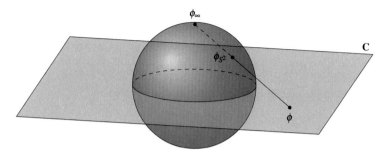

Fig. 6.1 Stereographic projection coordinates

The coordinate system and the Kähler potential discussed above are good for the entire S^2 except for the north pole. We therefore need to cover $\mathbb{C}\mathbb{P}^1$ by an alternative chart, with coordinate $z = \frac{1}{\phi}$ and Kähler potential $K_z/M_P^2 = N \log(1 + zz^*)$. As expected, in the overlapping patch, the two Kähler potentials are related by a Kähler transformation:

$$K_\phi - K_z = M_P^2\, N\, \log \frac{1 + \phi\phi^*}{1 + zz^*} = M_P^2\, N\, \log \frac{1 + \phi\phi^*}{1 + \frac{1}{\phi\phi^*}} \tag{6.98}$$

$$= M_P^2\, N\, \log \phi + M_P^2\, N\, \log \phi^* \tag{6.99}$$

$$= h(\phi) + h^*(\phi^*). \tag{6.100}$$

The problem of this transition function $h = M_P^2 N \log \phi$ is that it is multivalued on $\mathbb{C}$. However, in relating the two descriptions, what we really need is that the transition function for the fermions,

$$\exp\left[\frac{1}{4M_P^2}\left(h - h^*\right)\right] = \exp\left[\frac{N}{4}\log\frac{\phi}{\phi^*}\right],$$

be single valued. This is the case if and only if N is an even integer. Computing explicitly this expression at the equator

$$\exp\left[\frac{N}{4}\log\frac{\phi}{\phi^*}\right] = \exp\left[\frac{i}{2}N\alpha\right],$$

where α is the angle parametrizing the equator, we see that N has to be in $2\mathbb{Z}$ for making the fibers fit together when we complete a full revolution ($\alpha = 2\pi$ compared to $\alpha = 0$). Hence, while in global supersymmetry $N \in \mathbb{R}$, supergravity forces a quantization of its value and $N \in 2\mathbb{Z}$.

A curious fact is that now the quantization of N provides as a consequence the quantization of Newton's constant. In fact, when discussing scalar σ-models (ϕ is like a pion field), most of the physics is summarized in a dimensionful coupling constant, f^2, in front of the kinetic term of the scalars (which we generically reabsorbed in the fields). In cases where the σ-model is a coset manifold G/H, f is related to the vacuum expectation value of the field which breaks the global symmetry group G to a subgroup H in the effective theory. In any case, from the kinetic term of the σ-model under consideration, we have that

$$f^2 = N M_P^2 = \frac{N}{8\pi G_N}, \tag{6.101}$$

or, turning this around,

$$G_N = \frac{N}{8\pi f^2}. \tag{6.102}$$

We therefore see that Newton's constant is quantized in units of f.

Exercises

6.1. Verify that (6.25) and (6.26) are required by the consistency of the last of (6.39) and the holomorphicity of W.

6.2. Consider the same model considered in the globally supersymmetric case with three chiral multiplets and one vector multiplet (cf. Problem 5.4). This model has Kähler potential

$$K = |\phi|^2 + |A|^2 + |B|^2$$

and superpotential

$$W = \phi A B.$$

The three scalar fields are charged under a U(1) symmetry, which is gauged and generates a Fayet–Iliopoulos term. Their charges are q_ϕ, q_A, and q_B, not specified for the moment. The kinetic term for the vector field is canonical. Considering the coupling to gravity:

(a) Compute the scalar potential.
(b) Find the supersymmetric and non-supersymmetric vacua.
(c) Compute the boson masses at the vacua.
(d) Find out for which value of the charges we can get a potential of the form $V = \frac{g^2}{2}\eta^2$, where g is the gauge coupling constant.
(e) Discuss the type of supersymmetry breaking.

6.3. Find the Killing vectors of the isometries of the scalar manifold generated by

$$K = -\log\left[-\frac{i}{2}(s - \bar{s})((t - \bar{t})^2 - (u - \bar{u})^2)\right].$$

Compute its prepotentials and discuss the invariance of the Kähler potential under the action of these Killing vectors.

References

1. E. Cremmer, S. Ferrara, L. Girardello, A. Van Proeyen, Coupling supersymmetric Yang-Mills theories to supergravity. Phys. Lett. B **116**, 231–237 (1982)
2. E. Cremmer, S. Ferrara, L. Girardello, A. Van Proeyen, Yang-Mills theories with local supersymmetry: Lagrangian, transformation laws and SuperHiggs effect. Nucl. Phys. B **212**, 413 (1983)
3. J.A. Bagger, Coupling the gauge invariant supersymmetric nonlinear sigma model to supergravity. Nucl. Phys. B **211**, 302 (1983)
4. H. Elvang, D.Z. Freedman, B. Körs, Anomaly cancellation in supergravity with Fayet–Iliopoulos couplings. JHEP **11**, 068 (2006). [arXiv:hep-th/0606012 [hep-th]]
5. J. De Rydt, J. Rosseel, T.T. Schmidt, A. Van Proeyen, M. Zagermann, Symplectic structure of N=1 supergravity with anomalies and Chern-Simons terms. Class. Quant. Grav. **24**, 5201–5220 (2007). [0705.4216]
6. P. Binetruy, G. Dvali, R. Kallosh, A. Van Proeyen, Fayet–Iliopoulos terms in supergravity and cosmology. Class. Quant. Grav. **21**, 3137–3170 (2004). [hep-th/0402046]
7. R. D'Auria, S. Ferrara, On fermion masses, gradient flows and potential in supersymmetric theories. JHEP **05**, 034 (2001). [arXiv:hep-th/0103153 [hep-th]]
8. E. Witten, J. Bagger, Quantization of Newton's constant in certain supergravity theories. Phys. Lett. **B115**, 202 (1982)

Phenomenological Aspects 7

In previous chapters, we gave a complete picture of the Lagrangian and the supersymmetry rules for $\mathcal{N} = 1$ supergravity coupled to vector and chiral multiplets. In the present chapter, we now take a closer look at these theories from a phenomenological point of view. Our main focus will again be on those phenomenological aspects that differ in important ways between locally and globally supersymmetric theories. Most of these phenomenological differences can be traced back to:

- The different form of the scalar potential in global and local supersymmetry
- The presence of the gravitino
- The presence of gravity and other non-renormalizable interactions in supergravity

Our discussion includes the important topic of supersymmetry breaking, various aspects associated with the gravitino, as well as some cosmological consequences.

7.1 Spontaneous Supersymmetry Breaking

7.1.1 Vacua

Having a supersymmetric theory that involves gravity, it would now be very tempting to investigate the general interplay between supersymmetry, spacetime geometry, and the various matter fields. For many phenomenological questions, however, it is sufficient to restrict oneself to solutions of 4D, $\mathcal{N} = 1$ matter-coupled supergravity in which the spacetime metric is maximally symmetric, i.e., either Minkowski, de Sitter, or anti-de Sitter spacetime. We will refer to such solutions as "vacua" of the corresponding supergravity theory. The only non-trivial field configurations that could be consistent with the 4D homogeneity and isotropy of

© Springer-Verlag GmbH Germany, part of Springer Nature 2021
G. Dall'Agata, M. Zagermann, *Supergravity*, Lecture Notes in Physics 991,
https://doi.org/10.1007/978-3-662-63980-1_7

these spacetimes are the metric and constant scalar fields (or possibly also constant Lorentz invariant condensates of fermionic fields, such as gaugino condensates, which, however, we will not discuss here).[1]

Given a theory with n_S real scalar fields, $\varphi^i(x)$ $(i, j, \ldots = 1, \ldots, n_S)$, the scalar field equation with constant scalars $\varphi^i(x) = \varphi^i_c$ and all other non-gravitational fields set to zero reduces to

$$\partial_i V |_{\varphi^i = \varphi^i_c} \equiv \frac{\partial V(\varphi)}{\partial \varphi^i}\bigg|_{\varphi = \varphi^i_c} = 0, \tag{7.1}$$

i.e., the scalar fields have to be at a critical point of the scalar potential. In the following, we will, in a semi-classical sense, often refer to quantities that are evaluated at the critical point $\varphi^i = \varphi^i_c$ as the *vacuum expectation value (vev)* of that quantity and denote it with $\langle \ldots \rangle$, so that, e.g., (7.1) becomes $\langle \partial_i V \rangle = 0$.

The value $\langle V \rangle \equiv V(\varphi_c)$ of the potential at the critical point then is the vacuum energy density, ρ_{vac}, which enters the Einstein equation as a cosmological constant $\Lambda = \rho_{\text{vac}} M_P^{-2} = \langle V \rangle M_P^{-2}$, so that $\langle V \rangle = 0$, $\langle V \rangle > 0$ and $\langle V \rangle < 0$ correspond to, respectively, Minkowski, de Sitter, and anti-de Sitter spacetime.

A vacuum of the above type is stable against small fluctuations, $\varphi^i_c \to \varphi^i_c + \delta\varphi^i$, of the scalar field if the Hessian, $\langle \partial_i \partial_j V \rangle$, at the critical point is positive definite. If some eigenvalues of the Hessian are zero, the potential may have flat directions at the critical point, which correspond, for $\langle V \rangle \geq 0$, to marginal stability. Field directions along which the Hessian has a negative eigenvalue are called tachyonic field directions or tachyons and imply a perturbative instability of the vacuum along that direction if $\langle V \rangle \geq 0$. For $\langle V \rangle < 0$, on the other hand, the solution may be stable even for tachyonic field directions, as long as their mass eigenvalue satisfies the Breitenlohner–Freedman bound we discussed in (4.87).

In supergravity, we deal with complex scalar fields, $\phi^m(x)$ $(m, n, \ldots = 1, \ldots, n_C)$, and their complex conjugates, $\phi^{\bar{m}}(x)$. In terms of these, the critical point condition is correspondingly $\langle \partial_m V \rangle = 0$, and the stability is then dictated by

---

[1] Our definition of a "vacuum" here should not be confused with the notion of a "vacuum solution" in general relativity, which is more general and refers to solutions to Einstein's equation without contributions to the energy-momentum tensor $T_{\mu\nu}$ from any kind of matter excitation. Note that a *constant scalar field* $\phi(x) = \phi_0 = $ const. does not count as an excitation here and hence should be allowed in such a vacuum solution, because its contribution to $T_{\mu\nu}$ arises only from its constant potential $V(\phi(x)) = V(\phi_0)$, which is indistinguishable from a contribution due to a cosmological constant, Λ, in the Einstein equation. In other words, our maximally symmetric "vacua" are special cases of the "vacuum solutions" of general relativity, which, however, also encompass less symmetric matter-free solutions such as the Schwarzschild metric or a gravitational wave in empty space. Viewing the inhomogeneities of the metric in these solutions as gravitational excitations, one could also characterize our "vacua" as solutions that are free of matter *and* gravitational excitations and hence form the natural semi-classical analogues of the Poincaré-invariant vacua in conventional quantum field theories on Minkowski spacetime.

the eigenvalues of the matrix

$$\begin{pmatrix} \langle \partial_m \partial_n V \rangle & \langle \partial_m \partial_{\bar{n}} V \rangle \\ \langle \partial_{\bar{m}} \partial_n V \rangle & \langle \partial_{\bar{m}} \partial_{\bar{n}} V \rangle \end{pmatrix}. \tag{7.2}$$

The stability and many other phenomenologically important properties of a given vacuum crucially depend on whether supersymmetry is preserved or broken in this vacuum. We will therefore now introduce some basic terminology of spontaneous supersymmetry breaking in supergravity.

7.1.2 General Features of Spontaneous Supersymmetry Breaking

In this subsection, we analyze the conditions for spontaneous supersymmetry breaking in matter-coupled 4D, $\mathcal{N} = 1$ supergravity and explore some of their most immediate phenomenological consequences. As explained in the previous subsection, we will do this only for maximally symmetric spacetimes, but it should be mentioned that supersymmetry breaking can also be discussed for other physically interesting backgrounds such as black hole spacetimes, cosmic strings, domain wall solutions, or instantons with many interesting theoretical applications. For simplicity, we will also not consider non-trivial fermion condensates, so that the only fields that can have a non-trivial vev are the scalar fields (with constant vevs) and the metric.[2]

In matter-coupled 4D, $\mathcal{N} = 1$ supergravity, the scalar potential is the sum of a (Kähler covariantized) F-term contribution, V_F, a D-term contribution, V_D, and a new supergravity contribution, V_G:

$$V = V_F - V_G + V_D = e^{\frac{K}{M_P^2}} \left(|\mathscr{D}W|^2 - 3\frac{|W|^2}{M_P^2} \right) + \frac{1}{2}\mathrm{Re}\, f_{IJ} D^I D^J, \tag{7.3}$$

where, for convenience, in this section we introduced the splitting

$$V_F = e^{\frac{K}{M_P^2}} |\mathscr{D}W|^2, \tag{7.4}$$

$$V_D = \frac{1}{2}\mathrm{Re}\, f_{IJ}\, D^I D^J, \tag{7.5}$$

$$V_G = 3e^{\frac{K}{M_P^2}}\, \frac{|W|^2}{M_P^2}, \tag{7.6}$$

---

[2] Effects such as gaugino condensation, where strongly coupled gauge dynamics induce a non-trivial vev for gaugino bilinears, $\langle \bar{\lambda}\lambda \rangle \neq 0$, can, in an effective field theory below the condensation scale, often be described by an additional scalar field and contribute to spontaneous supersymmetry breaking in the effective scalar field sector. This effective scalar field dynamics would be contained in our analysis.

as this is useful in the following.[3] When the metric and the (constant) scalar fields are the only non-trivial fields, the supersymmetry rules can be written as

$$\delta\psi_{\mu L} = M_P \, \mathscr{D}_\mu \epsilon_L + \gamma_\mu \, S \, \epsilon_R, \tag{7.7}$$

$$\delta\chi_L^m = \mathscr{N}^m \epsilon_L, \tag{7.8}$$

$$\delta\lambda_L^I = N^I \epsilon_L. \tag{7.9}$$

Here, S is the fermionic shift in the gravitino transformation, which enters the Lagrangian also as a mass term for the gravitino (cf. also the next subsection),

$$S \equiv \frac{1}{2M_P} \, e^{\frac{K}{2M_P^2}} \, W, \tag{7.10}$$

whereas the shift of the chiralini is the covariantized F-term,

$$\mathscr{N}^m \equiv -\frac{1}{2} g^{m\bar{n}} e^{\frac{K}{2M_P^2}} \, \mathscr{D}_{\bar{n}} W^*, \qquad \mathscr{N}^{\bar{m}} \equiv (\mathscr{N}^m)^*, \tag{7.11}$$

and the shift of the gaugini is given by the D-terms,

$$N^I \equiv \frac{i}{2} D^I. \tag{7.12}$$

As we already explained in Sect. 6.1.6, supersymmetry gives an extremely useful relation between the "shifts" in the supersymmetry transformations of the fermions and the scalar potential. More concretely, the latter is simply given by the square of the shifts in (7.7)–(7.9), with appropriate signs and prefactors:

$$V = -12 \, S S^* + 4 \, g_{m\bar{n}} \, \mathscr{N}^m \mathscr{N}^{\bar{n}} + 2 \, \mathrm{Re}(f_{IJ}) \, N^I N^{*J}. \tag{7.13}$$

We see that the fermionic shifts from the matter fields always give positive semi-definite contributions, while the shifts from the gravitini give a negative semi-definite contribution. This relation between fermionic shifts in the supersymmetry transformations and the scalar potential is an important consequence of a Ward identity, similar to (4.63), which also generalizes to extended supersymmetry and other dimensions.

---

[3] Note that what we call $V_F - V_G$ here is usually called the "F-term potential" and would usually be denoted by V_F.

Having expressed the scalar potential in terms of the fermionic shifts, we can easily read off some important properties of the scalar potential and its vacua:

- The scalar potential can be positive, negative, or zero. This is the first striking difference with respect to the globally supersymmetric case. The reason is the negative contribution coming from the shift of the gravitino supersymmetry transformation rule.
- Supersymmetry is preserved if and only if the supersymmetry variations of the fermionic fields are vanishing in the background under consideration: $\langle \delta \psi_\mu \rangle = \langle \delta \chi^m \rangle = \langle \delta \lambda^I \rangle = 0$. This requires both (we recall that for $\langle W \rangle \neq 0$, the second condition follows from the first).

$$\langle \mathcal{N}^m \rangle = 0 \qquad \Leftrightarrow \qquad \langle \mathcal{D}_m W \rangle = 0 \tag{7.14}$$

and

$$\langle N^I \rangle = 0 \qquad \Leftrightarrow \qquad \langle D^I \rangle = 0. \tag{7.15}$$

On the other hand, there is no condition on the gravitino shift S, because a non-trivial S may be cancelled by a non-trivial $\mathcal{D}_\mu \epsilon$. Taking the covariant derivative of $M_P \mathcal{D}_\mu \epsilon_L = -\gamma_\mu S \epsilon_R$, however, one finds that this is possible only in anti-de Sitter space (see the discussion in Sect. 4.2 and the next item).

- For supersymmetric vacua the potential is negative (semi-)definite:

$$\langle V \rangle = -12 \langle S S^* \rangle = -3 \left\langle e^{\frac{K}{M_P^2}} \frac{|W|^2}{M_P^2} \right\rangle \leq 0. \tag{7.16}$$

This condition is also different from the globally supersymmetric case, where the vev of V had to vanish for unbroken supersymmetry. A supersymmetric Minkowski spacetime in supergravity is recovered only when the superpotential is vanishing on the vacuum: $\langle W \rangle = 0$. In this case, unbroken supersymmetry is equivalent to $\langle \partial_m W \rangle = \langle W \rangle = \langle D^I \rangle = 0$. By contrast, when the superpotential is not vanishing on the vacuum, $\langle W \rangle \neq 0$, the scalar potential has a negative value, and therefore we obtain a supersymmetric configuration with negative cosmological constant: the vacuum is an anti-de Sitter space.

- Conversely, having a negative or vanishing potential in the vacuum does not imply unbroken supersymmetry. This is also different from the globally supersymmetric case, where only a positive cosmological constant could break supersymmetry. For instance, we could have $\langle \mathcal{D}_m W \rangle \neq 0$ and $\langle W \rangle \neq 0$ at the same time in a way that they compensate in the potential so that $\langle V \rangle = 0$. A nice example for this are the *no-scale models*, which we introduce in Sect. 7.4.2. Also, it should be clear that supersymmetry breaking does not imply $\langle W \rangle \neq 0$, because we could have $\langle \mathcal{D}_m W \rangle \neq 0$, but $\langle W \rangle = 0$, which would imply a positive cosmological constant.

- In supergravity, D-terms and F-terms are related through (6.57). Hence, when both are present in the potential, there is no pure D-term breaking unless the superpotential vanishes on the vacuum. In this case, the relation (6.57) may yield a finite result even for vanishing $\langle \mathscr{D}_m W \rangle$. In any case, we cannot uplift supersymmetric AdS vacua (where necessarily $\langle W \rangle \neq 0$) to non-supersymmetric ones only by introducing a pure D-term breaking.

7.1.3 Mass Scales Related to Supersymmetry Breaking

The spontaneous breakdown of supersymmetry in a phenomenologically realistic model defines a number of important mass scales. In this subsection, we define these mass scales and exhibit the model-independent relation between some of them. Other relations are more model-dependent and will be discussed in the subsequent sections of this chapter, where we will focus on the sectors that define these mass scales in more detail.

7.1.3.1 The Supersymmetry Breaking Scale M_{susy}

As we have just seen, the relevant quantity that signals spontaneous supersymmetry breaking in supergravity are the vevs of the F-term shifts of the chiralini and the D-term shifts of the gaugini. Just as in global supersymmetry, we therefore define the supersymmetry breaking scale, M_{susy}, by

$$M_{\text{susy}}^4 \equiv \langle V_F \rangle + \langle V_D \rangle \tag{7.17}$$

as the natural order parameter of spontaneous supersymmetry breaking. In a generic supersymmetry breaking scenario, we expect M_{susy} to be larger than the electroweak scale, M_{ew}, because no sparticle has been observed yet.

7.1.3.2 The Vacuum Energy Scale M_{vac}

In spontaneously broken global supersymmetry, the supersymmetry breaking scale M_{susy} coincides with the mass scale, M_{vac}, of the tree-level vacuum energy density, $\rho_{\text{vac}} = M_{\text{vac}}^4$, which, in the presence of gravity, would be related to the cosmological constant Λ as $M_{\text{vac}}^4 = M_P^2 \Lambda$. In supergravity, however, M_{susy} and M_{vac} decouple, because in supergravity the latter also depends on the contribution, V_G, from the gravitino shift,

$$M_{\text{vac}}^4 = \langle V_F \rangle + \langle V_D \rangle - \langle V_G \rangle = M_{\text{susy}}^4 - \langle V_G \rangle. \tag{7.18}$$

This means that in supergravity a phenomenologically realistic value, $M_{\text{susy}} > 10^2$ GeV, can coexist with the tiny measured value of the vacuum energy density, $M_{\text{vac}}^4 \sim (10^{-3} \text{ eV})^4$, already at tree level. It is obvious from these considerations that it does not make much sense to compute the vacuum energy density in supersymmetry without taking into account gravity effects. An interesting class of models that have a tree-level Minkowski vacuum for arbitrary large M_{susy} are the no-scale models, which we discuss in Sect. 7.4.2.

7.1.3.3 The Gravitino Mass $M_{3/2}$

Another important mass scale is the mass of the gravitino after spontaneous supersymmetry breaking. It has a simple relation to M_{susy} if one neglects the small observed cosmological constant and approximates $M_{\text{vac}} \approx 0$. The (real) gravitino mass is then (omitting the brackets for vevs here)

$$M_{3/2} = e^{\frac{K}{2M_P^2}} \frac{|W|}{M_P^2} = \sqrt{\frac{V_G}{3M_P^2}} = \sqrt{\frac{M_{\text{susy}}^4 - M_{\text{vac}}^4}{3M_P^2}} \approx \frac{M_{\text{susy}}^2}{\sqrt{3}M_P} \tag{7.19}$$

We see that, for $M_{\text{susy}} > M_{\text{ew}} = 246\,GeV$, the gravitino mass could a priori be anywhere between the Planck and the milli-electronvolt scale. In order to constrain it further, one would have to feed in the desired mass scale for the superpartners of the Standard Model particles and to specify a particular supersymmetry breaking mechanism. We will discuss the gravitino sector in Sect. 7.2.

7.1.3.4 The Soft Masses M_{soft}

The abovementioned masses of the superpartners of the Standard Model particles are among the *soft supersymmetry breaking terms* that one can add to a supersymmetric extension of the Standard Model without introducing quadratic divergences to the Higgs mass. These "soft masses" must be above the energy scales that have already been probed by accelerators, i.e., at least near the TeV scale, with the precise values depending on the particle type and the assumed model.

In order to address naturalness issues of the Standard Model, the soft masses would have to be not too far above the electroweak scale either, which would lead to a typical scale, M_{soft}, of the soft masses somewhere near the TeV scale.

It should be emphasized, however, that M_{soft} is not a precise value for a given model but that the actual soft masses may of course cover a certain range. In models that address the hierarchy problem, this range cannot be too big, but if one abandons supersymmetry as a solution to the hierarchy problem, this mass range can be substantially bigger and in fact define several hierarchically different mass scales. This is in particular the case in models with split supersymmetry [1, 2], where the fermionic superpartners are assumed to be near the TeV scale, whereas the scalar masses could be substantially bigger without spoiling coupling unification or the existence of possible dark matter candidates among the fermionic superpartners.

In any case, the soft masses are related to the gravitino mass and the supersymmetry breaking scale in different ways depending on the particular scheme of supersymmetry breaking. This will be discussed in Sect. 7.3.

7.1.3.5 Moduli Masses M_{mod}

Moduli are scalar fields that couple to ordinary matter such as the Standard Model particles only with very weak, typically M_P-suppressed, interactions. Moduli are a typical ingredient of string compactifications but may also be introduced for purely phenomenological reasons, e.g., as a sector that realizes spontaneous supersymmetry breaking. As we will discuss, a natural scale for the moduli masses is the gravitino mass, but one can easily construct models where at least some moduli are much lighter or much heavier than the gravitino. We will discuss moduli masses in Sect. 7.4.1.

7.1.3.6 The Inflationary Energy Scale M_{inf}

This energy scale is not directly related to our present vacuum but refers to the potential energy density, $\rho_{inf} = M_{inf}^4$, of the inflaton during early Universe inflation. This potential energy must be high enough to enable the standard hot Big Bang scenario after reheating that leads to the present Universe. It therefore is also associated with a strong breaking of supersymmetry during the period of inflation so that we include it in this list. Some basic aspects of inflation in supergravity will be discussed in Sect. 7.4.4.

7.1.4   Rigid Limits

In our discussions above, we have seen that the question of vacuum energy can only be studied in a meaningful way when one also takes into account the effects from supergravity. This means that if one is interested in the limit of global supersymmetry, such a limit should only be taken after a specific vacuum has been chosen in the supergravity theory.

Once a vacuum has been identified in supergravity, it is clear that a rigid limit should involve sending $M_P \rightarrow \infty$. This limit kills all the supergravity related interactions in the Lagrangian and also modifies the supersymmetry transformation rules, as we discussed earlier. However, the supersymmetry breaking scale and the gravitino mass may also depend on M_P so that some care is required when this limit is performed and certain quantities are to be kept fixed. There are in fact two distinct procedures that could be taken in order to recover a proper rigid limit in the presence of supersymmetry breaking:

- **Keep $M_{3/2}$ fixed**.
  This means that the supersymmetry breaking scale is also taken to be high

$$M_{susy}^2 \sim M_{3/2} M_P \rightarrow \infty. \tag{7.20}$$

The resulting global theory features *explicitly broken supersymmetry* with tree-level soft terms of $\mathcal{O}(M_{3/2})$ and a decoupled goldstino (eaten up by the gravitino).[4]

---

[4] The couplings of the goldstino are proportional to inverse powers of M_{susy} and hence vanish in this limit.

- **Keep M_{susy} fixed**. This implies

$$M_{3/2} \sim \frac{M_{\text{susy}}^2}{M_P} \to 0. \tag{7.21}$$

Hence we obtain a scenario with *spontaneously broken supersymmetry* (with a goldstino interacting with matter) and decoupled gravitino.

One thing that has to be stressed in this procedure is that the rigid limit may affect the geometry of the scalar manifold in a non-trivial way. Indeed, the rigid and local geometries of the scalar σ-models are generically different, although for minimal supersymmetry the local case can be seen as a subcase of the rigid one and only concerns the global structure of the manifold (see Sect. 6.2). As we will discuss in Chap. 8, the difference between the geometries are more pronounced in extended supersymmetry and also concern the local structure. This phenomenon has not been thoroughly studied in the literature, and the phenomenological literature does not seem to be aware of the problem, mainly because usually only topologically trivial models are discussed. In general [3], only a subset of the scalars of the supergravity theory appears in the rigid limit for $\mathcal{N} > 1$, though for $\mathcal{N} = 1$ we can still keep all of them, as we would expect.

7.2 Gravitino, Goldstino, and Super-Higgs Mechanism

When supersymmetry is spontaneously broken in a Minkowski or AdS vacuum, the gravitino field acquires a Lagrangian mass term proportional to the (necessarily non-vanishing) expectation value of the superpotential, though in an AdS vacuum, as explained in Sect. 4.2.2, the Lagrangian mass term may not coincide with the actual mass definition. In a de Sitter vacuum, supersymmetry is always broken, and the vev of the superpotential may or may not be zero.

In any case, a massive gravitino should acquire new spin-$1/2$ polarizations with respect to the massless case, just as a gauge boson acquires an additional spin-0 degree of freedom when a gauge symmetry is spontaneously broken through the Higgs mechanism. Since we are now dealing with supersymmetry, this process has been named super-Higgs mechanism [4]. The goldstino, the massless fermion related to the spontaneous breaking of supersymmetry, is "eaten" by the gravitino, giving the latter the missing degrees of freedom to become a massive particle.

The details of this process depend on the type of vacuum one considers. We will discuss here in detail only the case of pure F-term breaking in a Minkowski vacuum. A non-trivial cosmological constant or the presence of D-terms would change some of the details of the computation, but the general scheme of the super-Higgs mechanism should become clear already in our restricted setup and can be worked out along the same lines [5].

We recall that in ordinary gauge theories the Goldstone bosons correspond to those directions in field space along which the broken symmetry generators act non-trivially in a given vacuum. Alternatively, they can be identified as the scalar field fluctuations that have a non-trivial coupling to the gauge fields of the broken symmetries in the vacuum. They are zero eigenvectors of the scalar mass matrix and can be absorbed by a redefinition of the vector fields of the broken symmetries. All these properties of the Goldstone boson have a close analogue for the goldstino.

7.2.1 The Goldstino in Global Supersymmetry

In order to exhibit the difference between global and local supersymmetry most clearly, let us briefly recapitulate the properties of the goldstino in a globally supersymmetric theory with n_C chiral multiplets and pure F-term supersymmetry breaking.

The fermionic part of the Lagrangian is

$$\mathscr{L}_F = -g_{m\bar{n}}\left[\overline{\chi}_L^m \slashed{\mathscr{D}} \chi_R^{\bar{n}} + \overline{\chi}_R^{\bar{n}} \slashed{\mathscr{D}} \chi_L^m\right]$$

$$- (\mathscr{D}_m \partial_n W)\overline{\chi}_L^m \chi_L^n - (\mathscr{D}_{\bar{m}} \partial_{\bar{n}} W^*)\overline{\chi}_R^{\bar{m}} \chi_R^{\bar{n}} \tag{7.22}$$

and the supersymmetry transformations of the chiralini are

$$\delta\chi_L^m = \frac{1}{2}\slashed{\partial}\phi^m \varepsilon_R - \frac{1}{2}F^m \varepsilon_L, \qquad \delta\chi_R^{\bar{m}} = \frac{1}{2}\slashed{\partial}\phi^{\bar{m}} \varepsilon_L - \frac{1}{2}F^{\bar{m}} \varepsilon_R, \tag{7.23}$$

where $F^m \equiv g^{m\bar{n}}\partial_{\bar{n}}W^*$ and $F^{\bar{m}} \equiv (F^m)^*$. Considering the fermions on a Minkowski background with constant scalar fields, we can replace $\mathscr{D}_\mu \to \partial_\mu$, and the supersymmetry transformations become

$$\delta\chi_L^m = -\frac{1}{2}F^m \varepsilon_L, \qquad \delta\chi_R^{\bar{m}} = -\frac{1}{2}F^{\bar{m}} \varepsilon_R. \tag{7.24}$$

Here and in the following, all scalar field-dependent quantities such as F^m, $g_{m\bar{n}}$, etc. are meant to denote their constant vevs (i.e., we drop the brackets for vevs for simplicity), which are then also inert under supersymmetry transformations. Obviously the supersymmetry transformation is non-trivial only along the direction defined by F^m and its complex conjugate in field space, and hence we define the goldstino field, ζ, as the projection of the chiralini along that direction,

$$\zeta_L := \frac{F_m \chi_L^m}{\sqrt{F_n F^n}} \tag{7.25}$$

and similarly for ζ_R, where the denominator ensures canonical normalization. The chiralini χ_L^m can then be decomposed into ζ_L and its orthogonal complement, $\chi_L^{m\perp}$, with respect to the vev of the Kähler metric,

$$\chi_L^m = \frac{F^m \zeta_L}{\sqrt{F_n F^n}} + \underbrace{\chi_L^m - \frac{F^m \zeta_L}{\sqrt{F_n F^n}}}_{:=\chi_L^{m\perp}}, \tag{7.26}$$

so that the kinetic term in (7.23) becomes

$$-[\overline{\zeta}_L \slashed{\partial} \zeta_R + \overline{\zeta}_R \slashed{\partial} \zeta_L] - g_{m\overline{n}}\left[\overline{\chi}_L^{m\perp} \slashed{\partial} \chi_R^{\overline{n}\perp} + \overline{\chi}_R^{\overline{n}\perp} \slashed{\partial} \chi_L^{m\perp}\right], \tag{7.27}$$

and

$$F_m \chi_L^{m\perp} = 0. \tag{7.28}$$

The supersymmetry transformations of ζ_L and $\chi_L^{m\perp}$ are

$$\delta\zeta_L = -\frac{1}{2}\sqrt{F_m F^m}\,\varepsilon_L, \tag{7.29}$$

$$\delta\chi_L^{m\perp} = 0, \tag{7.30}$$

so that indeed the goldstino ζ captures the full supersymmetry transformation.

To show that the goldstino is a massless fermion, we need the critical point condition, $\partial_l V = 0$, which can be written as

$$0 = \partial_l V = \mathscr{D}_l V = \mathscr{D}_l(\partial_m W g^{m\overline{n}} \partial_{\overline{n}} W^*) = (\mathscr{D}_l \partial_m W) g^{m\overline{n}} \partial_{\overline{n}} W^* = (\mathscr{D}_l \partial_m W) F^m, \tag{7.31}$$

where $\mathscr{D}_l g^{m\overline{n}} = 0$ and $\mathscr{D}_l \partial_{\overline{n}} W^* \equiv \partial_l \partial_{\overline{n}} W^* - \Gamma_{l\overline{n}}^{\overline{p}} \partial_{\overline{p}} W^* = 0$ have been used. As the matrix $(\mathscr{D}_l \partial_m W) \equiv (\partial_l \partial_m W - \Gamma_{lm}^p \partial_p W)$ is manifestly symmetric in l and m, we thus obtain

$$0 = (\mathscr{D}_l \partial_m W) F^m = (\mathscr{D}_m \partial_l W) F^m. \tag{7.32}$$

Using this equation, it is then easy to see that

$$-(\mathscr{D}_m \partial_n W)\overline{\chi}_L^m \chi_L^n = -(\mathscr{D}_m \partial_n W)\overline{\chi}_L^{m\perp} \chi_L^{n\perp} \tag{7.33}$$

i.e., the goldstino drops out of the fermionic mass term and hence is indeed massless.

7.2.2   The Goldstino and the Gravitino in Supergravity

As we will now discuss, the situation in local supersymmetry has many similarities to the above but also some important differences that have to do with the presence of the gravitino as the gauge field of local supersymmetry.

Before we start, we collect some useful relations for the Kähler invariant function, $\mathscr{G} \equiv K + M_P^2 \log(|W|^2/M_P^6)$, and its derivatives that will help us keeping the equations in a compact form. We define $\mathscr{G}_n \equiv \partial_n \mathscr{G}$, $\mathscr{G}^m \equiv g^{m\bar{n}} \partial_{\bar{n}} \mathscr{G}$, etc. so that

$$\mathscr{G}_m = \partial_m K + M_P^2 \frac{\partial_m W}{W} = M_P^2 \frac{\mathscr{D}_m W}{W}, \tag{7.34}$$

$$\partial_m \mathscr{G}_n = M_P^2 \partial_m \frac{\mathscr{D}_n W}{W} = \left[\frac{\partial_m \mathscr{D}_n W}{W} - \frac{\partial_m W \mathscr{D}_n W}{W^2} \right] M_P^2, \tag{7.35}$$

$$\mathscr{D}_m \mathscr{G}_n = \partial_m \mathscr{G}_n - \Gamma_{mn}^k \mathscr{G}_k = \left[\frac{\mathscr{D}_m \mathscr{D}_n W}{W} - \frac{\mathscr{D}_m W \mathscr{D}_n W}{W^2} \right] M_P^2, \tag{7.36}$$

where we also used the fact that we have full covariantization of the derivatives with respect to Kähler transformations and field redefinitions, like in (6.38). We recall the form of the scalar potential in terms of $\mathscr{G}$, as in (6.41)

$$V = e^{\frac{K}{M_P^2}} \left(g^{m\bar{n}} \mathscr{D}_m W \mathscr{D}_{\bar{n}} W^* - 3 \frac{|W|^2}{M_P^2} \right) = e^{\mathscr{G}/M_P^2} \left(M_P^2 \, g^{m\bar{n}} \mathscr{G}_m \mathscr{G}_{\bar{n}} - 3M_P^4 \right), \tag{7.37}$$

so that the Minkowski condition, $V = 0$, implies

$$\mathscr{G}_m \mathscr{G}^m = 3M_P^2. \tag{7.38}$$

Furthermore, the condition that we are at a critical point of the potential gives (by recalling that $\partial_m \mathscr{G}_{\bar{n}} = \partial_m \partial_{\bar{n}} K = g_{m\bar{n}}$ and $\partial_l g^{m\bar{n}} = -\Gamma_{lk}^m g^{k\bar{n}}$, as in (5.48)):

$$0 = \partial_n V|_{V=0} = e^{\frac{\mathscr{G}}{M_P^2}} M_P^2 \left[\mathscr{G}^m \mathscr{D}_n \mathscr{G}_m + \mathscr{G}_n \right], \tag{7.39}$$

which implies

$$(\mathscr{D}_n \mathscr{G}_m) \mathscr{G}^m = -\mathscr{G}_n. \tag{7.40}$$

The fermionic part of the Lagrangian in a Minkowski background with constant scalar fields and no vector multiplets is

$$
\mathcal{L}_F = -\frac{1}{2}\overline{\psi}_\mu \gamma^{\mu\nu\rho}\partial_\nu\psi_\rho - g_{m\bar{n}}\left[\overline{\chi}_L^m \slashed{\partial}\chi_R^{\bar{n}} + \overline{\chi}_R^{\bar{n}}\slashed{\partial}\chi_L^m\right]
$$
$$
+ \left[-\mathcal{M}_{mn}\overline{\chi_L}^m\chi_L^n + \frac{2S}{M_P^2}\overline{\psi}_{\mu R}\gamma^\mu\chi_L^m\mathcal{G}_m + \frac{S}{M_P}\overline{\psi}_{\mu R}\gamma^{\mu\nu}\psi_{\nu R} + \text{h.c.}\right],
$$

$$(7.41)$$

where

$$
\mathcal{M}_{mn} := e^{K/2M_P^2}(\mathcal{D}_m\mathcal{D}_n W) = \frac{2S}{M_P^3}\left[M_P^2\mathcal{D}_m\mathcal{G}_n + \mathcal{G}_m\mathcal{G}_n\right], \qquad (7.42)
$$

S denotes the gravitino shift (7.10), and we have dropped a mixing term between the gravitino and the chiralini involving derivatives of the (here still constant) scalars.

The supersymmetry transformations of the fermions in such a background are

$$
\delta\psi_{\mu L} = M_P\partial_\mu\varepsilon_L + S\gamma_\mu\varepsilon_R \qquad (7.43)
$$

$$
\delta\chi_L^m = -\frac{1}{2}e^{K/2M_P^2}g^{m\bar{n}}\mathcal{D}_{\bar{n}}W^*\varepsilon_L = -\frac{S^*}{M_P}\mathcal{G}^m\varepsilon_L, \qquad (7.44)
$$

We are now ready to discuss the goldstino and the super-Higgs effect. In the case of pure F-term breaking at $V = 0$, the goldstino is, just as for global supersymmetry, given by the linear combination of the chiralini along the direction of supersymmetry breaking, i.e., along $\mathcal{G}^m$,

$$
\zeta_L := \frac{\mathcal{G}_m\chi_L^m}{\sqrt{\mathcal{G}_n\mathcal{G}^n}} = \frac{\mathcal{G}_m\chi_L^m}{\sqrt{3}M_P}, \qquad (7.45)
$$

where we used (7.38) in the second equality.

Just as for global supersymmetry, we then decompose χ_L^m into ζ_L and its orthogonal complement, $\chi_L^{m\perp}$,

$$
\chi_L^m = \frac{\mathcal{G}^m}{\sqrt{\mathcal{G}_n\mathcal{G}^n}}\zeta_L + \underbrace{\left(\chi_L^m - \frac{\mathcal{G}^m}{\sqrt{\mathcal{G}_n\mathcal{G}^n}}\zeta_L\right)}_{=:\chi_L^{m\perp}}, \qquad (7.46)
$$

where the normalizations have again been chosen such that the kinetic term factorizes and is canonically normalized,

$$-g_{m\bar{n}}[\overline{\chi}_L^m \not{\partial} \chi_R^{\bar{n}} + \overline{\chi}_R^{\bar{n}} \not{\partial} \chi_L^m] = -[\overline{\zeta}_L \not{\partial} \zeta_R + \overline{\zeta}_R \not{\partial} \zeta_L]$$
$$- g_{m\bar{n}} \left[\overline{\chi}_L^{m\perp} \not{\partial} \chi_R^{\bar{n}\perp} + \overline{\chi}_R^{\bar{n}\perp} \not{\partial} \chi_L^{m\perp} \right], \tag{7.47}$$

and, moreover,

$$\mathcal{G}_m \chi_L^{m\perp} = 0. \tag{7.48}$$

The supersymmetry transformation (7.44) then becomes

$$\delta\zeta_L = -\sqrt{3}S^*\varepsilon_L \tag{7.49}$$
$$\delta\chi_L^{m\perp} = 0 \tag{7.50}$$

so that the goldstino indeed captures the entire supersymmetry transformation.

In the case of global supersymmetry, the goldstino dropped out of the mass term for the chiralini and was thus recognized as a massless fermion. In local supersymmetry, the situation is a bit more complicated due to the presence of the gravitino. In order to see this, we first note that Eqs. (7.38) and (7.40) now imply

$$\mathcal{M}_{mn}\mathcal{G}^n = \frac{4S}{M_P}\mathcal{G}_m, \tag{7.51}$$

which does not vanish, unlike the analogous expression (7.32) in global supersymmetry. This in turn implies that

$$- \mathcal{M}_{mn}\overline{\chi}_L^m \chi_L^n = - \left[\frac{4S}{M_P} \overline{\zeta}_L \zeta_L + \mathcal{M}_{mn}\overline{\chi}_L^{m\perp} \chi_L^{n\perp} \right], \tag{7.52}$$

i.e., although the goldstino decouples from its orthogonal complement in the naive mass matrix of the chiralini, it does get a non-vanishing mass contribution from it. This is different from the analogous expression (7.33), where the goldstino drops out. In contrast to global supersymmetry, however, this is not yet the end of the story, as in supergravity there is also a mixing term between the gravitino and the chiralini in (7.41), which upon using (7.46) becomes

$$\frac{2\sqrt{3}S}{M_P}\overline{\psi}_{\mu R}\gamma^\mu \zeta_L + \text{h.c.} \tag{7.53}$$

Putting everything together, the fermionic action (7.41) now becomes

$$
\mathcal{L}_F = -\frac{1}{2}\bar{\psi}_\mu \gamma^{\mu\nu\rho}\partial_\nu\psi_\rho - [\bar{\zeta}_L\partial\!\!\!/\zeta_R + \bar{\zeta}_R\partial\!\!\!/\zeta_L] - g_{m\bar{n}}\left[\overline{\chi_L}^{m\perp}\partial\!\!\!/\chi_R^{\bar{n}\perp} + \overline{\chi_R}^{\bar{n}\perp}\partial\!\!\!/\chi_L^{m\perp}\right]
$$

$$
+ \left[-\frac{4S}{M_P}\bar{\zeta}_L\zeta_L + \frac{2\sqrt{3}S}{M_P}\bar{\psi}_{\mu R}\gamma^\mu\zeta_L + \frac{S}{M_P}\bar{\psi}_{\mu R}\gamma^{\mu\nu}\psi_{\nu R} + \text{h.c.}\right] \quad (7.54)
$$

$$
- \left[\mathcal{M}_{mn}\overline{\chi_L}^{m\perp}\chi_L^{n\perp} + \text{h.c.}\right]
$$

In order to understand the mass spectrum, we have to get rid of the mixing term between the goldstino and the gravitino. This could be done by expressing everything in terms of a redefined gravitino field of the form $\psi_{\mu L} + \gamma_\mu\zeta_R/\sqrt{3}$, but this would introduce a mixing term in the kinetic terms. To also remove that one, one instead has to use the redefined gravitino

$$
\tilde{\psi}_{\mu L} \equiv \psi_{\mu L} + \frac{1}{\sqrt{3}S^*}\partial_\mu\zeta_L + \frac{1}{\sqrt{3}}\gamma_\mu\zeta_R, \quad (7.55)
$$

in terms of which (7.54) becomes

$$
\mathcal{L}_F = -\frac{1}{2}\bar{\tilde{\psi}}_\mu\gamma^{\mu\nu\rho}\partial_\nu\tilde{\psi}_\rho - g_{m\bar{n}}\left[\overline{\chi_L}^{m\perp}\partial\!\!\!/\chi_R^{\bar{n}\perp} + \overline{\chi_R}^{\bar{n}\perp}\partial\!\!\!/\chi_L^{m\perp}\right]
$$

$$
+ \left[\frac{S}{M_P}\bar{\tilde{\psi}}_{\mu R}\gamma^{\mu\nu}\tilde{\psi}_{\nu R} + \text{h.c.}\right] - \left[\mathcal{M}_{mn}\overline{\chi_L}^{m\perp}\chi_L^{n\perp} + \text{h.c.}\right]. \quad (7.56)
$$

We notice that the goldstino has not only decoupled from the gravitino but that it actually has disappeared completely from the action. Moreover, when we compare (7.55) with the supersymmetry transformation (7.43), we realize that the transition from ψ_μ to $\tilde{\psi}_\mu$ precisely takes the form of a local supersymmetry transformation with $\varepsilon_L = \zeta_L/(\sqrt{3}S^*)$, which in turn would precisely transform ζ_L to zero. We thus realize that in local supersymmetry, the goldstino becomes a pure gauge degree of freedom that can be entirely absorbed as the longitudinal mode of the massive gravitino. The gauge where $\zeta = 0$ is called the unitary gauge.

We can finally read off the gravitino mass from the above equation:

$$
M_{3/2} = \frac{2|S|}{M_P}. \quad (7.57)
$$

This is equivalent to the expression given in (7.19).

Sometimes it is easier to compute the fermion masses directly from the original χ_L^m basis, but taking into account the field redefinition (7.55) for the gravitino. In this case one diagonalizes a mass matrix that contains n_C fields, among which there is the goldstino. This matrix is

$$\widetilde{\mathscr{M}}_{mn} \equiv \mathscr{M}_{mn} - \frac{2}{3}\frac{S}{M_P^3}\mathscr{G}_m\mathscr{G}_n = \frac{2S}{M_P^3}\left[M_P^2\mathscr{D}_m\mathscr{G}_n + \frac{1}{3}\mathscr{G}_m\mathscr{G}_n\right], \tag{7.58}$$

which indeed satisfies

$$\widetilde{\mathscr{M}}_{mn}\mathscr{G}^n = 0. \tag{7.59}$$

7.2.3 Gravitino Couplings

So far, we considered constant scalar fields and a fixed Minkowski background, but the super-Higgs mechanism also works when all fields are dynamical. In order to understand a subtlety of the gravitino couplings to matter fields in theories with spontaneously broken supersymmetry, it is useful to briefly also sketch this general case.

To this end, we start from the general supergravity Lagrangian with n_C chiral multiplets,

$$e^{-1}\mathscr{L} = \frac{M_P^2}{2}R - \frac{1}{2}\overline{\psi}_\mu\gamma^{\mu\nu\rho}\mathscr{D}_\nu\psi_\rho - g_{m\bar{n}}\left[(\partial_\mu\phi^m)(\partial^\mu\phi^{\bar{n}}) + \overline{\chi}_L^m\slashed{\mathscr{D}}\chi_R^{\bar{n}} + \overline{\chi}_R^{\bar{n}}\slashed{\mathscr{D}}\chi_L^m\right]$$

$$- \left[e^{K/2M_P^2}(\mathscr{D}_m\mathscr{D}_n W)\overline{\chi}_L^m\chi_L^n + \text{h.c.}\right] - \frac{1}{M_P}\left[\overline{J}_R^\mu\psi_{\mu R} + \overline{J}_L^\mu\psi_{\mu L}\right] \tag{7.60}$$

$$+ \frac{1}{M_P}\left[S\overline{\psi}_{\mu R}\gamma^{\mu\nu}\psi_{\nu R} + S^*\overline{\psi}_{\mu L}\gamma^{\mu\nu}\psi_{\nu L}\right] - V + e^{-1}\mathscr{L}_{4F},$$

with the supercurrent

$$\overline{J}_L^\mu = -g_{m\bar{n}}\overline{\chi}_L^m\gamma^\mu\slashed{\partial}\phi^{\bar{n}} + \frac{2S^*}{M_P}\mathscr{G}_{\bar{n}}\overline{\chi}_R^{\bar{n}}\gamma^\mu \tag{7.61}$$

and the supersymmetry transformation laws

$$\delta e_\mu^a = \frac{1}{2M_P}\overline{\varepsilon}\gamma^a\psi_\mu, \tag{7.62}$$

$$\delta\psi_{\mu L} = M_P\mathscr{D}_\mu\varepsilon_L + S\gamma_\mu\varepsilon_R, \tag{7.63}$$

$$\delta\phi^m = \overline{\varepsilon}_L\chi_L^m, \tag{7.64}$$

$$\delta\chi_L^m = \frac{1}{2}\slashed{\partial}\phi^m\varepsilon_R - \frac{S^*}{M_P}\mathscr{G}^m\varepsilon_L. \tag{7.65}$$

As the scalar fields are now dynamical, we have to distinguish carefully again the dynamical fields $\phi^m(x)$ and their vevs $\langle \phi^m \rangle$ and define the shifted fields

$$\Delta \phi^m(x) := \phi^m(x) - \langle \phi^m \rangle \tag{7.66}$$

and similarly for $\Delta \mathcal{G}^m$, ΔS, etc. The goldstino, ζ, and its orthogonal complements $\chi^{m\perp}$ are then given by

$$\zeta_L := \frac{\langle \mathcal{G}_m \rangle}{\sqrt{\langle \mathcal{G}_m \mathcal{G}^m \rangle}} \chi_L^m = \frac{\langle \mathcal{G}^m \rangle}{\sqrt{3} M_P} \chi_L^m, \tag{7.67}$$

$$\chi_L^{m\perp} := \chi_L^m - \frac{\langle \mathcal{G}^m \rangle}{\sqrt{\langle \mathcal{G}_m \mathcal{G}^m \rangle}} \zeta_L. \tag{7.68}$$

For the goldstino we then have the following supersymmetry variation

$$\delta \zeta_L = \frac{\langle \mathcal{G}_m \rangle}{\sqrt{\langle \mathcal{G}_m \mathcal{G}^m \rangle}} \left[\frac{1}{2} \partial\!\!\!/\phi^m \varepsilon_R - \frac{S^*}{M_P} \mathcal{G}^m \varepsilon_L \right] \tag{7.69}$$

$$= -\frac{\langle \mathcal{G}_m \mathcal{G}^m S^* \rangle}{M_P \sqrt{\langle \mathcal{G}_m \mathcal{G}^m \rangle}} \varepsilon_L + \underbrace{\frac{\langle \mathcal{G}_m \rangle}{\sqrt{\langle \mathcal{G}_m \mathcal{G}^m \rangle}} \left[\frac{1}{2} \partial\!\!\!/\phi^m \varepsilon_R - \frac{\Delta(S^* \mathcal{G}^m)}{M_P} \varepsilon_L \right]}_{=: \delta' \zeta_L} \tag{7.70}$$

$$= -\sqrt{3} \langle S^* \rangle \varepsilon_L + \delta' \zeta_L. \tag{7.71}$$

In the case of constant scalar fields, one has $\delta' \zeta_L = 0$, and one can simply choose $\varepsilon_L = \zeta_L / (\sqrt{3} \langle S^* \rangle)$ to gauge away the goldstino. In the general case with non-vanishing $\delta' \zeta^m$, the gaugino can still be gauged away, but the explicit form of the ε that achieves this is different. However, if, as we will assume from now on, the scalar field fluctuations and gradients are much smaller than the supersymmetry breaking scale, $M_{\text{susy}} = [\langle \mathcal{G}_m \mathcal{G}^m |S|^2 \rangle / M_P^2]^{1/4}$, i.e., if

$$\left| \frac{\Delta(S^* \mathcal{G}^m)}{M_P} \right| \ll \left| \frac{\langle S^* \mathcal{G}^m \rangle}{M_P} \right| \sim M_{\text{susy}}^2, \tag{7.72}$$

$$\left| \partial_\mu \phi^m \right| \ll \left| \frac{\langle S^* \mathcal{G}^m \rangle}{M_P} \right| \sim M_{\text{susy}}^2, \tag{7.73}$$

then $\delta' \zeta_L$ is only a small correction to the leading term,

$$\delta \zeta_L = \underbrace{-\sqrt{3} \langle S^* \rangle \varepsilon_L}_{\mathcal{O}(M_{\text{susy}}^2)} + \underbrace{\delta' \zeta_L}_{\ll M_{\text{susy}}^2}. \tag{7.74}$$

The ε necessary to gauge away ζ can then be expanded in powers of the small quantities $\partial_\mu \phi^m / M_{\text{Susy}}^2$ and $\Delta(\mathscr{G}^m S^*)/(M_P M_{\text{Susy}}^2)$ with the leading term given by the value $\zeta_L/(\sqrt{3}\langle S^* \rangle)$ for constant scalar fields, i.e.,

$$\varepsilon_L = \frac{\zeta_L}{\sqrt{3}\langle S^* \rangle}[1+x], \tag{7.75}$$

where x are small corrections of order $\partial_\mu \phi^m / M_{\text{Susy}}^2$ and $\Delta(\mathscr{G}^m S^*)/(M_P M_{\text{Susy}}^2)$ that could be determined explicitly in an iterative procedure but won't be needed here. Using this ε, we can then again eliminate the goldstino from the original supergravity action by expressing it in terms of the transformed fields $\widetilde{e}_\mu^a = e_\mu^a + \delta e_\mu^a$, $\widetilde{\psi}_\mu = \psi_\mu + \delta\psi_\mu$, etc. As the original action is supersymmetric, the action with the transformed fields takes the same form as the original action, but now with $\widetilde{\zeta} \equiv 0$ everywhere. In particular, we now have the Noether coupling

$$-\frac{1}{M_P}\left[\widetilde{J}_L^\mu|_{\widetilde{\zeta}_L=0}\widetilde{\psi}_{\mu L} + \widetilde{J}_R^\mu|_{\widetilde{\zeta}_R=0}\widetilde{\psi}_{\mu R}\right], \tag{7.76}$$

where, as indicated, we have to set $\widetilde{\zeta} = 0$ in the supercurrents. This coupling describes the main coupling of the physical gravitino to the remaining matter fields $\widetilde{\phi}^m$ and $\widetilde{\chi}_L^{m\perp}$. These couplings have the schematic form $\frac{1}{M_P}\partial\widetilde{\phi}^m\widetilde{\chi}_L^{m\perp}\widetilde{\psi}_\mu$ and $\frac{1}{M_P}\widetilde{\chi}_R^{m\perp}\Delta\left(\frac{S^*\mathscr{G}_m}{M_P}\right)\widetilde{\psi}_\mu$. For large enough momenta, the first term usually dominates and gives a contact interaction of strength E/M_P, which is highly suppressed at accessible energies $E \ll M_{\text{Susy}} \ll M_P$.

For energies much higher than the gravitino mass, i.e., in the regime $M_{3/2} \ll E \ll M_{\text{Susy}}$, however, an enhanced gravitino coupling is often used based on the equivalence theorem, which states that the matter couplings of the spin-1/2 polarization states of the gravitino are effectively given by the corresponding goldstino couplings.

To understand what is meant here, we start with the transformed gravitino field that has eaten the goldstino (ignoring small corrections of order $\partial_\mu \phi^m / M_{\text{Susy}}^2$ or $\Delta(S\mathscr{G}^m)/(M_P M_{\text{Susy}}^2)$),

$$\widetilde{\psi}_{\mu L} \cong \psi_{\mu L} + \frac{M_P}{\sqrt{3}\langle S^* \rangle}\mathscr{D}_\mu\zeta_L + \frac{1}{\sqrt{3}}\gamma_\mu\zeta_R = \psi_{\mu L} + \frac{2}{\sqrt{3}M_{3/2}}\mathscr{D}_\mu\zeta_L + \frac{1}{\sqrt{3}}\gamma_\mu\zeta_R, \tag{7.77}$$

which, for energies $E \gg M_{3/2}$, is dominated by the derivative term. The dominant coupling of the gravitino to the supercurrent then takes the schematic form

$$\frac{1}{M_P M_{3/2}}\partial\widetilde{\phi}^m\widetilde{\chi}_L^{m\perp}\mathscr{D}_\mu\zeta_L \sim \frac{1}{M_{\text{Susy}}^2}\partial\widetilde{\phi}^m\widetilde{\chi}_L^{m\perp}\mathscr{D}_\mu\zeta_L, \tag{7.78}$$

which, for $E \gg M_{3/2}$, scales as

$$\frac{E^2}{M_{\text{Susy}}^2} \sim \frac{E^2}{M_P M_{3/2}} \gg \frac{E}{M_P}, \tag{7.79}$$

i.e., it is enhanced by a factor $E/M_{3/2} \gg 1$ relative to the naive coupling. This would be particularly relevant for scenarios with small gravitino mass such as, e.g., scenarios with gauge-mediated supersymmetry breaking (see the next section).

It should be noted, however, that there is a subtlety with the above coupling when one takes the rigid limit $M_P \to \infty$, $M_{3/2} \to 0$ with M_{Susy} fixed. In this limit, the above coupling between the derivative of the goldstino and the supercurrent survives, and one would conclude that in global supersymmetry, there is a goldstino coupling of the form

$$\frac{1}{M_{\text{Susy}}^2} \overline{J}_L^\mu |_{\zeta_L=0} \partial_\mu \zeta_L \tag{7.80}$$

which, in particular, contains couplings of the schematic form $\frac{1}{M_P^2}(\partial \widetilde{\phi}^m \widetilde{\chi}_L^{m\perp}) \partial_\mu \zeta_L$. In the original Lagrangian (5.50), however, there is no analogous coupling with two spacetime derivatives and two fermions.[5] The resolution of this is that the above coupling still refers to the transformed matter fields $\widetilde{\phi}^m$ and $\widetilde{\chi}_L^{m\perp}$, which are related to the original fields, ϕ^m and $\chi_L^{m\perp}$, by a local supersymmetry transformation with supersymmetry parameter (7.75). While this is a symmetry in supergravity, it is not a symmetry in global supersymmetry so that the form of the action does change when one switches from the original fields ϕ^m, $\chi_L^{m\perp}$ to the transformed fields $\widetilde{\phi}^m$, $\widetilde{\chi}_L^{m\perp}$. In fact, we know precisely how the action changes from our discussion in Chap. 2, where we found that a local supersymmetry transformation in a globally supersymmetric theory leads to an uncancelled variation of the form $\overline{J}_L^\mu \partial_\mu \varepsilon_L$, which for an ε_L as in (7.75) precisely reproduces (7.80). Thus, the above goldstino coupling to the supercurrent only occurs in global supersymmetry upon a field redefinition to fields that are adapted to the vacuum of the corresponding supergravity Lagrangian but that are not particularly natural in global supersymmetry.

7.2.4 Generalizations

When both D- and F-terms are present, the combination ζ giving the goldstino can be found in an analogous way as above by examining the supersymmetry transformation laws in the vacuum. A simpler way is to read it off directly from

---

[5] In fact, it is easy to construct globally supersymmetric theories in which the goldstino completely decouples from the $\chi^{m\perp}$.

the mixing terms with the gravitino in the Lagrangian (6.48):

$$\zeta_L = M_P \frac{\mathscr{D}_m W}{W} \chi_L^m + \frac{e^{-\frac{\kappa}{2M_P^2}} M_P}{W} \frac{i}{2} \lambda_L^I \mathscr{P}_I. \tag{7.81}$$

Again, one finds that after an analogous redefinition of the gravitino, this field disappears from the action. The gravitino now also has Noether couplings to the gauge fields and the gaugini, and similar remarks apply to these couplings as in the case of chiral multiplets.

As mentioned earlier, we will not discuss in detail this case and the ones with a non-trivial cosmological constant, which however can be found in [5].

7.3 Mass Sum Rules and Mediation Mechanisms

In this section, we discuss how the soft supersymmetry breaking terms in supersymmetric extensions of the Standard Model might be generated and how the scale of the soft masses, M_{soft}, would be related to the fundamental supersymmetry breaking scale, M_{susy}, and the gravitino mass, $M_{3/2}$, in some popular scenarios.

7.3.1 Mass Sum Rules, Hidden Sectors, and Mediation Mechanisms

In our discussion of spontaneous supersymmetry breaking in the global case in Sect. 5.4, we mentioned a phenomenological problem with the corresponding mass sum rule $Str \mathscr{M}^2 = 0$ that is valid for renormalizable couplings at tree level. If we restrict to canonical Kähler potentials, constant gauge kinetic functions, and no gauging (i.e., no D-terms), the corresponding tree-level expression in supergravity is [6]

$$Str \mathscr{M}^2 = \sum_J (-1)^{2J} (2J+1) \mathscr{M}_J^2 = 2(n_c - 1) M_{3/2}^2, \tag{7.82}$$

where n_c is the number of chiral multiplets, and we considered pure F-term breaking on a Minkowski vacuum. This modified mass sum rule indicates that soft masses generically get a supergravity contribution of order $M_{3/2}$ that is absent in the globally supersymmetric case. On the other hand, this contribution may alleviate the mass splitting problem encountered in global supersymmetry provided $M_{3/2}$ turns out large enough. In the following, we discuss some of the most popular supersymmetry breaking scenarios with special attention to the relation between the soft masses, M_{soft}, and the gravitino mass, $M_{3/2}$, resp. the supersymmetry breaking scale M_{susy} in those scenarios. As explained in Sect. 5.4, we will assume that supersymmetry is broken in a hidden sector without renormalizable couplings at tree level to the Standard Model sector and mostly restrict ourselves to the case of soft masses near the TeV scale.

Gravity Mediation

The non-renormalizable tree-level interactions present in supergravity Lagrangians generically transmit effects of supersymmetry breaking from the hidden sector to the visible fields, even if these have no renormalizable couplings to one another. When this is the dominant source for the soft masses in the visible sector, one calls this gravity-mediated supersymmetry breaking [7].[6] Due to the special role played by supergravity in this scenario, we will discuss it in more detail in Sect. 7.3.2. In the simplest version, the soft masses are of order $M_{\text{soft}} \sim M_{\text{susy}}^2/M_P$ and hence of the order the gravitino mass, as may also be estimated from the mass sum rule (7.82). For soft masses around the TeV scale, the gravitino mass would then also be in this regime, whereas the fundamental supersymmetry breaking scale M_{susy} would be around 10^{11} GeV, depending on the details.

Anomaly Mediation

For special forms of the Kähler and superpotential, the abovementioned gravity mediated tree-level contribution to the soft masses may be suppressed. This happens, for example, in the no-scale models to be discussed below or in so-called sequestered models. A set of loop corrections that can also be related to certain quantum anomalies can then give the dominant contribution to soft masses and generate gaugino and sfermion masses of order $\frac{g^2}{16\pi^2} M_{3/2}$, where g represents a Standard Model gauge coupling. This mechanism is generally referred to as anomaly-mediated supersymmetry breaking [8]. Due to the loop suppression factors, the gravitino mass in pure anomaly mediation scenarios is one or two orders of magnitude larger than the soft masses in the Standard Model sector. For TeV scale soft masses, this would imply $M_{3/2} \sim \mathcal{O}(10^2 \text{ TeV})$, corresponding to a SUSY breaking scale around 10^{12} or 10^{13} GeV.

Gauge Mediation

Another important scenario in which the tree-level terms from gravity mediation (as well as from anomaly mediation) are subdominant (for a different reason than in anomaly mediation, see below) is gauge-mediated supersymmetry breaking [9]. In gauge mediation, the dominant soft terms in the Standard Model sector are generated by loop diagrams that involve messenger fields, a set of new fields charged under the Standard Model gauge group and with renormalizable couplings to the hidden sector where the fundamental supersymmetry breaking takes place. The soft masses are typically of order $\frac{g^2}{16\pi^2} M_{\text{mess}}$, where g again represents a Standard Model gauge coupling, and M_{mess} denotes the mass scale of the messenger particles.[7] If M_{susy} and M_{mess} turn out to be of the same order, TeV scale soft

---

[6] This name is a bit misleading, as it is not exactly gravity that induces the soft terms but other Planck-suppressed contact interactions that are (in part) related to gravity by supersymmetry; see Sect. 7.3.2.

[7] The gaugino masses arise at one-loop, whereas the squares of the sfermion masses are due to a two-loop correction.

masses would be generated for $M_{\text{mess}} \sim M_{\text{susy}} \sim \mathscr{O}(10^2 \text{ TeV})$, which would correspond to a very small gravitino mass in the eV regime. Making M_{susy} larger while keeping the soft masses fixed, one can reach larger gravitino masses in the keV or even MeV regime at the expense of introducing larger and larger hierarchies between M_{susy} and M_{mess}. Thus, while in scenarios dominated by gravity or anomaly mediation the gravitino mass is comparable to the soft masses or slightly larger, it is generically much smaller than the soft masses for gauge mediation. This is self-consistent in the sense that it is only because of this small gravitino mass that one can neglect the effects of gravity and anomaly mediation (which are generically present) relative to the gauge mediation effects.

The above describes the most commonly discussed scenarios of supersymmetry breaking in their simplest form. If supersymmetry is indeed realized in nature, it might of course be broken in a way that cannot be embedded unambiguously in one of the above frameworks. It could also be that supersymmetry breaking works in a way that makes use of an effective field theory that seems non-generic from a low-energy point of view but may look more natural from a top-down approach such as string theory. Likewise the simple picture of having all soft masses in the TeV region may be wrong as, e.g., in the various incarnations of split supersymmetry, or simply because supersymmetry could turn out to be irrelevant for low energy physics.

However, one may hope that if signs of supersymmetry are to be found in the future and a particular susy breaking scheme emerges as favored, we can learn something about the physics at even higher energy scales, possibly way outside the reach of direct accelerator experiments.

7.3.2 Gravity-Mediated Supersymmetry Breaking and the Polonyi Model

7.3.2.1 The Polonyi Model

The simplest model for generating a hidden sector breaking supersymmetry at vanishing cosmological constant was suggested by Polonyi in an unpublished preprint [10]. This model has a canonical Kähler potential for a single scalar field

$$K_h = \phi\phi^* \tag{7.83}$$

and a superpotential with a constant contribution and an arbitrary mass scale fixed by m and $\alpha \in \mathbb{R}$,

$$W_h = mM_P\left(\phi + \alpha M_P\right). \tag{7.84}$$

In terms of the dimensionless field, $Z := \frac{\phi}{M_P} = v + iw$, the resulting scalar potential is

$$V = m^2 M_P^2 e^{ZZ^*} \left[|1 + Z^*(Z + \alpha)|^2 - 3|Z + \alpha|^2 \right]. \tag{7.85}$$

The critical point condition $\partial_Z V|_{V=0} = 0$ admits the solution $w = 0$, to which we restrict ourselves in the following. The critical point condition $\partial_Z V|_{V=0} = 0$ then becomes

$$(2v + \alpha)(1 + v(v + \alpha)) - 3(v + \alpha) = 0, \tag{7.86}$$

whereas the Minkowski condition, $V = 0$, implies

$$1 + v(v + \alpha) = \pm\sqrt{3}(v + \alpha). \tag{7.87}$$

Inserting (7.87) into (7.86) implies $(v+\alpha) = \pm\sqrt{3} - v$, which can be re-inserted into (7.87) to give the four solutions $\alpha = -2 \mp \sqrt{3}$, $v = 1 \pm \sqrt{3}$, and $\alpha = 2 \pm \sqrt{3}$, $v = -1 \pm \sqrt{3}$, out of which, however, only two are local minima,

$$\begin{aligned} \alpha = -2 + \sqrt{3}, &\qquad \phi = M_P(1 - \sqrt{3}), \\ \alpha = 2 - \sqrt{3}, &\qquad \phi = M_P(-1 + \sqrt{3}), \end{aligned} \tag{7.88}$$

The scale of supersymmetry breaking in this model then turns out to be

$$M_{\text{susy}}^2 = \sqrt{3} m M_P e^{2-\sqrt{3}}, \tag{7.89}$$

and the mass of the gravitino is

$$M_{3/2} = e^{K_h/(2M_P^2)} \frac{W_h}{M_P^2} = e^{2-\sqrt{3}} m. \tag{7.90}$$

Gravity mediation would then lead to soft masses in the TeV range if m is chosen to be in the TeV range. This simple analysis shows two unpleasant features of this model: the parameter α has to be extremely fine-tuned, and also m needs to be chosen appropriately to obtain interesting phenomenology. We can, however, consider this setup as a simple toy model for a hidden sector that can serve as the basis to illustrate the effects of gravity mediation.

7.3.2.2 Illustration of Gravity Mediation: Scalar Soft Terms

The idea of gravity mediation is that the effects of supersymmetry breaking in the hidden sector are transmitted to the visible sector by the M_P-suppressed extra interactions that distinguish supergravity from globally supersymmetric field theories. There are different ways to realize such a setup depending on the details of the hidden sector dynamics and how exactly the hidden and visible sectors appear in the Kähler and superpotential. We discuss here only one of the simplest realizations in which the hidden sector is modeled by a single chiral multiplet with complex scalar ϕ that develops a Planck scale vev and breaks supersymmetry spontaneously. For the sake of concreteness, we realize this with the above Polonyi model,

$$K_h = \phi\phi^*, \qquad W_h = mM_P(\phi + \alpha M_P), \tag{7.91}$$

but the general conclusions do not depend on this choice. More crucial is the coupling of this model to the observable sector (the MSSM, for instance), which is done through a separable Kähler potential and superpotential

$$K = |\phi|^2 + \sum_i |y^i|^2, \tag{7.92}$$

$$W = W_h(\phi) + W_v(y^i). \tag{7.93}$$

Here, y^i denotes the scalars from the visible sector. Note that in global supersymmetry, such a separable K and W would imply a complete decoupling of the hidden sector and the visible sector. Hence, no matter how badly supersymmetry is broken in the hidden sector, the visible sector would not feel this in global supersymmetry in this model. Let us now investigate how this changes in supergravity.

We first write down the F-term potential,

$$V = e^{\sum_i \frac{|y^i|^2}{M_P^2} + \frac{|\phi|^2}{M_P^2}} \left[\left| \mathcal{D}_\phi W_h + \frac{\phi^*}{M_P^2} W_v \right|^2 \right.$$

$$\left. + \sum_i \left| \frac{\partial W_v}{\partial y^i} + \frac{y^{i*}}{M_P^2} + \frac{y^{i*}}{M_P^2} W_h \right|^2 - 3 \left| \frac{W_h + W_v}{M_P} \right|^2 \right]. \tag{7.94}$$

We assume that the vevs of W_v and $\frac{\partial W_v}{\partial y^i}$ are negligible compared to the vevs of, respectively, W_h and $\frac{\partial W_h}{\partial \phi}$ so that the dynamics in the hidden sector is essentially unaffected by the presence of the visible sector and yields essentially the same vevs in that sector as without the visible sector.[8] We thus still have to a very good

---

[8] This is not a strong assumption for the MSSM, because if $SU(3)_c$ and $U(1)_{em}$ as well as R-parity are to be unbroken, only the neutral Higgs fields can get a vev in the visible sector, and this vev is around the electroweak scale. The only term in W_v that can then get a vev is the quadratic term in the Higgs fields, the μ-term, which will then also be near the electroweak scale. The vev of W_h in the Polonyi model, by contrast, is of order $M_P^2 M_{3/2}$, which is many orders of magnitude higher. Similar remarks apply to the derivatives of the superpotentials.

approximation

$$\left\langle e^{\frac{|\phi|^2}{2M_P^2}} |\mathscr{D}_\phi W_h| \right\rangle = M_{\text{susy}}^2 = \sqrt{3} M_P M_{3/2} = \sqrt{3} \left\langle e^{\frac{|\phi|^2}{2M_P^2}} \frac{|W_h|}{M_P} \right\rangle. \tag{7.95}$$

The mass of ϕ is of order m, i.e., of the same order as the gravitino mass, $M_{3/2}$, which, as we will see in a moment, also sets the scale of the gravity-mediated soft terms in this model. We can thus not integrate out ϕ while keeping the other fields based on simple energy considerations. However, we can split $\phi = \langle \phi \rangle + \Delta \phi$ and use the fact that the vev of ϕ is of order M_P, so that M_P-suppressed terms involving ϕ may have significant contributions to the dynamics of the visible sector due to the contribution from $\langle \phi \rangle$, whereas the contributions from the fluctuation $\Delta \phi$ will just give rise to M_P-suppressed interactions with the visible sector fields. We thus ignore these interactions and replace ϕ, and consequently $\mathscr{D}_\phi W_h$ and W_h, by their vevs using the above relations to obtain

$$V = e^{\sum_i \frac{|y^i|^2}{M_P^2}} \left[\left| \sqrt{3} M_P M_{3/2} + \frac{\langle \phi^* \rangle}{M_P^2} \widehat{W}_v \right|^2 \right.$$

$$\left. + \sum_i \left| \frac{\partial \widehat{W}_v}{\partial y^i} + \frac{y^{i*}}{M_P^2} \widehat{W}_v + y^{i*} M_{3/2} \right|^2 - 3 \left| M_P M_{3/2} + \frac{\widehat{W}_v}{M_P} \right|^2 \right], \tag{7.96}$$

where

$$\widehat{W}_v := \left\langle e^{\frac{|\phi|^2}{2M_P^2}} \right\rangle W_v \tag{7.97}$$

is an appropriately rescaled superpotential for the visible sector. Taking now the rigid limit $M_P \to \infty$ with $M_{3/2}$ and $\langle \phi \rangle / M_P$ fixed, we obtain

$$V = \sum_i \left| \frac{\partial \widehat{W}_v}{\partial y^i} \right|^2 + \sum_i |y^i|^2 M_{3/2}^2$$

$$+ M_{3/2} \left[\left(\frac{\sqrt{3} \langle \phi \rangle}{M_P} - 3 \right) \widehat{W}_v + \sum_i \frac{\partial \widehat{W}_v}{\partial y^i} y^i + \text{h.c.} \right]. \tag{7.98}$$

The first term is the usual F-term potential for the fields y^i in global supersymmetry. The remaining terms are soft terms, with the second term being a mass term for the sfermions and the terms in brackets corresponding to bilinear and trilinear soft terms if $\widehat{W}_v$ contains only quadratic and cubic terms as in the MSSM. We note that the universal sfermion masses, $M_{\text{sfermion}} = M_{3/2}$, are a consequence of the simple Kähler potential (7.92). Non-universal sfermion masses may, e.g., be obtained from more general Kähler potentials such as

$$K = |\phi|^2 + \sum_i h_i(\phi, \phi^*)|y^i|^2. \tag{7.99}$$

7.3.2.3 Gaugino Masses

Gaugino masses can also be obtained if the gauge kinetic function, f_{IJ}, has a non-trivial dependence on the hidden sector fields, because the general supergravity action contains the term

$$\frac{1}{4} e^{K/2M_P^2} \mathcal{D}_m W g^{m\bar{n}} (\partial_{\bar{n}} f_{IJ}^*) \bar{\lambda}_R^I \lambda_R^J + \text{h.c.} \tag{7.100}$$

If $\langle \partial_{\bar{n}} f_{IJ}^* \rangle$ is of order $1/M_P$, this term would yield a gaugino mass, $M_{1/2}$, of order $M_{3/2}$, so that all soft masses would be comparable to the gravitino mass.

Note, however, that the above coupling (7.100) is already present in the most general globally supersymmetric theory (5.120). Thus the gaugino masses are in this sense not really gravity mediated, and the mass scale of $\partial_{\bar{n}} f_{IJ}^*$ need not necessarily be related to M_P but could a priori be related to any mass scale $\mu < M_P$ so that $\langle \partial_{\bar{n}} f_{IJ}^* \rangle \sim 1/\mu$. A simple example would, e.g., be $f(\phi) = 1 + \frac{1}{\mu}(\phi - \langle \phi \rangle)$, which satisfies $\langle f(\phi) \rangle = 1$, but $\langle \partial_\phi f \rangle = 1/\mu$, which would be much bigger than $1/M_P$ if μ is much smaller than M_P. In at least two important cases, however, a different mass scale $\mu \ll M_P$ in f does not enter the physical gaugino masses:

(i) $f(\phi) = \frac{\phi}{\mu} = \frac{\langle \phi \rangle}{\mu} + \frac{1}{\mu}(\phi - \langle \phi \rangle)$;
(ii) $f(\phi) = 1 + g^2 \log\left(\frac{\phi}{\mu}\right)$.

In case (i), $\langle \partial_\phi f \rangle = 1/\mu$ naively seems to result in a gaugino mass of the form $M_{3/2} M_P/\mu \gg M_{3/2}$, but this is not correct, because $\langle f \rangle = \langle \phi \rangle/\mu \gg 1$ so that the kinetic term for the gaugini and the gauge fields would not be canonically normalized. Expressing the mass term in terms of the canonically normalized gaugini $\tilde{\lambda} = \sqrt{M_P/\mu}\lambda$ then results again in a mass term of order $M_{3/2}$ if $\langle \phi \rangle$ is of order M_P.

In case (ii), one again splits off the vev of f,

$$f = \underbrace{1 + g^2 \log\left(\frac{\langle \phi \rangle}{\mu}\right)}_{=:X} + g^2 \log\left(\frac{\phi}{\langle \phi \rangle}\right) \tag{7.101}$$

and switches to canonically normalized fields $\tilde{\lambda} = \sqrt{X}\lambda$ so that the new gauge kinetic function becomes[9]

$$\tilde{f} = 1 + \frac{g^2}{X} \log\left(\frac{\phi}{\langle\phi\rangle}\right). \tag{7.102}$$

If g is a gauge coupling, the canonical normalization of the gauge fields also rescales g^2 to $\tilde{g}^2 = g^2/X$ so that

$$\frac{\partial \tilde{f}}{\partial \phi} = \frac{g^2}{X}\frac{1}{\phi} = \tilde{g}^2\frac{1}{\phi} \tag{7.103}$$

which results in gaugino masses of order $\tilde{g}^2 M_{3/2}$ if $\langle\phi\rangle$ is of order M_P. Thus, up to a possible slight suppression due to the factor $\tilde{g}^2$, one again obtains gaugino masses of the same order as for the sfermions.

It should be noted, however, that there is a difference between the gaugino masses and the sfermion masses also in the above two cases. Namely, in the case of the sfermion masses, the mass due to gravity mediation does not rely on having $\langle\phi\rangle$ of order M_P. In the derivation of the sfermion masses, we only used that the vevs of W_h and $\mathscr{D}_\phi W_h$ are much larger than the vevs of W_v and $\partial_{y^i} W_v$, which would be the case for vevs of ϕ that are large compared to the electroweak scale, but not necessarily of order M_P. The gaugino masses in the above two examples, on the other hand, are of order $M_{3/2}$ only for $\langle\phi\rangle \sim M_P$.

7.4 Moduli Stabilization, de Sitter Vacua, and Inflation

In this section, we derive some model-independent constraints on masses of scalar, in particular modulus-like, fields and discuss some issues associated with the cosmologically interesting case of positive potential energy.

7.4.1 Moduli Stabilization and Moduli Masses

An interesting possible ingredient of supersymmetric extensions of the Standard Model is moduli or modulus-like fields. These are scalar fields that are uncharged with respect to the gauge interactions and couple to ordinary matter only very weakly, typically with interactions suppressed by a very high mass scale such as M_P. Moduli can be useful for realizing spontaneous supersymmetry breaking, and they are a natural ingredient of phenomenologically realistic string compactifica-

---

[9] The canonical normalization factor X is not parametrically large due to the logarithm, so unlike in case (i), the canonical normalization does not change the result by at most an order of magnitude.

tions, where they correspond to deformations of the compactification background that cost little energy.[10]

A phenomenologically very important parameter is the mass of a modulus field, because if the moduli are too light, they can give rise to various phenomenological problems. Depending on the exact mass range, these may include, e.g., conflicts with the successful predictions of Big Bang Nucleosynthesis (BBN), problems with the measured energy density of the Universe or tensions with fifth force experiments. The simplest way to avoid all these problems is to ensure that the moduli have masses above about 30 TeV.[11] The implementation of mechanisms that generate such masses is generally referred to as moduli stabilization and forms an important part of modern string phenomenology.

In many models, a natural mass scale of the moduli is the gravitino mass, $M_{3/2}$, but moduli may also have masses considerably larger or smaller than $M_{3/2}$, depending on the details of the model. Moreover, some moduli may develop a negative mass squared, which means that they are tachyonic directions of the scalar potential that indicate an instability of that particular vacuum.

7.4.1.1 The sgoldstini

In this subsection,[12] we restrict ourselves mainly to a few rather general statements one can make based on the general form of the scalar potential. A crucial role in this discussion is played by the sgoldstini, the two scalar superpartners of the goldstino. Unlike the goldstino itself, the sgoldstini are not eaten by any other field and hence are part of the physical scalar spectrum.

To define the sgoldstini, it is convenient to shift all scalar fields, ϕ^m, by their vevs such that we have $\langle \phi^m \rangle = 0$. In a given vacuum, the sgoldstini are then the two real components of the complex scalar field

$$\Sigma := \frac{\langle \mathcal{G}_m \rangle}{\sqrt{\langle \mathcal{G}_n \mathcal{G}^n \rangle}} \phi^m. \tag{7.104}$$

---

[10] In the original sense, moduli denote massless scalar fields, usually related to constant energy deformations of the internal space in string compactifications or to exactly marginal deformations of conformal quantum field theories. As described above, in phenomenologically realistic string compactifications, such scalar fields should have a certain mass, i.e., the corresponding deformations of the internal space should cost some energy (e.g., due to the presence of p-form fluxes in the internal space or as a consequence of non-perturbative quantum corrections). Nevertheless, these massive scalar fields are still called moduli.

[11] This bound is an approximate estimate based on the following assumptions [11]: (1) Moduli couple with M_P-suppressed interactions to other matter with a resulting decay width $\Gamma \sim \mathcal{O}(M_{\mathrm{Mod}}^3/M_P^2)$; (2) the Hubble scale after inflation is larger than M_{Mod}, and when it drops to $H \approx M_{\mathrm{Mod}}$, the modulus starts oscillating around its potential minimum; (3) the moduli so produced decay before Big Bang Nucleosynthesis (BBN).

[12] This subsection is based on [12].

With respect to the scalar field metric in the vacuum, $\langle g_{m\bar{n}} \rangle$, its orthogonal complement is

$$\phi^{m\perp} := \phi^m - \frac{\langle \mathcal{G}^m \rangle}{\sqrt{\langle \mathcal{G}_l \mathcal{G}^l \rangle}} \Sigma, \tag{7.105}$$

which satisfies

$$\langle \mathcal{G}_m \rangle \phi^{m\perp} = 0 \tag{7.106}$$

so that indeed

$$- \langle g_{m\bar{n}} \rangle (\partial_\mu \phi^m)(\partial^\mu \phi^{\bar{n}}) = - \langle g_{m\bar{n}} \rangle (\partial_\mu \phi^{m\perp})(\partial^\mu \phi^{\bar{n}\perp}) - (\partial_\mu \Sigma)(\partial^\mu \Sigma^*), \tag{7.107}$$

and Σ and $\phi^{m\perp}$ are the superpartners of, respectively, the goldstino, ζ_L, and its orthogonal complement, $\chi_L^{m\perp}$, in the given vacuum,

$$\delta\Sigma = \bar{\varepsilon}_L \zeta_L, \qquad \delta\phi^{m\perp} = \bar{\varepsilon}_L \chi_L^{m\perp}. \tag{7.108}$$

Several interesting conclusions can be drawn from the behavior of the scalar potential in the complex plane given by Σ. In order to probe the scalar potential along arbitrary real straight lines through the origin of this plane, we introduce the rotated field

$$Z := e^{i\theta} \Sigma, \tag{7.109}$$

where $e^{i\theta}$ is a constant phase factor, and split it into canonically normalized real and imaginary parts, X and Y,

$$Z = \frac{1}{\sqrt{2}}(X + iY). \tag{7.110}$$

Studying V along X for all possible θ is then equivalent to studying V along all possible straight real lines through the origin of the Σ-plane.

If all other fields are frozen, the canonically normalized mass of X would be

$$\begin{aligned}
m_X^2 = \partial_X \partial_X V &= \frac{1}{2}(V_{ZZ} + 2V_{ZZ^*} + V_{Z^*Z^*}) \\
&= \frac{1}{2}(e^{-2i\theta} V_{\Sigma\Sigma} + 2V_{\Sigma\Sigma^*} + e^{2i\theta} V_{\Sigma^*\Sigma^*}) \\
&= V_{\Sigma\Sigma^*} + \mathrm{Re}[e^{-2i\theta} V_{\Sigma\Sigma}], \tag{7.111}
\end{aligned}$$

where $V_{ZZ} \equiv \partial_Z \partial_Z V$, etc., and all quantities are meant to be evaluated at the critical point.

Setting now $V_{\Sigma\Sigma} = Ae^{i\alpha}$ with $A := |V_{\Sigma\Sigma}|$ and α being a possible phase, we can write

$$m_X^2 = V_{\Sigma\Sigma^*} + A\cos(\alpha - 2\theta). \tag{7.112}$$

Thus, for the particular field direction given by $\theta = \alpha/2 - \pi/4$, the angular term can be eliminated, and we obtain

$$m_X^2 = V_{\Sigma\Sigma^*} = \frac{1}{\mathcal{G}_n \mathcal{G}^n} \mathcal{G}^l V_{l\bar{k}} \mathcal{G}^{\bar{k}}, \tag{7.113}$$

where again all quantities are meant to be constant vevs.

In Appendix 7.A at the end of this chapter, we show that, at a critical point of the potential, (7.113) can be written as [12]

$$m_X^2 = M_P^2 e^{\mathcal{G}/M_P^2} \left[2 - \frac{R_{\bar{k}l\bar{p}m} \mathcal{G}^{\bar{k}} \mathcal{G}^l \mathcal{G}^{\bar{p}} \mathcal{G}^m}{\mathcal{G}_n \mathcal{G}^n} \right]. \tag{7.114}$$

A number of interesting conclusions regarding moduli mass scales and the stability of the vacuum can be drawn from this equation.

7.4.1.2 A Constraint on the Lightest Modulus Mass

In this subsection, we discuss how, under certain assumptions, Eq. (7.114) provides a bound on the mass of the lightest modulus-like field [12]. To this end, we use $M_{3/2}^2 = M_P^2 e^{\mathcal{G}/M_P^2}$ and neglect the small cosmological constant in the present Universe, setting $\mathcal{G}_n \mathcal{G}^n = 3M_P^2$. Equation (7.114) then becomes

$$M_X^2 = \partial_X \partial_X V = M_{3/2}^2 [2 - r], \tag{7.115}$$

where

$$r := \frac{1}{3M_P^2} R_{\bar{k}l\bar{n}m} \mathcal{G}^{\bar{k}} \mathcal{G}^l \mathcal{G}^{\bar{n}} \mathcal{G}^m. \tag{7.116}$$

Now, M_X^2 is not necessarily an eigenvalue of the mass matrix, because the Hessian of V in general contains non-trivial off-diagonal terms. However, $M_X^2 = \partial_X \partial_X V$ is a diagonal element of the mass matrix, and as the smallest eigenvalue of a diagonalizable real matrix cannot be bigger than any of its diagonal elements, we can conclude that the smallest scalar mass eigenvalue, M_{min}^2, is bounded from above by M_X^2,

$$M_{min}^2 \leq M_X^2 = M_{3/2}^2 [2 - r]. \tag{7.117}$$

Thus, if the curvature of the scalar manifold along the sgoldstino directions is at most of order M_P^{-2}, such that $|r| = \mathcal{O}(1)$ or smaller, Eq. (7.117) implies that the lightest modulus-like field cannot be parametrically heavier than the gravitino.

Thus, if one wants the moduli masses to be above 30 TeV, the gravitino mass could then likewise not be parametrically smaller than this mass scale. This in turn would be a challenge for scenarios with small gravitino masses such as gauge mediation, and it might be part of an explanation for why the superpartners of the Standard Model particles might be somewhat heavier than what simple naturalness arguments suggest (see, e.g., [13]).

7.4.1.3 Possible Caveats

While it would be a fascinating possibility, there are also a number of caveats to the above argument. More precisely, there are two types of possible caveats. The first type concerns all the assumptions that underlie the phenomenological 30 TeV estimate for the lightest modulus mass (see footnote 11). If any of these assumptions is too strong, the moduli masses might be smaller without conflicting observations, which would then also relax the above-described constraints on the gravitino mass from Eq. (7.117). The second type of caveats concerns the conclusion that the lightest modulus cannot be parametrically heavier than the gravitino due to Eq. (7.117). We focus here only on this latter type of caveats, which include the following:

- The smallest eigenvalue of the scalar mass matrix need not necessarily be a modulus-like field but could in principle be another type of scalar, e.g., a charged scalar field. In that case, the bound (7.117) would not constrain the lightest modulus-like field but this other type of scalar field. However, if there is no significant mixing between the moduli and these other scalar fields and if supersymmetry is predominantly broken in the moduli sector, one can, to a very good approximation, repeat the above analysis in the moduli sector alone and would then again conclude that the lightest modulus is indeed constrained by an equation of the form (7.117).
- We did not consider D-terms in our analysis. If they play a significant role in supersymmetry breaking and/or moduli stabilization, the above analysis would have to be re-done with D-terms.
- If the scalars, ϕ^m, are not canonically normalized with respect to the vev of the scalar field metric in the kinetic term, i.e., if $\langle g_{m\bar{n}} \rangle \neq \delta_{m\bar{n}}$, the eigenvalues of the Hessian of V do not give the physical masses of the canonically normalized fields. The transition to canonically normalized scalars would instead rescale the physical mass eigenvalues. In the case at hand, however, this would not change (7.117) and the conclusions drawn from it. The reason is that the real sgoldstino fields X and Y are already canonically normalized and orthogonal to the other scalars $\phi^{m\perp}$. Thus a redefinition of the $\phi^{m\perp}$ to make them also canonically normalized would not change the diagonal element $\partial_X \partial_X V$ of the mass matrix but at most the terms involving at least one derivative with respect to $\phi^{m\perp}$. Thus, after canonical normalization, the smallest eigenvalue of the new

mass matrix would still be bounded by the unchanged diagonal element (7.117) with r being expressed in terms of the new canonically normalized fields. But as r is a scalar under scalar field redefinitions, it does not change its value under these redefinitions, and hence also the canonically normalized fields satisfy the bound (7.117).

- If r is negative and $|r|$ is extremely large, i.e., if $-r \gg 1$, the lightest modulus may be much heavier than the gravitino, and the above bound would allow for much smaller gravitino masses. A special case of this would be the limit $r \to \infty$, which would typically correspond to a particular realization of non-linear supersymmetry, which is thus also not directly covered by the argument above.

7.4.1.4 A Counterexample with Large Curvature

Let us briefly illustrate the last item in the above list with a simple toy model [14] given by a single chiral multiplet with scalar ϕ and

$$K = \phi\phi^* - \frac{\phi^2\phi^{*2}}{\Lambda^2}, \qquad W = \mu^2(\phi + cM_P), \tag{7.118}$$

where μ and Λ are mass parameters with $\Lambda \ll M_P$, and c is a numerical constant. For suitably fine-tuned c, the resulting F-term potential has a non-supersymmetric Minkowski minimum with a very small vev of ϕ. This may be seen by a perturbative expansion in the small parameter $\Lambda/M_P \ll 1$, which, after some algebra, leads to

$$c = \frac{1}{\sqrt{3}}\left(1 + \mathcal{O}\left(\frac{\Lambda^2}{M_P^2}\right)\right), \tag{7.119}$$

$$\langle\phi\rangle = \langle\phi^*\rangle = M_P\left(\frac{1}{2\sqrt{3}}\frac{\Lambda^2}{M_P^2}\right)\left(1 + \mathcal{O}\left(\frac{\Lambda^2}{M_P^2}\right)\right) \ll M_P. \tag{7.120}$$

The curvature parameter r at this vacuum is

$$r = -12\frac{M_P^2}{\Lambda^2}\left(1 + \mathcal{O}\left(\frac{\Lambda^2}{M_P^2}\right)\right) \ll -1 \tag{7.121}$$

so that the bound (7.117) allows for moduli masses parametrically larger than $M_{3/2}$. And indeed, one finds that, to lowest order in the small parameter Λ^2/M_P^2, the two eigenvalues of the scalar mass matrix are, using $M_{3/2}^2 = \mu^4/3M_P^2$,

$$M_1^2 = M_2^2 = \frac{4\mu^4}{\Lambda^2}\left(1 + \mathcal{O}\left(\frac{\Lambda^2}{M_P^2}\right)\right) = 12M_{3/2}^2\frac{M_P^2}{\Lambda^2}\left(1 + \mathcal{O}\left(\frac{\Lambda^2}{M_P^2}\right)\right) \gg M_{3/2}^2. \tag{7.122}$$

In the limit $\Lambda \to 0$, the curvature parameter r and with it the scalar masses diverge, and the low-energy theory becomes effectively a specific version of a local non-linearly realized supersymmetry with the metric and a massive gravitino as the only light fields, which one may also describe with the help of nilpotent superfields [14, 15].

Interestingly, even though some of the above caveats might in principle apply, many models derived from string theory do indeed have a lightest modulus that is not parametrically heavier than the gravitino [12].

7.4.2 No Scale Models

In the counterexample above, we saw that the sgoldstini can in principle be parametrically heavier than the gravitino when the curvature in the sgoldstino plane takes on extreme (negative) values. There is, however, also the opposite possibility that an exact or approximate cancellation of the two terms in (7.117) could make the sgoldstini and hence the lightest modulus much lighter than the gravitino, or even massless. An extreme example for this are no-scale models, where scalars that do not appear in W are exactly massless at tree level, while the gravitino mass can take on arbitrarily high values. At the same time, the tree-level cosmological constant remains exactly zero, no matter how high the scale of supersymmetry breaking is [16]. Subleading corrections in general lift this vacuum degeneracy and could generate a hierarchically small supersymmetry breaking scale [17]. Unfortunately, the smallness of the cosmological constant relative to the supersymmetry breaking scale in general does not survive these corrections. Nevertheless, no-scale models play an important role as a first step in model building due to their natural emergence in string theory compactifications.

7.4.2.1 The Simplest Example
The simplest version of a no-scale model is based on one chiral multiplet with dimensionless complex scalar, T, and

$$K = -3M_P^2 \log(T + T^*), \qquad W = \alpha M_P^3 \qquad (7.123)$$

with $\alpha \in \mathbb{R}$. In that case, one has $\partial_T W = 0$ and

$$\partial_T K = -\frac{3M_P^2}{T + T^*}, \qquad g_{TT^*} = \frac{3M_P^2}{(T + T^*)^2} \qquad \Rightarrow \qquad K^T K^{T^*} g_{TT^*} = 3M_P^2 \qquad (7.124)$$

so that the F-term contribution from T to the scalar potential satisfies

$$g^{TT^*} \mathscr{D}_T W \mathscr{D}_{T^*} W^* = 3\frac{|W|^2}{M_P^2}, \qquad (7.125)$$

which exactly cancels the negative term coming from the gravitino contribution.

As a result, the potential is completely flat, $V \equiv 0$, so that we have an infinite set of degenerate Minkowski vacua parameterized by the vev of an unfixed massless modulus T. The supersymmetry breaking scale and the gravitino mass, however, are not zero but depend on the vev of T,

$$M_{\text{SUSY}}^2 = \frac{\sqrt{3}\alpha M_P^2}{\langle (T + T^*)^{3/2} \rangle}, \qquad M_{3/2} = \frac{\alpha M_P}{\langle (T + T^*)^{3/2} \rangle}. \qquad (7.126)$$

7.4.2.2 Generalizations

The crucial property of the above example that leads to the cancellation of the negative term in the potential is (apart from $\partial_T W = 0$) the geometric property

$$K_T K^T = 3M_P^2. \qquad (7.127)$$

A straightforward generalization of this for n complex dimensionless scalars t^α ($\alpha, \beta, \ldots = 1, \ldots, n$) is obtained when K and W satisfy

$$K^\alpha K_\alpha = 3M_P^2, \qquad \partial_\alpha W = 0, \qquad (7.128)$$

where, as usual, $K_\alpha \equiv \partial_\alpha K$, etc. This would then again lead to $V \equiv 0$.

For a concrete realization of such Kähler potentials, we split the n scalars t^α into n_1 complex scalars, T^A ($A, B, \ldots = 1, \ldots n_1$), and n_2 complex scalars, φ^a ($a, b, \ldots = 1, \ldots, n_2$), and take the Kähler potential to be of the following form

$$K = -M_P^2 \log[F(J^A)], \qquad J^A := T^A + T^{\overline{A}} - N^A(\varphi^a, \varphi^{\overline{a}}), \qquad (7.129)$$

where F is a homogeneous function of degree three of the composite variables J^A,

$$J^A \frac{\partial F}{\partial J^A} = 3F, \qquad (7.130)$$

and N is an arbitrary real function of the fields φ^a.

To verify the no-scale property,

$$K^\alpha K_\alpha = K^A K_A + K^a K_a = 3M_P^2, \qquad (7.131)$$

we first note that

$$K_A \equiv \frac{\partial K}{\partial T^A} = \frac{\partial K}{\partial T^{\overline{A}}} = \frac{\partial K}{\partial J^A} = -\frac{M_P^2}{F} \frac{\partial F}{\partial J^A} \qquad (7.132)$$

so that, with (7.130),

$$J^A K_A = -3M_P^2. \qquad (7.133)$$

With this, we compute

$$0 = \partial_{\overline{\alpha}}(J^A K_A) = (\partial_{\overline{\alpha}} J^A) \underbrace{\frac{\partial K}{\partial T^A}}_{\frac{\partial K}{\partial J^A}} + J^A K_{\overline{\alpha} A} = \partial_{\overline{\alpha}} K + g_{\overline{\alpha} A} J^A, \tag{7.134}$$

or, raising the index $\overline{\alpha}$ with $g^{\alpha \overline{\alpha}}$,

$$K^\alpha = -\delta^\alpha_A J^A, \tag{7.135}$$

so that (7.133) indeed implies $K_\alpha K^\alpha = 3M_P^2$.

Frequently encountered examples are

- $K = -M_P^2 \log \left(T + T^* - c\delta_{a\overline{b}} \varphi^a \varphi^{\overline{b}} \right)^3 = -3M_P^2 \log \left(T + T^* - c\delta_{a\overline{b}} \varphi^a \varphi^{\overline{b}} \right)$.
  For the special case $c = 1/3$, this is the symmetric space $SU(1, 1+n_2)/S(U(1) \times U(1 + n_2))$.
- $K = -M_P^2 \log \left[d_{ABC}(T^A + T^{\overline{A}})(T^B + T^{\overline{B}})(T^C + T^{\overline{C}}) \right]$, where d_{ABC} is a real symmetric tensor.

Although the above form of the Kähler potential looks quite peculiar, it should be stressed that no-scale models are in fact not uncommon in string theory compactifications and other types of dimensional reductions. Upon reducing a theory from D to $d < D$ dimensions, one finds, in particular, that the volume modulus, and more generally the Kähler moduli, behaves like the T^A in the leading order Kähler potential.

7.4.2.3 Adding Scalars Without No-Scale Property

If one adds scalars, z^i ($i, j, \ldots = 1, \ldots, m$), to a no-scale model with scalars, t^α, as above such that

$$K = \tilde{K}(t^\alpha, t^{\overline{\alpha}}) + \widehat{K}(z^i, z^{\overline{i}}), \qquad W = W(z^i) \tag{7.136}$$

with $\tilde{K}_\alpha \tilde{K}^\alpha = 3M_P^2$, the F-term of the t^α still cancels the negative gravitino contribution in the potential, which then becomes positive semi-definite,

$$V = \exp \left(\frac{K(t, z, t^*, z^*)}{M_P^2} \right) g^{i\overline{j}} \mathscr{D}_i W \mathscr{D}_{\overline{j}} W^* \geq 0, \tag{7.137}$$

where i, j run only over the z^i fields, and the only dependence on t^α is in the exponential factor. This implies that the Minkowski vacua with $\langle \mathscr{D}_i W \rangle = 0$ are the global minima of the potential at which the z^i are in general stabilized. This degenerate valley of Minkowski minima is therefore parameterized by the vevs of the unstabilized t^α, because the t^α-dependence of the potential drops out in the valley, where $V=0$. Just as in the simple example (7.123), for fixed z^i, the vevs of the t^α determine the supersymmetry breaking scale and the gravitino mass, which can take on arbitrary values even though the cosmological constant vanishes at tree level.

In general, however, the flatness of the potential along the no-scale moduli t^α is lifted by quantum corrections, and the t^α cannot be chosen at will but need to be evaluated at a suitable minimum of the corrected potential (if such a minimum exists). As they are generated only at subleading order, the masses of the t^α may be hierarchically smaller than the masses of the z^i. If that is the case, one can integrate out the heavier z^i and work with an effective theory of the t^α alone. A well-known example for this is the KKLT scenario in type IIB string theory [18], where the z^i are complex structure moduli that are stabilized at tree level by fluxes of antisymmetric tensor fields in the compact space, and the t^α correspond to Kähler moduli with smaller masses generated by non-perturbative quantum corrections to W.

7.4.2.4 A D-Term Analogue

While the above examples of no-scale models are all based on pure F-term potentials, there is also a construction involving D-terms that shares some of the features of the pure F-term models [19]. To this end, we couple $\mathcal{N} = 1$ supergravity to one chiral multiplet with the following Kähler potential and superpotential

$$K = -2\,M_P^2\,\log(\phi + \phi^*), \qquad W = a\,M_P^3, \tag{7.138}$$

where a is a real constant and ϕ is dimensionless. This is obviously not of the no-scale type discussed above and by itself would not lead to a flat potential. To achieve this, however, we can now also couple one vector multiplet (A_μ, λ) to the above theory with a gauge kinetic function $f(\phi)$ given by

$$f(\phi) = \frac{1}{g^2} \tag{7.139}$$

where g is a real constant (i.e., the ϕ dependence is trivial here). Obviously, the Kähler metric on the scalar manifold spanned by the complex field ϕ only depends on the real part of ϕ so that the field transformation

$$\phi \rightarrow \phi + \alpha\,\xi^\phi = \phi + i\,\alpha \tag{7.140}$$

is an isometry on the scalar manifold with symmetry parameter α and Killing vector $\xi^\phi = i$. Gauging this isometry then leads to a D-term potential given by

$$V_D = \frac{1}{2}(\mathrm{Re}\,f)^{-1}\,\mathscr{P}\,\mathscr{P}, \tag{7.141}$$

where $\mathscr{P} \equiv i\xi^\phi \partial_\phi K$ is the Killing prepotential of ξ^ϕ. The total potential in the special case $2g^2 = a^2 \neq 0$ is then again flat, $V \equiv 0$, and the supersymmetry breaking scale can take on any value depending on the undetermined vev of ϕ.

Also in this case, one can imagine adding additional scalars and vector fields without spoiling the no-scale property of the resulting potential, provided certain requirements are satisfied. The Kähler potential and superpotentials can be extended to

$$
K = -2\,M_P^2\,\log(\phi + \phi^*) + \Delta K(\phi, \phi^*, z^i, z^{\bar i}), \qquad W = a\,M_P^3 + \Delta W(z^i),
\tag{7.142}
$$

and the gauge kinetic functions generalized to

$$
f_{ab} = f_{ab}^{(0)}(z^i) + f_{ab}^{(1)}(z^i)\phi,
\tag{7.143}
$$

provided $\langle \Delta K \rangle = \langle \Delta W \rangle = 0$ at the minimum and the supersymmetric conditions $\mathscr{D}_i W = 0 = D_a$ are satisfied for the z^i directions and for the new gauge fields A_μ^a.

7.4.3 Dark Energy and de Sitter Vacua

In Sect. 7.4.1, we were primarily concerned with the absolute value of moduli masses in a Minkowski vacuum and how these masses compared to the mass of the gravitino based on the bound (7.117). What we did not yet discuss is the possibility that the sign of some squared moduli masses might actually be *negative* at a given critical point. In fact, (7.114) indicates that for scalar manifolds with sufficiently strong *positive* curvature in the sgoldstino plane, the curvature term in (7.114) may overcompensate the positive first term and imply that the lightest modulus must have a negative mass squared. In other words, the critical point would not be a local minimum of the potential but would have a tachyonic direction in field space indicating a perturbative instability of that vacuum.

The danger of the sgoldstino bound becoming negative is generally smaller for AdS than for Minkowski or even de Sitter vacua. To understand this, we go back to the original equation (7.114), which is valid for any value and sign of the vacuum energy, and write it as[13]

$$
m_{\min}^2 \leq m_X^2 = \partial_X \partial_X V = M_P^2\,\frac{e^{\mathscr{G}/M_P^2}}{\mathscr{G}_n \mathscr{G}^n}\left[2\mathscr{G}_l \mathscr{G}^l - R_{\bar{k}l\bar{p}m}\,\mathscr{G}^{\bar k}\mathscr{G}^l \mathscr{G}^{\bar p}\mathscr{G}^m\right].
\tag{7.144}
$$

---

[13] This subsection is based on [20].

Obviously, a tachyon is implied if the term in brackets becomes negative, i.e., if the quartic curvature term overcompensates the first term quadratic in $\mathscr{G}^m$. For sub-Planckian curvatures and small $|\mathscr{G}^m| \equiv \sqrt{\mathscr{G}_l \mathscr{G}^l} \ll M_P^2$, however, the first term should always dominate and make M_X^2 positive so that a negative $M_{\min}^2$ is not necessarily implied.[14] Remembering $V = M_P^2 e^{\mathscr{G}/M_P^2}[\mathscr{G}_m \mathscr{G}^m - 3M_P^2]$, however, the case $|\mathscr{G}^m| \ll M_P^2$ is seen to correspond to negative vacuum energy, i.e., an AdS vacuum.[15]

For a Minkowski vacuum, by contrast, we have instead $|\mathscr{G}^m| = \sqrt{3}M_P$ so that a (positive) curvature $R_{\bar{k}l\bar{p}m}$ of order $1/M_P^2$ would make the quartic term in (7.144) comparable to the quadratic term so that a negative M_X^2 becomes possible. For a de Sitter vacuum, one even has $|\mathscr{G}^m| > \sqrt{3}M_P$ so that a positive curvature can imply a negative M_X^2 more easily the larger $|\mathscr{G}^m|$ and hence M_{vac} is. In fact, for some models one can even rule out a dS minimum no matter how small the vacuum energy is, as we now describe.

To this end, we rewrite the quantity in brackets in (7.144) as

$$
\begin{aligned}
\lambda &:= 2\mathscr{G}_m \mathscr{G}^m - R_{\bar{k}l\bar{p}m} \mathscr{G}^{\bar{k}} \mathscr{G}^l \mathscr{G}^{\bar{p}} \mathscr{G}^m \\
&= -\frac{2}{3}(\mathscr{G}_n \mathscr{G}^n - 3)\mathscr{G}_l \mathscr{G}^l + \left[\frac{2}{3}(\mathscr{G}_l \mathscr{G}^l)^2 - R_{\bar{k}l\bar{p}m} \mathscr{G}^{\bar{k}} \mathscr{G}^l \mathscr{G}^{\bar{p}} \mathscr{G}^m \right] \qquad (7.145) \\
&= -\frac{2}{3}\widehat{V}(\widehat{V} + 3) + \sigma,
\end{aligned}
$$

where $\widehat{V} \equiv (\mathscr{G}_n \mathscr{G}^n - 3)$ and

$$
\sigma := \left[\frac{1}{3}(g_{l\bar{k}} g_{m\bar{p}} + g_{l\bar{p}} g_{m\bar{k}}) - R_{l\bar{k}m\bar{p}} \right] \mathscr{G}^l \mathscr{G}^{\bar{k}} \mathscr{G}^m \mathscr{G}^{\bar{p}}. \qquad (7.146)
$$

Thus, in a dS vacuum, with $\widehat{V} > 0$, a tachyon would necessarily be present if

$$
\sigma \le 0. \qquad (7.147)
$$

Note that the sign of σ only depends on the orientation of the $\mathscr{G}^m$ as well as on purely geometric quantities determined by K, but not on the absolute value $|\mathscr{G}^m|$ and hence also not on the magnitude of the (positive) cosmological constant.

We conclude this subsection with a few examples, where, for simplicity, we set $M_P = 1$ and work with dimensionless fields.

---

[14] This does of course not guarantee that there is really no tachyon, because (7.144) is just an upper bound.

[15] As we have discussed in Chap. 4, stable AdS vacua would even be consistent with tachyonic scalars as long as they satisfy the Breitenlohner–Freedman bound.

- $K = \phi\phi^* \Rightarrow R_{\phi\phi^*\phi\phi^*} = 0 \Rightarrow \sigma = \frac{2}{3}(\mathscr{G}_l \mathscr{G}^l)^2 > 0$. Hence, the bound (7.144) does not necessarily imply the existence of a de Sitter tachyon for this case. In fact, the Polonyi model with appropriately chosen constant term in the superpotential would give a simple example with a dS minimum.
- $K = -3\log(T + T^*) \Rightarrow \sigma = 0$. Hence, unless K receives corrections, this no-scale model does not allow for a dS minimum with a pure F-term potential, no matter how W is chosen (see also [21]).
- $K = -3\log(T + T^* - 1/3\sum_i |\phi^i|^2) \Rightarrow \sigma = 0$. Hence, in this model, there are no dS minima possible if one just uses an F-term potential and K does not receive corrections.

In this section, we focused on the stability of a given de Sitter extremum of an F-term potential and found that there are some dangers of developing a tachyonic instability in the sgoldstino plane. A somewhat related problem occurs when one starts from no-scale potentials with Minkowski vacua and tries to deform them into a de Sitter vacuum with a small cosmological constant by introducing small corrections to the no-scale potential. As shown in [22], many of such corrections may lead to a tachyon at the de Sitter extremum, which is not in the sgoldstino plane but along a field direction which only aligns with one of the sgoldstino directions in the Minkowski limit.

Of course tachyons can occur in other field directions as well, and as there is no analogue of a Breitenlohner–Freedman bound for de Sitter spacetime, any tachyon would lead to an instability, making this a general issue for de Sitter vacua.

Moreover, many F-term potentials that descend from tree-level dimensional reductions of higher-dimensional supergravity theories do not even have de Sitter critical points due to surprisingly simple no-go theorems [23]. These no-go theorems may be evaded, e.g., by introducing objects of negative energy density such as orientifold planes in type II string theories and/or by taking into account quantum or stringy corrections or by allowing for more exotic branes and fluxes.

7.4.4 Inflation and the Supergravity η-Problem

Another interesting cosmological problem that can be addressed in supergravity is inflation [24]. Inflation denotes a postulated period of accelerated cosmic expansion in the very early Universe that could solve various cosmological naturalness problems such as the flatness and homogeneity problem and provides a mechanism for the generation of density fluctuations. The simplest realization of inflation is via a scalar field, ϕ, ("inflaton") that slowly rolls down a relatively flat potential, $V(\phi)$, with its momentary potential energy driving the accelerated cosmic expansion. The flatness of the potential is usually expressed in terms of the slow-roll parameters ϵ and η, which have to satisfy

$$\epsilon \equiv \frac{1}{2} M_P^2 \left(\frac{V'}{V} \right)^2 \ll 1, \tag{7.148}$$

$$\eta \equiv M_P^2 \left| \frac{V''}{V} \right| \ll 1. \tag{7.149}$$

If one embeds this model in a supersymmetric theory, one might hope that supersymmetry could protect the inflaton direction from getting a large slope or curvature. However, in supergravity, this is surprisingly difficult to achieve, even at tree level. In order to see this, we assume that the inflaton potential during slow-roll inflation is dominated by the F-term potential,

$$V_{\text{infl}} \cong V_F = e^{K/M_P^2} \left(|\mathscr{D}W|^2 - 3 \frac{|W|^2}{M_P^2} \right) \tag{7.150}$$

Assuming now a canonical Kähler potential, $K = K_{\text{can}} = \phi\phi^*$, one finds that the derivatives of the exponential of K produce a term of the form

$$\partial_\phi \partial_{\phi^*} V_{\text{infl}} = \frac{1}{M_P^2} V_{\text{infl}} + \ldots \tag{7.151}$$

where the ellipsis denotes terms with derivatives acting also on the other terms. This then implies

$$\eta = 1 + \ldots, \tag{7.152}$$

i.e., a generic contribution of order one to the eta parameter.[16] This would have to be cancelled by the terms denoted by the ellipsis, which usually requires very special superpotentials and/or fine-tunings. This is the supergravity η-problem of F-term inflation[17] [25]. Possible solutions are the following:

- W is of a special form such that $\eta \ll 1$ is achieved. An example is given in the exercises.
- The Kähler potential might have a shift symmetry along, say, the imaginary part of ϕ [26]. More precisely, define $\varphi \equiv \text{Im}(\phi)$ and let $K = \frac{(\phi+\phi^*)^2}{2}$, which would still give rise to a canonical kinetic term for the scalars. In this case, the

---

[16] Note that we have been a bit sloppy here with the fact that ϕ in supergravity is complex and that in single field inflation the inflaton is real. One thus has to go over to the diagonalized mass matrix of the real and imaginary part of ϕ first.

[17] The inflaton mass term can alternatively be viewed as due to gravity-mediated spontaneous supersymmetry breaking during slow-roll inflation.

exponential of K would not depend on φ, and, taking φ as the inflaton, no η-problem would arise at this level. Note that this new Kähler potential is related to the canonical one by a Kähler transformation. In global supersymmetry, we can therefore always bring the Kähler potential to this shift symmetric form, but in supergravity, the Kähler transformation would also act on W, which would then pick up an exponential of the Kähler transformation, thereby maintaining the η-problem (it would just be shifted from K to W).

- Simply accept an $\mathcal{O}(10^{-2})$ fine-tuning (eta is constrained only to order 10^{-2}).
- Use D-term inflation, i.e., an inflaton potential dominated by D-terms [27, 28]. The problem with standard versions of D-term inflation, however, is that they usually produce cosmic strings after inflation. Moreover, in models in which moduli fields have to be stabilized (e.g., in the generic string theory low-energy effective actions), the stabilization usually works with F-term potentials, which can re-introduce the η-problem also in D-term inflation models [29].
- Use a potential that is not a supergravity potential or one with non-linearly realized supersymmetry [30, 31].

7.A   Appendix: Proof of Eq. (7.114)

In this Appendix, we show that, at a critical point of an F-term potential in $\mathcal{N} = 1$ supergravity, one has

$$\mathcal{G}^{\bar{k}}\mathcal{G}^{l} V_{\bar{k}l} = M_P^2\, e^{\mathcal{G}/M_P^2}\, \mathcal{G}_p\mathcal{G}^p \left[2 - \frac{R_{\bar{k}l\bar{n}m}\,\mathcal{G}^{\bar{k}}\mathcal{G}^{l}\mathcal{G}^{\bar{n}}\mathcal{G}^{m}}{\mathcal{G}_q\mathcal{G}^q} \right], \qquad (7.153)$$

which then implies Eq. (7.114) upon division by $\mathcal{G}_p\mathcal{G}^p$.

As a preparation, we first define

$$\widehat{V} := \mathcal{G}_m\mathcal{G}^m - 3M_P^2 \qquad (7.154)$$

so that

$$V = M_P^2 e^{\mathcal{G}/M_P^2}\, \widehat{V} \qquad (7.155)$$

and rewrite the critical point condition in a number of useful ways. Using (7.155), we first have

$$\partial_l V = M_P^2 e^{\mathcal{G}/M_P^2} \left[\frac{\mathcal{G}_l}{M_P^2}\, \widehat{V} + \widehat{V}_l \right] = 0 \qquad (7.156)$$

$$\Leftrightarrow \widehat{V}_l = -\frac{\mathcal{G}_l \widehat{V}}{M_P^2}, \qquad (7.157)$$

where $\mathcal{G}_l \equiv \partial_l \mathcal{G}$, etc. and all quantities are meant to be vevs.

To compute $\widehat{V}_l$ directly from the definition (7.154), we use $\mathscr{D}_l g^{m\bar{n}} = 0$ and $\mathscr{D}_l \mathscr{G}_{\bar{n}} = g_{l\bar{n}}$ to obtain

$$\widehat{V}_l = \partial_l \widehat{V} = \mathscr{D}_l \widehat{V} = (\mathscr{D}_l \mathscr{G}_m)\mathscr{G}^m + \mathscr{G}_m g^{m\bar{n}}(\mathscr{D}_l \mathscr{G}_{\bar{n}}) = \mathscr{G}^m(\mathscr{D}_l \mathscr{G}_m) + \mathscr{G}_l \qquad (7.158)$$

Hence, (7.157) becomes

$$\mathscr{G}^m \mathscr{D}_l \mathscr{G}_m = -\mathscr{G}_l \left(\frac{\widehat{V}}{M_P^2} + 1 \right). \qquad (7.159)$$

We are now ready to prove (7.153) and first compute

$$V_{\bar{k}l} = \partial_{\bar{k}}(\partial_l V) \overset{(7.156)}{=} \partial_{\bar{k}} \left[e^{\mathscr{G}/M_P^2}(\mathscr{G}_l \widehat{V} + M_P^2 \widehat{V}_l) \right] \qquad (7.160)$$

$$= \frac{\mathscr{G}_{\bar{k}}}{M_P^2} \underbrace{\partial_l V}_{=0} + e^{\mathscr{G}/M_P^2} \left[\underbrace{(\partial_{\bar{k}}\mathscr{G}_l)}_{g_{\bar{k}l}} \widehat{V} + \underbrace{\mathscr{G}_l \widehat{V}_{\bar{k}}}_{\overset{(7.157)}{=} -\mathscr{G}_l \mathscr{G}_{\bar{k}} \widehat{V}/M_P^2} + M_P^2 \widehat{V}_{l\bar{k}} \right] \qquad (7.161)$$

$$= V \left[\frac{g_{l\bar{k}}}{M_P^2} - \frac{\mathscr{G}_l \mathscr{G}_{\bar{k}}}{M_P^4} \right] + M_P^2 e^{\mathscr{G}/M_P^2} \widehat{V}_{l\bar{k}}. \qquad (7.162)$$

In $\widehat{V}_{l\bar{k}}$, one can replace $\partial_{\bar{k}}$ by $\mathscr{D}_{\bar{k}}$, because the Christoffel symbols with mixed indices vanish, so that

$$\widehat{V}_{l\bar{k}} = \mathscr{D}_{\bar{k}} \widehat{V}_l \overset{(7.158)}{=} (\mathscr{D}_{\bar{k}}\mathscr{G}_{\bar{n}})(\mathscr{D}_l \mathscr{G}_m)g^{m\bar{n}} + \mathscr{G}^m(\mathscr{D}_{\bar{k}}\mathscr{D}_l \mathscr{G}_m) + g_{\bar{k}l} \qquad (7.163)$$

$$= (\mathscr{D}_{\bar{n}}\mathscr{G}_{\bar{k}})(\mathscr{D}_m \mathscr{G}_l)g^{m\bar{n}} + \mathscr{G}^m(\mathscr{D}_{\bar{k}}\mathscr{D}_l \mathscr{G}_m) + g_{\bar{k}l} \qquad (7.164)$$

where $\mathscr{D}_l \mathscr{G}_m \equiv \partial_l \mathscr{G}_m - \Gamma^p_{lm}\mathscr{G}_p = \mathscr{D}_m \mathscr{G}_l$ and the complex conjugate thereof was used in the last step. Using now $\partial_l g_{\bar{k}m} = \Gamma^p_{lm} g_{\bar{k}p}$ and $\partial_{\bar{k}} \Gamma^p_{lm} = R_{\bar{k}l}{}^p{}_m$, one finds $\mathscr{D}_{\bar{k}}\mathscr{D}_l \mathscr{G}_m = -R_{\bar{k}l\bar{p}m}\mathscr{G}^{\bar{p}}$ so that

$$\widehat{V}_{l\bar{k}} = (\mathscr{D}_{\bar{n}}\mathscr{G}_{\bar{k}})(\mathscr{D}_m \mathscr{G}_l)g^{m\bar{n}} - R_{\bar{k}l\bar{p}m}\mathscr{G}^m \mathscr{G}^{\bar{p}} + g_{\bar{k}l}. \qquad (7.165)$$

Inserting this into (7.162) and contracting with $\mathscr{G}^{\bar{k}}\mathscr{G}^l$ then gives

$$
\mathscr{G}^l\mathscr{G}^{\bar{k}}V_{l\bar{k}} = V\left[\frac{\mathscr{G}_l\mathscr{G}^l}{M_P^2} - \frac{(\mathscr{G}_l\mathscr{G}^l)^2}{M_P^4}\right]
$$

$$
+M_P^2 e^{\mathscr{G}/M_P^2}\left[(\underbrace{\mathscr{G}^{\bar{k}}\mathscr{D}_{\bar{n}}\mathscr{G}_{\bar{k}}}_{-\mathscr{G}_{\bar{n}}\left(\frac{\hat{V}}{M_P^2}+1\right)})\ g^{m\bar{n}}\ \underbrace{(\mathscr{G}^l\mathscr{D}_m\mathscr{G}_l)}_{-\mathscr{G}_m\left(\frac{\hat{V}}{M_P^2}+1\right)} - R_{\bar{k}l\bar{p}m}\mathscr{G}^{\bar{k}}\mathscr{G}^l\mathscr{G}^{\bar{p}}\mathscr{G}^m + \mathscr{G}_l\mathscr{G}^l\right].
$$

(7.166)

Factoring out $\mathscr{G}_l\mathscr{G}^l$ and replacing $\mathscr{G}_n\mathscr{G}^n$ by $(\hat{V}+3M_P^2)$ in the remaining term, one finds, after some algebra, that all $\hat{V}$-dependent terms cancel and (7.153) is obtained.

Exercises

7.1. Compute the masses of the fermions at the non-supersymmetric vacuum of the model with Kähler potential

$$
K = |\phi|^2 + |A|^2 + |B|^2
$$

and superpotential

$$
W = \phi A B.
$$

Discuss the super-Higgs mechanism at this vacuum.

7.2. Consider an $\mathcal{N} = 1$ supersymmetric theory of three chiral multiplets with complex scalars S, ϕ_+, ϕ_- and the following Kähler and superpotentials:

$$
K = |S|^2 + |\phi_+|^2 + |\phi_-|^2
$$

(7.167)

$$
W = S\left(\kappa\phi_+\phi_- - \mu^2\right).
$$

(7.168)

Here, μ is a mass parameter and κ denotes a dimensionless coupling constant.

(a) Show that in *global* supersymmetry the resulting classical F-term potential has positive mass terms for ϕ_+ and ϕ_- if $|S| > S_c = \mu/\sqrt{\kappa}$.
(b) For $|S| > S_c$, one can thus assume $\langle\phi_\pm\rangle = 0$. Show, again in *global* supersymmetry, that the classical F-term potential is flat along S for vanishing

vevs of ϕ_+ and ϕ_- and that supersymmetry is spontaneously broken in this valley.

(c) The abovementioned classical flatness along S is lifted by two effects. One is due to the spontaneously broken supersymmetry, which induces a logarithmic loop correction. This correction is welcome as it may induce a gentle slope of the potential along S, which may then serve as an inflaton. If S slowly rolls down its potential and reaches the critical point $|S| = S_c$, one of the scalars ϕ_+ and ϕ_- becomes tachyonic and condenses, and the scalar potential drops to zero. This would then be a supersymmetric realization of a hybrid inflation model.

This naive picture, however, could be ruined by the second type of corrections, namely, the classical supergravity corrections to the F-term potential, i.e., all the M_P-suppressed terms in the supergravity F-term potential that are absent in global supersymmetry. As discussed above, these generically give rise to a too large inflaton mass/eta parameter. Compute the classical supergravity F-term potential along the line $\phi_+ = \phi_- = 0$ and show that, for this particular model, the supergravity η-problem is actually *absent*, i.e., that the quadratic terms in S induced by K and W precisely cancel.

References

1. N. Arkani-Hamed, S. Dimopoulos, Supersymmetric unification without low energy supersymmetry and signatures for fine-tuning at the LHC. JHEP **06**, 073 (2005). [arXiv:hep-th/0405159 [hep-th]]
2. G.F. Giudice, A. Romanino, Split supersymmetry. Nucl. Phys. B **699**, 65–89 (2004). [arXiv:hep-ph/0406088 [hep-ph]]
3. M. Billo, F. Denef, P. Fre, I. Pesando, W. Troost, A. Van Proeyen, D. Zanon, The rigid limit in special Kahler geometry: From K3 fibrations to special Riemann surfaces: a detailed case study'. Class. Quant. Grav. **15**, 2083–2152 (1998). [arXiv:hep-th/9803228 [hep-th]]
4. E. Cremmer, B. Julia, J. Scherk, P. van Nieuwenhuizen, S. Ferrara, L. Girardello, Super-higgs effect in supergravity with general scalar interactions. Phys. Lett. B **79**, 231–234 (1978)
5. S. Ferrara, A. Van Proeyen, Mass formulae for broken supersymmetry in curved space-time. Fortsch. Phys. **64**(11–12), 896–902 (2016). [arXiv:1609.08480 [hep-th]]
6. M.T. Grisaru, M. Rocek, A. Karlhede, The superhiggs effect in superspace. Phys. Lett. B **120**, 110–118 (1983)
7. H.P. Nilles, Phys. Lett. B **115**, 193 (1982); A.H. Chamseddine, R. Arnowitt, P. Nath, Phys. Rev. Lett. **49**, 970 (1982); R. Barbieri, S. Ferrara, C.A. Savoy, Phys. Lett. B **119**, 343 (1982); L.E. Ibáñez, Phys. Lett. B **118**, 73 (1982); L.J. Hall, J.D. Lykken, S. Weinberg, Phys. Rev. D **27**, 2359 (1983); N. Ohta, Prog. Theor. Phys. **70**, 542 (1983)
8. L. Randall, R. Sundrum, Out of this world supersymmetry breaking. Nucl. Phys. B **557**, 79–118 (1999); G.F. Giudice, M.A. Luty, H. Murayama, R. Rattazzi, Gaugino mass without singlets. JHEP **12**, 027 (1998); A. Pomarol, R. Rattazzi, Sparticle masses from the superconformal anomaly. JHEP **05**, 013 (1999)
9. M. Dine, W. Fischler, M. Srednicki, Nucl. Phys. B **189**, 575 (1981); M. Dine, W. Fischler, Phys. Lett. B **110**, 227 (1982); M. Dine, A.E. Nelson, Phys. Rev. D **48**, 1277 (1993) [arXiv:hep-ph/9303230]; M. Dine, A.E. Nelson, Y. Shirman, Phys. Rev. D **51**, 1362 (1995) [arXiv:hep-ph/9408384]
10. J. Polonyi, *Generalization of the Massive Scalar Multiplet Coupling to the Supergravity*, Hungary Central Inst. Res. KFKI-77-93 preprint (1977, rec. July 1978), 5 pages, unpublished

11. G.D. Coughlan, W. Fischler, E.W. Kolb, S. Raby, G.G. Ross, Phys. Lett. B **131**, 59–64 (1983); J.R. Ellis, D.V. Nanopoulos, M. Quiros, Phys. Lett. B **174**, 176–182 (1986); B. de Carlos, J.A. Casas, F. Quevedo, E. Roulet, Phys. Lett. B **318**, 447–456 (1993); T. Banks, D.B. Kaplan, A.E. Nelson, Phys. Rev. D **49**, 779–787 (1994)

12. B.S. Acharya, G. Kane, E. Kuflik, Bounds on scalar masses in theories of moduli stabilization. Int. J. Mod. Phys. A **29**, 1450073 (2014)

13. M.R. Douglas, The string landscape and low energy supersymmetry (2012). https://doi.org/10.1142/9789814412551_0012. [arXiv:1204.6626 [hep-th]]

14. R. Casalbuoni, S. De Curtis, D. Dominici, F. Feruglio, R. Gatto, Nonlinear realization of supersymmetry algebra from supersymmetric constraint. Phys. Lett. B **220**, 569–575 (1989)

15. G. Dall'Agata, E. Dudas, F. Farakos, On the origin of constrained superfields. JHEP **05**, 041 (2016) [arXiv:1603.03416 [hep-th]]

16. E. Cremmer, S. Ferrara, C. Kounnas, D.V. Nanopoulos, Phys. Lett. B **133**, 61 (1983); J.R. Ellis, C. Kounnas, D.V. Nanopoulos, Nucl. Phys. B **247**, 373–395 (1984); S. Ferrara, C. Kounnas, F. Zwirner, Nucl. Phys. B **429**, 589 (1994). Erratum: [Nucl. Phys. B 433 (1995) 255] [hep-th/9405188]

17. J. Polchinski, L. Susskind, Phys. Rev. D **26**, 3661 (1982); H.-P. Nilles, M. Srednicki, D. Wyler, Phys. Lett. B **124**, 337 (1983); J.R. Ellis, A.B. Lahanas, D.V. Nanopoulos, K. Tamvakis, Phys. Lett. B **134**, 429 (1984); J.R. Ellis, C. Kounnas, D.V. Nanopoulos, Nucl. Phys. B **241**, 406 (1984) and B 247, 373 (1984); C. Kounnas, F. Zwirner, I. Pavel, Phys. Lett. B **335**, 403 (1994) [hep-ph/9406256]

18. S. Kachru, R. Kallosh, A.D. Linde, S.P. Trivedi, De Sitter vacua in string theory. Phys. Rev. D **68**, 046005 (2003). [arXiv:hep-th/0301240 [hep-th]]

19. G. Dall'Agata, F. Zwirner, New class of $N = 1$ no-scale supergravity models. Phys. Rev. Lett. **111**(25), 251601 (2013). [arXiv:1308.5685 [hep-th]]

20. L. Covi, M. Gomez-Reino, C. Gross, J. Louis, G.A. Palma, C.A. Scrucca, de Sitter vacua in no-scale supergravities and Calabi-Yau string models. JHEP **06**, 057 (2008). [arXiv:0804.1073 [hep-th]]

21. R. Brustein, S.P. de Alwis, Moduli potentials in string compactifications with fluxes: Mapping the discretuum. Phys. Rev. D **69**, 126006 (2004). [arXiv:hep-th/0402088 [hep-th]]

22. D. Junghans, M. Zagermann, A universal tachyon in nearly no-scale de Sitter compactifications. JHEP **07**, 078 (2018) [arXiv:1612.06847 [hep-th]]

23. G.W. Gibbons, Aspects of supergravity theories, in *Supersymmetry, Supergravity and Related Topics*, ed. by F. del Aguila, J.A. de Azcárraga and L.E. Ibáñez; B. de Wit, D.J. Smit, N.D. Hari Dass, Residual supersymmetry of compactified D=10 supergravity. Nucl. Phys. B **283**, 165 (1987); J.M. Maldacena, C. Nunez, Supergravity description of field theories on curved manifolds and a no go theorem. Int. J. Mod. Phys. A **16**, 822–855 (2001). [arXiv:hep-th/0007018 [hep-th]]

24. A.H. Guth, The inflationary universe: a possible solution to the horizon and flatness problems. Phys. Rev. D **23**, 347–356 (1981); A.D. Linde, A new inflationary universe scenario: a possible solution of the horizon, flatness, homogeneity, isotropy and primordial monopole problems. Phys. Lett. B **108**, 389–393 (1982); A. Albrecht, P.J. Steinhardt, Cosmology for grand unified theories with radiatively induced symmetry breaking. Phys. Rev. Lett. **48**, 1220–1223 (1982)

25. E.J. Copeland, A.R. Liddle, D.H. Lyth, E.D. Stewart, D. Wands, False vacuum inflation with Einstein gravity. Phys. Rev. D **49**, 6410–6433 (1994). [arXiv:astro-ph/9401011 [astro-ph]]

26. M. Kawasaki, M. Yamaguchi, T. Yanagida, Natural chaotic inflation in supergravity. Phys. Rev. Lett. **85**, 3572–3575 (2000). [arXiv:hep-ph/0004243 [hep-ph]]

27. P. Binetruy, G.R. Dvali, D term inflation. Phys. Lett. B **388**, 241–246 (1996). [arXiv:hep-ph/9606342 [hep-ph]]

28. P. Binetruy, G. Dvali, R. Kallosh, A. Van Proeyen, Fayet–Iliopoulos terms in supergravity and cosmology. Class. Quant. Grav. **21**, 3137–3170 (2004). [arXiv:hep-th/0402046 [hep-th]]

29. L. McAllister, An Inflaton mass problem in string inflation from threshold corrections to volume stabilization. JCAP **02**, 010 (2006). [arXiv:hep-th/0502001 [hep-th]]

30. S. Ferrara, R. Kallosh, A. Linde, Cosmology with nilpotent superfields. JHEP **10**, 143 (2014). https://doi.org/10.1007/JHEP10(2014)143. [arXiv:1408.4096 [hep-th]]
31. G. Dall'Agata, F. Zwirner, On sgoldstino-less supergravity models of inflation. JHEP **12**, 172 (2014). [arXiv:1411.2605 [hep-th]]

Part III

Extended, Gauged and Higher-Dimensional Supergravity

Extended Supergravities

8

Whereas in $\mathcal{N} = 1$ supersymmetry scalar fields are only mapped to certain spin-1/2 fermions, the larger multiplets of extended supersymmetry may link scalars also to fields of spin greater than 1/2, in particular vector fields. As we will see in the following, this heavily constrains the structure and couplings of extended supergravities. The main reason for this is that in a theory without charged fields, the field equations of Abelian vector fields in four spacetime dimensions exhibit the phenomenon of electric–magnetic duality. In extended supersymmetry, this electric–magnetic duality structure of the vector field sector is then linked to the scalar field geometry by supersymmetry and constrains the scalar manifolds of most supermultiplets. Moreover, this same duality structure is going to be at the basis of the consistent construction of deformations that make some of the fields charged under the gauge group. We therefore first discuss in Sect. 8.1 the main features of electric–magnetic duality in preparation for our discussion of extended supergravity. In the subsequent sections of this chapter, we then explain how electric–magnetic duality and the structure of the R-symmetry group constrain the geometries of the corresponding scalar manifolds. In Sect. 8.2, this will lead us to special Kähler and quaternionic manifolds for the scalar fields in, respectively, the vector and hypermultiplets in $\mathcal{N} = 2$ supergravity. The scalar manifolds for supergravity theories with $\mathcal{N} \geq 3$ supergravity are then discussed in Sect. 8.3. The appendix to this chapter contains computational details on the scalar field geometries in $\mathcal{N} = 2$ supergravity.

8.1 Electric–Magnetic Duality

It is well-known that Maxwell's equations in the vacuum are invariant under the exchange of the electric and magnetic vector fields: $\vec{E} \rightarrow \vec{B}$ and $\vec{B} \rightarrow -\vec{E}$. In the Lorentz covariant notation, this duality relation is expressed as the exchange of the

© Springer-Verlag GmbH Germany, part of Springer Nature 2021
G. Dall'Agata, M. Zagermann, *Supergravity*, Lecture Notes in Physics 991,
https://doi.org/10.1007/978-3-662-63980-1_8

gauge field strength, $F = 1/2\, dx^\mu \wedge dx^\nu\, F_{\mu\nu}$, and its Poincaré dual,

$$\widetilde{F} = \star F = \frac{1}{2} dx^\mu \wedge dx^\nu\, \widetilde{F}_{\mu\nu}, \tag{8.1}$$

where

$$\widetilde{F}_{\mu\nu} \equiv \frac{1}{2} \epsilon_{\mu\nu\rho\sigma} F^{\rho\sigma}. \tag{8.2}$$

In fact, Maxwell's equations in vacuum can be written as

$$\begin{cases} dF = 0, \\ d \star F = 0, \end{cases} \tag{8.3}$$

where the first equation is usually interpreted as a Bianchi identity for the curvature $F = dA$ of the gauge potential, A, and the second is its equation of motion. However, by the introduction of the dual gauge field strength, $\widetilde{F}$, these equations can be written in the more symmetric form

$$\begin{cases} dF = 0, \\ d\widetilde{F} = 0, \end{cases} \tag{8.4}$$

which is invariant under $F \to \star F = \widetilde{F}$ and $\widetilde{F} \to \star \widetilde{F} = (\star)^2 F = -F$. Actually, one could mix the two curvatures with general linear transformations without changing the content of (8.4). This duality transformation is obviously violated by minimal electric couplings and/or non-Abelian gauge field strengths as these contain direct couplings of the gauge potentials that cannot be written in terms of the Abelian field strengths alone. However, we can introduce a new generalized form of duality transformation for Abelian vector fields (and their field strengths) that is valid also in the presence of non-minimal matter couplings.

The first thorough analysis of the consequences of the existence of electric–magnetic duality for generic field theories containing vector fields was performed in [1]. This work also constitutes the basis of the discussion for their generalization, namely, U-duality symmetries, which we will give in the next chapter. Although we borrow a lot from that paper, in the following, we provide an original presentation, clarifying some aspects that are especially relevant for the construction of supergravity theories.

Consider a generic 2-derivative Lagrangian containing n_V *Abelian* vectors, A^I ($I = 1, \ldots, n_V$), through their field strengths, $F^I = dA^I$, and arbitrary couplings to other fields φ^i (denoting together bosonic and fermionic fields),

$$e^{-1} \mathcal{L}(F^I, \varphi^i, \partial_\mu \varphi^i) = \frac{1}{4} \mathscr{I}_{IJ} F^I_{\mu\nu} F^{J\mu\nu} + \frac{1}{4} \mathscr{R}_{IJ} F^I_{\mu\nu} \widetilde{F}^{J\mu\nu} + \frac{1}{2} \mathcal{O}^{\mu\nu}_I F^I_{\mu\nu}$$

$$+ e^{-1} \mathcal{L}_{\text{rest}}. \tag{8.5}$$

Here, $\mathscr{I}_{IJ}$ and $\mathscr{R}_{IJ}$ are symmetric matrices that may depend on the scalar fields, with $\mathscr{I}_{IJ}$ being negative definite to ensure unitarity, $\mathcal{O}^{\mu\nu}_I$ is a generic tensor function of the other fields containing at most a single derivative, and $\mathcal{L}_{\text{rest}}$ contains all the terms that do not depend on the vector field strengths. By definition, the vector field curvatures are $F^I = dA^I$ and hence

$$dF^I = 0 \tag{8.6}$$

describe the Bianchi identities of the n_V vector fields, A^I. The equations of motion for such fields then follow as usual from setting

$$\nabla^\mu \frac{\partial \mathcal{L}}{\partial F^{\mu\nu I}} = 0. \tag{8.7}$$

These equations can also take the form of Bianchi identities if we introduce dual variables

$$\widetilde{G}_{I\mu\nu} = 2 \frac{\partial \mathcal{L}}{\partial F^{I\mu\nu}} \tag{8.8}$$

and $\widetilde{G}_{I\mu\nu} = \frac{1}{2} \epsilon_{\mu\nu\rho\sigma} G^{\rho\sigma}_I$, so that (8.7) can be written as

$$dG_I = 0. \tag{8.9}$$

The system of Bianchi identities and equations of motion is a priori invariant under constant $\mathrm{GL}(2n_V, \mathbb{R})$ transformations,

$$\mathbb{F}' = S\mathbb{F}, \qquad \mathbb{F} \equiv \begin{pmatrix} F^I \\ G_J \end{pmatrix} \tag{8.10}$$

with $S \in \mathrm{GL}(2n_V, \mathbb{R})$. However, in order to preserve the definition of the G_I tensors in terms of F^I and via (8.8) for the transformed Lagrangian, we have to further constrain the matrix S.

In order to do so, notice that (8.9) implies the (local) existence of n_V dual 1-forms A_I such that $G_I = dA_I$. Obviously, the Lagrangian will depend only on one of the two, but we can trade one for the other by a Legendre-like transformation, introducing a total derivative term to the action of the form

$$S' = \frac{1}{2} \int F^I \wedge dA_I = \frac{1}{2} \int F^I \wedge G_I. \qquad (8.11)$$

Varying the action $S + S'$ with respect to A_I, one obtains the usual Bianchi identity $dF^I = 0$, while varying with respect to F^I, one obtains the definition of $G_I = dA_I$ and hence one could rewrite the Lagrangian in terms of the dual potentials by plugging the solution to these equations back into the action. Let us then perform a Legendre-like transformation from the F to the G variables and introduce the dual Lagrangian

$$e^{-1}\mathscr{L}_D = \left[e^{-1}\mathscr{L} - \frac{1}{2} F^I_{\mu\nu} \widetilde{G}^{\mu\nu}_I \right]_{F=F(\widetilde{G})}, \qquad (8.12)$$

where, after adding (8.11) to the original action, we replace each instance of the original field strengths, F^I, with their expression in terms of the dual ones, G_I. We call this transformation Legendre-like because the dual Lagrangian is defined by the difference between the Lagrangian itself and the product of the variables on which the original Lagrangian depends (the F^I field strengths, which we treat as black boxes) and the dual variables (the G_I forms we introduced above), like an actual Legendre transformation. This analogy can be actually extended to the point that the original variables are recovered by varying the dual Lagrangian with respect to the dual variables:

$$F^I_{\mu\nu} = -2\frac{\partial \mathscr{L}_D}{\partial \widetilde{G}^{\mu\nu}_I}. \qquad (8.13)$$

In fact, we can expand the dual Lagrangian in terms of the dual field strengths as follows:

$$e^{-1}\mathscr{L}_D = \frac{1}{4} \mathscr{A}^{IJ} \widetilde{G}_{I\mu\nu} \widetilde{G}^{\mu\nu}_J + \frac{1}{4} \mathscr{B}^{IJ} \widetilde{G}_{I\mu\nu} G^{\mu\nu}_J - \frac{1}{2} \mathscr{O}^{I\,\mu\nu} \widetilde{G}_{J\mu\nu} + \mathscr{L}'_{\text{rest}}, \qquad (8.14)$$

where, again, $\mathscr{A}^{IJ}$ and $\mathscr{B}^{IJ}$ are symmetric matrices, $\mathscr{O}^{I\mu\nu}$ is a tensor that does not contain the $\widetilde{G}_{I\mu\nu}$ fields, and $\mathscr{L}'_{\text{rest}}$ is the part of the Lagrangian that does not depend at all on the $\widetilde{G}_{I\mu\nu}$ fields. The fact that (8.14) has the form presented above can be argued by the linearity of the relation (8.8) between the original vector field strengths and their duals.

Now that we have an explicit form for both $\mathscr{L}(F^I)$ and $\mathscr{L}_D(G_I)$, we can recover the explicit relations following from applying (8.8) and (8.13). The results are collected in the following expressions:

$$\widetilde{G}_{I\,\mu\nu} = \mathscr{I}_{IJ}\, F^J_{\mu\nu} + \mathscr{R}_{IJ}\, \widetilde{F}^J_{\mu\nu} + \mathcal{O}_{I\,\mu\nu}\,,$$

$$G_{I\,\mu\nu} = -\mathscr{I}_{IJ}\, \widetilde{F}^J_{\mu\nu} + \mathscr{R}_{IJ}\, F^J_{\mu\nu} - \widetilde{\mathcal{O}}_{I\,\mu\nu}\,,$$

$$F^I_{\mu\nu} = -\mathscr{A}^{IJ}\, \widetilde{G}_{J\,\mu\nu} - \mathscr{B}^{IJ}\, G_{J\,\mu\nu} + \mathcal{O}^I_{\mu\nu}\,,$$

$$\widetilde{F}^I_{\mu\nu} = \mathscr{A}^{IJ}\, G_{J\,\mu\nu} - \mathscr{B}^{IJ}\, \widetilde{G}_{J\,\mu\nu} + \widetilde{\mathcal{O}}^I_{\mu\nu}\,,$$

(8.15)

where we explicitly wrote the relations involving the $\mathbb{F}_{\mu\nu}$ curvatures and their Hodge duals $\widetilde{\mathbb{F}}_{\mu\nu} = \frac{1}{2}\epsilon_{\mu\nu\rho\sigma}\mathbb{F}^{\rho\sigma}$, for completeness. Consistency of these conditions then gives the following conditions:

$$\mathscr{A} = -(\mathscr{I} + \mathscr{R}\mathscr{I}^{-1}\mathscr{R})^{-1}\,,$$

(8.16)

$$\mathscr{B} = \mathscr{A}\mathscr{R}\mathscr{I}^{-1} = \mathscr{I}^{-1}\mathscr{R}\mathscr{A}\,,$$

(8.17)

$$\mathcal{O}^I = \mathscr{A}^{IJ}\mathcal{O}_J - \mathscr{B}^{IJ}\widetilde{\mathcal{O}}_J\,,$$

(8.18)

$$\mathscr{L}'_{\text{rest}} = \mathscr{L}_{\text{rest}} + \frac{1}{4}\mathcal{O}^I\mathcal{O}_I\,,$$

(8.19)

which come from inserting the expressions (8.15) for F^I and $\widetilde{F}^I$ into the ones for G_I and $\widetilde{G}_I$. The outcome is that we can express $\mathscr{L}_D$ fully in terms of the quantities appearing in the original $\mathscr{L}$. Moreover, using these relations, we can then explicitly check that (8.12) is indeed identically satisfied as it should.

At this point, we can impose the consistency constraints on the duality transformations (8.10) by applying them to one of the explicit relations (8.15) and imposing its consistency. Let us see how. At the infinitesimal level, the duality transformation (8.10) can be written in terms of four real matrices

$$\delta F^I = A^I{}_J F^J + B^{IJ} G_J\,,$$

$$\delta G_I = C_{IJ} F^J + D_I{}^J G_J\,.$$

(8.20)

If we now apply this to the first line in (8.15), we see that

$$\delta\widetilde{G}_{I\,\mu\nu} = -\delta\mathscr{I}_{IJ}F^J_{\mu\nu} + \delta\mathscr{R}_{IJ}\widetilde{F}^J_{\mu\nu} - \mathscr{I}_{IJ}\delta F^J_{\mu\nu} + \mathscr{R}_{IJ}\delta\widetilde{F}^J_{\mu\nu} + \delta\mathcal{O}_{I\,\mu\nu}\,, \quad (8.21)$$

which gives

$$C_{IJ}\tilde{F}^J_{\mu\nu} + D_I{}^J\tilde{G}_{I\,\mu\nu} = -\,\delta\mathscr{I}_{IJ}F^J_{\mu\nu} + \delta\mathscr{R}_{IJ}\tilde{F}^J_{\mu\nu}$$

$$-\,\mathscr{I}_{IJ}\left(A^J{}_K F^K_{\mu\nu} + B^{JK}G_{K\,\mu\nu}\right) \tag{8.22}$$

$$+\,\mathscr{R}_{IJ}\left(A^J{}_K \tilde{F}^K_{\mu\nu} + B^{JK}\tilde{G}_{K\,\mu\nu}\right) + \delta\mathscr{O}_{I\,\mu\nu}.$$

This should be an identity once we express all G_I in terms of F^I or the opposite, using again (8.15). Once we do so, we find the following transformation rules for the matter couplings

$$\delta\mathscr{I}_{IJ} = D_I{}^K\mathscr{I}_{KJ} - \mathscr{I}_{IK}A^K{}_J - \mathscr{I}_{IK}B^{KL}\mathscr{R}_{LJ} - \mathscr{R}_{IK}B^{KL}\mathscr{I}_{LJ},$$

$$\delta\mathscr{R}_{IJ} = C_{IJ} + D_I{}^K\mathscr{R}_{KJ} + \mathscr{I}_{IK}B^{KL}\mathscr{I}_{LJ} - \mathscr{R}_{IK}A^K{}_J - \mathscr{R}_{IK}B^{KL}\mathscr{R}_{LJ},$$

$$\delta\mathscr{O}_I = D_I{}^J\mathscr{O}_J - \mathscr{I}_{IJ}B^{JL}\tilde{\mathscr{O}}_L - \mathscr{R}_{IJ}B^{JL}\mathscr{O}_L. \tag{8.23}$$

Moreover, consistency of $\delta\mathscr{I}^T = \delta\mathscr{I}$ and $\delta\mathscr{R}^T = \delta\mathscr{R}$ implies that

$$C = C^T, \quad B = B^T, \quad A = -D^T, \tag{8.24}$$

which are the conditions on the infinitesimal duality transformation to be in the algebra $\mathfrak{sp}(2n_v, \mathbb{R})$. In fact (8.24) is the statement that $S = \exp\begin{pmatrix} A & B \\ C & D \end{pmatrix}$ satisfies

$$S^T \Omega S = \Omega, \quad \Omega = \begin{pmatrix} 0 & \mathbb{1}_{n_V} \\ -\mathbb{1}_{n_V} & 0 \end{pmatrix}, \tag{8.25}$$

or, at the infinitesimal level, (8.24). In the following, we will call a $2n_V$-component object, Y, that transforms under electric–magnetic duality transformations in the same way as the field strength vector $\mathbb{F}_{\mu\nu}$,

$$Y \mapsto S \cdot Y, \tag{8.26}$$

a *symplectic vector* and define the symplectically invariant inner product between two symplectic vectors, Y and Z, as

$$\langle Y, Z \rangle := Y^T \Omega Z. \tag{8.27}$$

Before moving on, let us notice two important facts. The first one is that the invariance of the system $d\mathbb{F} = 0$ does not imply invariance of the Lagrangian.

Indeed, applying an infinitesimal duality transformation, the Lagrangian transforms as

$$\delta\mathcal{L} = \frac{1}{4}FC\tilde{F} + \frac{1}{4}GB\tilde{G} + \delta\mathcal{L}_{\text{rest}}. \tag{8.28}$$

Even if $\delta\mathcal{L}_{\text{rest}} = 0$, as we will prove later, this implies that the Lagrangian is invariant, up to a total derivative (the $FC\tilde{F}$ term), only if $B = 0$. This case corresponds to perturbative transformations, which in the quantum theory have to be restricted to $\mathrm{Sp}(2n_V, \mathbb{Z})$. The second one is that the full electric–magnetic duality group acts with fractional transformations on the scalar matrices and θ couplings of the original action. This becomes clear if we introduce the complex kinetic matrix

$$\mathcal{N}_{IJ} = \mathcal{R}_{IJ} + i\,\mathcal{I}_{IJ} \tag{8.29}$$

and the self-dual combination[1]

$$\theta^+ = \frac{1}{2}(\theta - i\,\tilde{\theta}), \tag{8.30}$$

which indeed satisfies $\tilde{\theta}^+ = i\theta^+$. If we act with an infinitesimal duality transformation (8.23) on these objects, we find

$$\delta\mathcal{N} = C + D\mathcal{N} - \mathcal{N}A - \mathcal{N}B\mathcal{N}, \tag{8.31}$$

and

$$\delta\theta^+ = \theta^+(D^T - B\mathcal{N}) = \theta^+(-A - B\mathcal{N}), \tag{8.32}$$

or (always at first order in A, B, C, and D)

$$\mathcal{N}' = C + (\mathbb{1} + D)\mathcal{N}(\mathbb{1} - A - B\mathcal{N}), \tag{8.33}$$

$$\theta^{+\prime} = \theta^+(\mathbb{1} - A - B\mathcal{N}). \tag{8.34}$$

It is straightforward to see that these are the expansion of

$$\mathcal{N}' = (\hat{C} + \hat{D}\mathcal{N})(\hat{A} + \hat{B}\mathcal{N})^{-1}, \tag{8.35}$$

---

[1] We define the self-dual and anti-self-dual tensor field combinations

$$T^{\pm}_{\mu\nu} \equiv \frac{1}{2}\left(T_{\mu\nu} \mp \frac{i}{2}\epsilon_{\mu\nu\rho\sigma}T^{\rho\sigma}\right),$$

which satisfy $\frac{1}{2}\epsilon_{\mu\nu\rho\sigma}T^{\rho\sigma\pm} = \pm i\,T^{\pm}_{\mu\nu}$.

and

$$\mathcal{O}^{+\prime} = \mathcal{O}^{+}(\hat{A} + \hat{B}\mathcal{N})^{-1}, \tag{8.36}$$

which are the fractional transformation of the kinetic couplings and the remaining matter couplings once we identify

$$S = \begin{pmatrix} \hat{A} & \hat{B} \\ \hat{C} & \hat{D} \end{pmatrix}, \tag{8.37}$$

and we recover the infinitesimal expansion $S = \mathbb{1} + \begin{pmatrix} A & B \\ C & D \end{pmatrix}$. This now also explains

why $\hat{B} \neq 0$ corresponds to non-perturbative duality transformations, because from (8.35) we see that $\hat{B} \neq 0$ involves inversions of the components of $\mathcal{N}$, which may change weak couplings to strong couplings or vice versa.

Finally, we close this section by noting that the introduction of self-dual and anti-self-dual field tensors allows the rewriting of the kinetic Lagrangian in the form

$$e^{-1}\mathscr{L}_{\text{kin}} = \frac{1}{4}(\text{Im}\mathscr{N}_{IJ})F^I_{\mu\nu}F^{\mu\nu J} - \frac{1}{8}(\text{Re}\mathscr{N}_{IJ})\epsilon^{\mu\nu\rho\sigma}F^I_{\mu\nu}F^J_{\rho\sigma} \tag{8.38}$$

$$= \frac{1}{2}\text{Im}\left[\mathscr{N}_{IJ}\, F^{I+}_{\mu\nu}\, F^{\mu\nu J+}\right], \tag{8.39}$$

so that the duality action $\mathbb{F}' = S\mathbb{F}$ becomes

$$F^{+\prime}_{\mu\nu} = (\hat{A} + \hat{B}\mathscr{N})F^+_{\mu\nu}, \qquad G^{+\prime}_{\mu\nu} = (\hat{C} + \hat{D}\mathscr{N})F^+_{\mu\nu}, \tag{8.40}$$

which is compatible with the transformation (8.35) for the kinetic matrix $\mathscr{N}$. Notice also that

$$G^+_{I\mu\nu} = \mathscr{N}_{IJ}F^{J+}_{\mu\nu}. \tag{8.41}$$

8.2 $\mathscr{N} = 2$ Supergravity

We have seen in Chaps. 5 and 6 that the scalar manifold, $\mathscr{M}_{\text{scalar}}$, of $\mathscr{N} = 1$ chiral multiplets must be a Kähler manifold in 4D global and local $\mathscr{N} = 1$ supersymmetry. The coupling to supergravity introduces a global restriction, which leads to the subclass of Kähler-Hodge manifolds, as discussed in Sect. 6.2, but the Kähler manifolds in $\mathscr{N} = 1$ supersymmetry are otherwise arbitrary.

In $\mathscr{N} = 2$ supersymmetry, matter can reside in $\mathscr{N} = 2$ vector or hypermultiplets, which, at the linearized level, can be thought of as being composed of one $\mathscr{N} = 1$ vector and one chiral multiplet, or two chiral multiplets, respectively. The presence of global or local $\mathcal{N} = 2$ supersymmetry, however, has quite distinct consequences for the possible scalar geometries of these two types of $\mathcal{N} = 2$ multiplets and leads to a factorization of the scalar manifold,

$$\mathcal{M}_{\text{scalar}}^{\mathcal{N}=2} = \mathcal{M}_{\text{vec}} \times \mathcal{M}_{\text{hyper}}, \tag{8.42}$$

because the superpartners of the scalars in vector and hypermultiplets involve different spins and hence are not related by any symmetry, unless one has $\mathcal{N} \geq 2$ supersymmetry.

As we will now discuss, the scalar fields of the $\mathcal{N} = 2$ vector multiplets parameterize a *Special Kähler manifold*, which comes in two different versions, depending on whether one has global or local $\mathcal{N} = 2$ supersymmetry. The hypermultiplet scalars, on the other hand, are constrained to form a hyper-Kähler manifold in global supersymmetry or a quaternionic-Kähler manifold in supergravity. Apart from quaternionic-Kähler manifolds, all the above geometries are particular subclasses of Kähler manifolds. For the globally supersymmetric theories, this is clear, because any field theory with $\mathcal{N}$-extended global supersymmetry is only a special case of the field theories with $\mathcal{N}' < \mathcal{N}$ global supersymmetry, whereas an analogous statement is in general not true in supergravity due to the missing gravitini in theories with lower $\mathcal{N}$.

The above $\mathcal{N} = 2$ scalar field geometries feature prominently in various applications of string and field theory and define interesting mathematical structures in their own right. We therefore devote an extra section to their structure and explain how this structure is imposed by supersymmetry.

For the rest of this chapter, we set

$$M_P = 1, \tag{8.43}$$

unless stated otherwise.

8.2.1 $\mathcal{N} = 2$ Vector Multiplets and Special Kähler Geometry

As $\mathcal{N} = 2$ supersymmetry interpolates between the complex scalar and the vector field of an $\mathcal{N} = 2$ vector multiplet, the symplectic duality covariance of the vector field sector discussed in Sect. 8.1 should leave an imprint on the scalar manifold $\mathcal{M}_{\text{vec}}$. This imprint is expected to be different for global and local supersymmetry, because the latter involves an additional vector field from the $\mathcal{N} = 2$ supergravity multiplet. In the following, we will explain how precisely the resulting special Kähler geometry arises for global and local supersymmetry and discuss their differences.

8.2.1.1 Rigid ("Affine") Special Kähler Geometry
The special Kähler geometry that arises in $\mathcal{N} = 2$ global supersymmetry is called "rigid" or "affine" special Kähler geometry (for a review see [2, 31]). The field

theories with global $\mathcal{N} = 2$ supersymmetry form a subclass of the field theories with global $\mathcal{N} = 1$ supersymmetry, defined by the invariance under the larger $\mathcal{N} = 2$ Poincaré superalgebra. The essential features of rigid special Kähler geometry can be easily derived by imposing this larger invariance on the general $\mathcal{N} = 1$ field theories of Chap. 5. In fact, it is sufficient to just impose invariance under the discrete R-symmetry [3]

$$Q^{(1)} \to Q^{(2)}, \qquad Q^{(2)} \to -Q^{(1)}, \tag{8.44}$$

which exchanges the two prospective supersymmetry generators $Q^{(1)}$ and $Q^{(2)}$. This R-symmetry is an element of the full R-symmetry group $U(2)_R$.

In terms of $\mathcal{N} = 1$ language, an $\mathcal{N} = 2$ vector multiplet is decomposed of an $\mathcal{N} = 1$ vector multiplet (A_μ, λ) and an $\mathcal{N} = 1$ chiral multiplet (χ, ϕ). In a theory with only n_V $\mathcal{N} = 2$ vector multiplets, one therefore has to restrict oneself to an equal number, n_V, of $\mathcal{N} = 1$ vector and chiral multiplets, which we therefore label by a common index $I = 1, \ldots, n_V$. To make the equations look more natural, we furthermore slightly adjust the names and normalizations of the fields and write the field content as

$$(A_\mu^I, \lambda^{I(1)}, \lambda^{I(2)}, X^I) = \left(A_\mu^I, \chi^I, \frac{\lambda^I}{2}, \phi^I \right), \tag{8.45}$$

where the right-hand side contains the original $\mathcal{N} = 1$ fields, and we have combined the χ^I and $\lambda^I/2$ into the $\mathcal{N} = 2$ gaugini $\lambda^{I(i)}$ $(i, j, \ldots = 1, 2)$ as indicated. Also, in the following, to adhere to usual $\mathcal{N} = 2$ notation, we will denote the complex conjugation on scalars with a bar $\overline{X}^I = (X^I)^*$.

As the gaugini $\lambda^{I(i)}$ are obtained by acting with $Q^{(i)}$ on the vector fields A_μ^I, the discrete R-symmetry (8.44) now exchanges the gaugini $\lambda^{I(1)}$ and $\lambda^{I(2)}$,

$$\lambda^{I(1)} \to \lambda^{I(2)}, \qquad \lambda^{I(2)} \to -\lambda^{I(1)} \tag{8.46}$$

and leaves the other fields invariant.

This means, in particular, that the kinetic terms of the fermions (cf. the $\mathcal{N} = 1$ Lagrangian (5.120)),

$$A := -g_{I\bar{J}} \left[\overline{\lambda}_L^{I(1)} \, \overleftrightarrow{\mathcal{D}} \lambda_R^{\bar{J}(1)} + \overline{\lambda}_R^{\bar{J}(1)} \, \overleftrightarrow{\mathcal{D}} \lambda_L^{I(1)} \right] \tag{8.47}$$

$$B := -2(\operatorname{Re} f_{IJ}) \overline{\lambda}^{I(2)} \, \overleftrightarrow{\mathcal{D}} \lambda^{J(2)} + i \partial_\mu (\operatorname{Im} f_{IJ}) \overline{\lambda}^{I(2)} \gamma_5 \gamma^\mu \lambda^{J(2)}, \tag{8.48}$$

where we integrated by parts in the last term, have to transform into each other under (8.46). Using $\mathscr{D}_\mu \chi_L^J = \partial_\mu \chi_L^J + (\partial_\mu X^I)\Gamma_{IK}^J \chi_L^K$ and $\partial_\mu(\text{Im } f_{IJ}) = (\partial_\mu X^K)\partial_K \text{Im } f_{IJ} + (\partial_\mu \overline{X}^{\overline{K}})\partial_{\overline{K}} \text{Im } f_{IJ}$, this then leads to the constraints

$$g_{I\overline{J}} = g_{J\overline{I}} = 2\,\text{Re } f_{IJ} \tag{8.49}$$

$$2i\,\partial_K(\text{Im } f_{IJ}) = \Gamma_{KI\overline{J}} = \Gamma_{KJ\overline{I}}. \tag{8.50}$$

The holomorphicity of f_{IJ} and the Kähler relation $\Gamma_{KI\overline{J}} = \partial_K g_{I\overline{J}}$ imply (8.50) from (8.49), which is thus the only non-trivial condition. Due to the holomorphicity of f_{IJ} and the Kähler relation $g_{I\overline{J}} = \partial_I \partial_{\overline{J}} K$, (8.49) is (up to Kähler transformations) solved by

$$K = i\left(X^I \overline{F_I} - \overline{X}^I F_I\right), \tag{8.51}$$

where $F_I(X)$ are holomorphic functions of the scalars X^I. In this expression, the factor i is introduced in order to have a relative minus sign between the two terms, which renders the Kähler potential manifestly $\text{Sp}(2n_V, \mathbb{R})$ invariant if we identify X^I and F_J as components of a symplectic vector,

$$\mathscr{V} := \begin{pmatrix} X^I \\ F_J \end{pmatrix}. \tag{8.52}$$

Indeed, after we introduce the symplectic invariant matrix Ω, the Kähler potential can be written as

$$K = i\langle \mathscr{V}, \overline{\mathscr{V}} \rangle = i\mathscr{V}^T \Omega \overline{\mathscr{V}}, \tag{8.53}$$

which is manifestly $\text{Sp}(2n_V, \mathbb{R})$ invariant under

$$\mathscr{V}' = S\mathscr{V}, \tag{8.54}$$

for any $S \in \text{Sp}(2n_V, \mathbb{R})$.

From (8.51) and (8.49), one further obtains

$$g_{I\overline{J}} = \partial_I \partial_{\overline{J}} K = -i(\partial_I F_J - \overline{\partial_J F_I}) = (f_{IJ} + \overline{f_{IJ}}), \tag{8.55}$$

so that $f_{IJ} = -i\partial_I F_J$. The symmetry of f_{IJ} then implies $\partial_{[I} F_{J]} = 0$, i.e.,

$$F_I = \partial_I F, \tag{8.56}$$

where $F(X)$ is a holomorphic function of the scalars, and hence

$$f_{IJ} = -i\,\partial_I \partial_J F. \tag{8.57}$$

The holomorphic function $F(X)$ is called the *holomorphic prepotential*. It determines the Kähler potential K and the gauge kinetic matrix f_{IJ} by its first and second derivatives, respectively. As in Sect. 8.1, the gauge kinetic matrix is often expressed in terms of its complex conjugate

$$\mathcal{N}_{IJ} := -i\,\overline{f_{IJ}} = \overline{\partial_I \partial_J F} \equiv \overline{F_{IJ}}, \tag{8.58}$$

in terms of which (8.55) becomes

$$\text{Im}\mathcal{N}_{IJ} = -\frac{1}{2}g_{I\bar{J}}. \tag{8.59}$$

Equations (8.52), (8.53), (8.56), and (8.58) capture the essence of rigid special Kähler geometry, which describes the scalar field geometry of $\mathcal{N} = 2$ vector multiplets in field theories with global $\mathcal{N} = 2$ supersymmetry.[2] From a mathematical point of view, however, the above equations are still unsatisfactory as a basis for an intrinsically geometric and coordinate independent definition of rigid special Kähler geometry. The core of this problem lies in Eq. (8.54), which states that the vector $\mathcal{V}$ transforms under a symplectic duality transformation $S \in Sp(2n_V, \mathbb{R})$ in the same way as the vector field strengths $F_{\mu\nu}^I$ and their duals $G_{\mu\nu I}$. This identification of $\mathcal{V}$ with a symplectic vector, however, then raises an important technical question: How is the general holomorphic reparameterization invariance of the scalars, $X^I \mapsto \tilde{X}^I(X)$, on a Kähler manifold compatible with the symplectic covariance of the theory, when the X^I are restricted to belong to a symplectic vector that only admits *linear* reparametrizations?

To understand this, we consider the supersymmetry transformations of the fermions with respect to the original $\mathcal{N} = 1$ supersymmetry, which we take to be parameterized by $\epsilon^{(1)}$ (cf. Eqs. (5.123)),

$$\delta^{(1)}\lambda_L^{I(1)} = \frac{1}{2}(\partial\!\!\!/ X^I)\epsilon_R^{(1)} \tag{8.60}$$

$$\delta^{(1)}\lambda_L^{I(2)} = \frac{1}{8}\gamma^{\mu\nu}F_{\mu\nu}^I\epsilon_L^{(1)} = \frac{1}{8}\gamma^{\mu\nu}F_{\mu\nu}^{I-}\epsilon_L^{(1)}, \tag{8.61}$$

where we have used $\epsilon_L^{(1)} = P_L\epsilon^{(1)}$ and $\gamma^{\mu\nu} = (i/2)\epsilon^{\mu\nu\rho\sigma}\gamma_{\rho\sigma}\gamma_5$ to convert $F_{\mu\nu}^I$ to $F_{\mu\nu}^{I-}$. The transformation (8.60) suggests, just as in $\mathcal{N} = 1$ supersymmetry, that $\lambda_L^{I(1)}$ transforms as a holomorphic tangent vector under scalar reparameterizations $X^I \mapsto \tilde{X}^I(X)$, i.e., $\lambda_L^{I(1)} \mapsto (\partial\tilde{X}^I/\partial X^J)\lambda_L^{J(1)}$. The R-symmetry (8.46) then implies that also $\lambda_L^{I(2)}$ has to transform in this way, so that (8.61) then ultimately would imply that also the field strengths transform with the Jacobian, $F_{\mu\nu}^{I-} \mapsto (\partial\tilde{X}^I/\partial X^J)F_{\mu\nu}^{J-}$. If the transformation $X^I \mapsto \tilde{X}^I(X)$ is nonlinear, however, the

---

[2] In the literature these equations often appear with different normalizations.

Jacobian is X^I-dependent, so that the field equations and Bianchi identities (8.6) and (8.7) would no longer hold for the transformed field strengths and instead get modified by spacetime derivatives of scalar fields.

The most natural resolution of this tension between symplectic duality and scalar reparameterization invariance is achieved by separating these two types of transformations. To this end, one considers the symplectic vector $\mathcal{V}(z)$ as an abstract holomorphic $2n_V$-dimensional vector field with components labeled by $I, J, \ldots = 1, \ldots, n_V$ on the scalar manifold $\mathcal{M}_{\text{vec}}$, which itself is locally parameterized by n_V complex coordinates, z^m, $(m, n, \ldots = 1, \ldots, n_V)$. The electric–magnetic duality transformations then only act on the components of $\mathcal{V}$ with a symplectic matrix $S \in Sp(2n_V, \mathbb{R})$ as in (8.54), but not on the coordinates z^m of the scalar manifold. In other words, $\mathcal{V}(z)$ is a section in a holomorphic vector bundle with fiber $\mathbb{C}^{2n_V}$ and structure group $Sp(2n_V, \mathbb{R})$. As the symplectic matrix S does not depend on the coordinates z^m, this bundle is topologically trivial.

The physical scalar fields are then the z^m, which in general are not equal to the components $X^I(z)$. The special coordinate systems in which the coordinates z^m are equal to the components $X^I(z)$ (i.e., $z^1 = X^1(z)$, etc.) are called *special coordinates*. Under a general holomorphic scalar field reparameterization $z^m \mapsto \tilde{z}^m(z)$, the components of $\mathcal{V}$ do not get rotated, but behave as ordinary functions on the scalar manifold. In other words, the symplectic vector bundle is not identified with or related to the tangent bundle of the scalar manifold.

Just as in $\mathcal{N} = 1$ supersymmetry, the left- and right-handed components of the fermions transform under scalar reparameterizations $z^m \mapsto \tilde{z}^m(z)$ as, respectively, holomorphic and antiholomorphic tangent vectors on the scalar manifold and are consequently also denoted with indices $m, n, \ldots$:

$$\lambda_L^{m(j)} \mapsto \left(\frac{\partial \tilde{z}^m}{\partial z^n}\right) \lambda_L^{n(j)}, \qquad \lambda_R^{\overline{m}(j)} \mapsto \left(\frac{\partial \tilde{z}^{\overline{m}}}{\partial \overline{z}^{\overline{n}}}\right) \lambda_R^{\overline{n}(j)}. \tag{8.62}$$

Just as the scalar fields, z^m, the fermions are then considered invariant with respect to symplectic duality transformations.

A covariant form of the supersymmetry transformations that respects all the above symmetries and reduces to (8.60) and (8.61) for special coordinates is then given by

$$\delta^{(1)} \lambda_L^{m(1)} = \frac{1}{2} (\partial\!\!\!/ z^m) \epsilon_R^{(1)} \tag{8.63}$$

$$\delta^{(1)} \lambda_L^{m(2)} = -\frac{1}{4} g^{m\overline{n}} \overline{\partial_n X^I} (\operatorname{Im}\mathcal{N}_{IJ}) F_{\mu\nu}^{J-} \gamma^{\mu\nu} \epsilon_L^{(1)}, \tag{8.64}$$

where $g^{m\overline{n}}(z, \overline{z})$ denotes the inverse of the metric on the scalar manifold. For (8.63) this is easy to see, as both sides are symplectically invariant and transform as tangent vectors with respect to scalar reparameterizations with an obvious reduction to (8.60) upon choosing special coordinates. As for (8.64), both sides are again manifestly transforming as holomorphic tangent vectors with respect to scalar field reparameterizations. The left-hand side is also manifestly symplectically invariant. It thus remains to show that the right-hand side is also symplectically invariant and reduces to the right-hand side of (8.61) upon switching to special coordinates.

To show the symplectic invariance, we use $G_{\mu\nu I-} = \overline{\mathcal{N}_{IJ}} F_{\mu\nu}^{J-}$ and compute

$$\overline{\partial_n X^I} (\operatorname{Im} \mathcal{N}_{IJ}) F_{\mu\nu}^{J-} = \frac{1}{2i} \overline{\partial_{\overline{n}} X^I} (\mathcal{N}_{IJ} - \overline{\mathcal{N}_{IJ}}) F_{\mu\nu}^{J-} \tag{8.65}$$

$$= \frac{1}{2i} \left(\overline{\partial_{\overline{n}} X^I} \mathcal{N}_{IJ} F_{\mu\nu}^{J-} - \overline{\partial_{\overline{n}} X^I} G_{\mu\nu I-} \right). \tag{8.66}$$

This is the symplectic inner product, $\frac{1}{2i} \langle \mathbb{F}_{\mu\nu}^-, \partial_{\overline{n}} \overline{\mathcal{V}} \rangle$, of the two symplectic vectors $\partial_{\overline{n}} \overline{\mathcal{V}} = \begin{pmatrix} \overline{\partial_{\overline{n}} X^I} \\ \overline{\partial_{\overline{n}} F_J} \end{pmatrix}$ and $\mathbb{F}_{\mu\nu}^- = \begin{pmatrix} F_{\mu\nu}^{I-} \\ G_{\mu\nu J-} \end{pmatrix}$ provided that

$$\overline{\partial_{\overline{n}} X^I} \mathcal{N}_{IJ} = \overline{\partial_{\overline{n}} F_J}. \tag{8.67}$$

This relation, however, simply follows from $\mathcal{N}_{IJ} = \partial_I \partial_J F$, $F_J = \partial_J F$ and the chain rule, showing that the right-hand side of (8.64) really is symplectically invariant.

The equivalence to (8.61) in special coordinates, finally, follows from $\partial_n X^I = \delta_n^I$ and (8.59), which are valid if the z^m are identified with the X^I.

We finally note that $(\partial_n X^I)$ is the Jacobian of the change from general coordinates z^n to special coordinates and hence must be invertible. This implies that (8.67) together with the relation $F_I = \partial_I F$ fixes the gauge kinetic matrix to

$$\mathcal{N}_{IJ} = (\overline{\partial_n X^I})^{-1} \cdot (\overline{\partial_n F_J}), \tag{8.68}$$

or equivalently

$$\mathcal{N}_{JI} = (\overline{\partial_n F_J}) \cdot (\overline{\partial_n X^I})^{-1}, \tag{8.69}$$

because $\mathcal{N}_{IJ} = \mathcal{N}_{JI}$. In this notation, we can now prove that $\mathcal{N}_{IJ}$ transforms under symplectic duality transformations in the way it should (cf. Eq. (8.35)). In fact, from the action of (8.37) on the symplectic sections $\mathcal{V}$, we inherit

$$\mathcal{N}_{JI}' = \left(\widehat{C}_{JK} \overline{\partial_n X^K} + \widehat{D}_J{}^K \overline{\partial_n F_K} \right) \cdot \left(\widehat{A}^I{}_L \overline{\partial_n X^L} + \widehat{B}^{IL} \overline{\partial_n F_L} \right)^{-1}, \tag{8.70}$$

which can also be written as

$$\mathcal{N}'_{JI} = \left(\widehat{C}_{JK} + \widehat{D}_J{}^L \overline{\partial_m F_L} \cdot \left(\overline{\partial_m X^K}\right)^{-1}\right)^{-1} \left(\partial_n X^K\right) \cdot$$

$$\cdot \left(\overline{\partial_n X^L}\right)^{-1} \left(\widehat{A}^I{}_L + \widehat{B}^{IM} \overline{\partial_m F_M} \cdot \left(\overline{\partial_m X^L}\right)^{-1}\right)^{-1},$$

(8.71)

which finally gives (8.35), once (8.69) is used.

Notice that (8.69) is invariant under a constant phase redefinition of the symplectic sections $\mathcal{V}$, as well as under a constant shift $\mathcal{V} \to \mathcal{V} + c$, with $c \in \mathbb{C}^{2n_V}$. In fact it is easy to see that the entire Lagrangian and the supersymmetry laws are invariant under such symmetries, provided that the gaugini $\lambda^{m(i)}$ and the supersymmetry parameter $\epsilon^{(i)}$ transform as well with a suitable chiral phase factor. Because of these symmetries, when we define a special Kähler manifold, we have to admit that on different coordinate patches the symplectic sections may differ by constant phases as well as constant complex shifts. This leads to the following general definition of rigid special Kähler geometry [2, 4]:

Definition. A *rigid special Kähler manifold* is a n-dimensional Kähler manifold of restricted type:

- it is equipped with a tensor bundle $\mathcal{H}$ given by the product of a flat U(1) bundle and a vector bundle with an inhomogeneous symplectic structure group, so that, on each patch, U_A, of a good cover, a section of $\mathcal{H}$ is described by a symplectic vector

$$\mathcal{V} := \begin{pmatrix} X^I \\ F_J \end{pmatrix}, \qquad I = 1, \ldots, n,$$

(8.72)

such that the transition functions between two different local trivializations of $\mathcal{H}$ on U_A and on U_B have the form

$$\mathcal{V}_A = e^{i\phi_{AB}} S_{AB} \mathcal{V}_B + c_{AB},$$

(8.73)

where S_{AB} is a constant matrix in $Sp(2n, \mathbb{R})$, $\phi_{AB} \in \mathbb{R}$, $c_{AB} \in \mathbb{C}^{2n}$;
- the Kähler potential is given by

$$K = i\langle \mathcal{V}, \overline{\mathcal{V}} \rangle = i\mathcal{V}^T \Omega \overline{\mathcal{V}},$$

(8.74)

where $\langle \cdot, \cdot \rangle$ is a Hermitian metric on $\mathcal{H}$;

(continued)

- the sections satisfy

$$\langle \partial_m \mathcal{V}, \partial_n \mathcal{V} \rangle = 0. \tag{8.75}$$

This last equation guarantees the existence of a prepotential and the symmetry of the kinetic vector matrix $\mathcal{N}_{IJ}$.

The transition functions are subject to the usual consistency conditions on triple overlaps:

$$\phi_{AB}\phi_{BC}\phi_{CA} = 1, \quad S_{AB}S_{BC}S_{CA} = 1. \tag{8.76}$$

The covariant transformation laws (8.63)–(8.64) and the relation (8.67) will have important analogues in $\mathcal{N} = 2$ supergravity to which we turn next.

8.2.1.2 Local ("Projective") Special Kähler Geometry

Local or "projective" special Kähler geometry arises when $\mathcal{N} = 2$ vector multiplets are coupled to $\mathcal{N} = 2$ supergravity. As we have just discussed, an $\mathcal{N} = 2$ vector multiplet contains one vector field, $A_\mu(x)$, two gaugini, $\lambda^{(i)}(x)$ $(i = 1, 2)$, and one complex scalar field, $z(x)$. The $\mathcal{N} = 2$ supergravity multiplet, on the other hand, consists of the vierbein, $e_\mu^a(x)$; two gravitini, $\psi_\mu^{(i)}(x)$ $(i, j, \ldots = 1, 2)$; and another vector field, $A_\mu'(x)$, often called the "graviphoton." The index i corresponds to the fundamental representation of the R-symmetry subgroup $SU(2)_R$ of the total R-symmetry group $U(2)_R$.

Thus, coupling n_V $\mathcal{N} = 2$ vector multiplets to $\mathcal{N} = 2$ supergravity gives rise to a theory with the field content

$$\{e_\mu^a, \psi_\mu^{(i)}, A_\mu^I, \lambda^{m(i)}, z^m\}, \tag{8.77}$$

where $I, J, \ldots = 0, 1, \ldots, n_V$, and $m, n, \ldots = 1, \ldots, n_V$. Note that the vector fields, A_μ^I, now include the graviphoton so that there are $(n_V + 1)$ vector fields and only n_V physical complex scalar fields, z^m. In global supersymmetry, by contrast, one has as many vector fields as physical complex scalars so that one can in principle use the same index type to label them, as we first did at the beginning of Sect. 8.2.1.1. As we saw there, however, even for global supersymmetry the clearest notation that makes all symmetries manifest is to distinguish the index I for the vector fields and the index m for the complex scalars and the gaugini. In $\mathcal{N} = 2$ supergravity, the mismatch in the number of vector and scalar fields makes this distinction even more natural, so that we will use it here right from the beginning.

We are thus immediately led to the interpretation that the $(n_V + 1)$ Abelian field strengths, $F_{\mu\nu}^{I+}$, and their magnetic duals, $G_{\mu\nu I+} = \mathcal{N}_{IJ} F_{\mu\nu}^{J+}$, combine into a $(2n_V + 2)$-component symplectic vector, $\mathbb{F}_{\mu\nu}^+$, that transforms in the fundamental representation of the electric–magnetic duality group $\mathrm{Sp}(2n_V + 2, \mathbb{R})$,

$$\mathbb{F}^+_{\mu\nu} = \begin{pmatrix} F^{I+}_{\mu\nu} \\ G_{\mu\nu J+} \end{pmatrix} \mapsto S \cdot \begin{pmatrix} F^{I+}_{\mu\nu} \\ G_{\mu\nu J+} \end{pmatrix} = S \cdot \mathbb{F}^+_{\mu\nu}, \qquad S \in \mathrm{Sp}(2n_V+2, \mathbb{R}). \quad (8.78)$$

The matrix $\mathcal{N}_{IJ}$ in the above expression for $G_{\mu\nu I+}$ is the scalar field-dependent gauge kinetic function of the theory, which thus transforms as in (8.35) under symplectic duality transformations. Its precise dependence on the scalar fields z^m is of a different form than in global supersymmetry, however, and will be derived further below.

The scalars z^m and the gaugini $\lambda^{m(i)}$, by contrast, are symplectically invariant, but transform under holomorphic scalar field reparameterizations as

$$z^m \mapsto \widetilde{z}^m(z), \qquad \lambda^{m(i)} \mapsto \left(\frac{\partial \widetilde{z}^m}{\partial z^n}\right) \lambda^{n(i)}, \quad (8.79)$$

with the vector fields A^I_μ being invariant.

As the coupling to supergravity does not involve new gaugini or new scalars, the basic supersymmetry relation between the gaugini and the scalars should again link left-handed fermion components with the z^m and right-handed fermion components with the complex conjugates $\overline{z}^{\overline{m}}$. Just as discussed in Sect. 5.2.4, the holonomy group of the scalar manifold of the z^m should then again be contained in $U(n_V)$, i.e., the scalar manifold should again be a (Hodge-)Kähler manifold. In parallel with the situation in global supersymmetry, however, we expect that the interplay with the symplectic duality of the vector fields and the extended supersymmetry will lead to restrictions on the possible Kähler geometries, and hence to another version of "special Kähler geometry." As the number of vector fields and the rank of the symplectic duality group are now increased, it is plausible that the type of special Kähler geometry will be different from the rigid special Kähler geometry discussed in the previous subsection. This is indeed true, and the essential differences with the rigid case turn out to be [2, 5–8]:

1. Just as in rigid $\mathcal{N} = 2$ supersymmetry, there is a holomorphic symplectic section $\mathcal{V}(z) = \begin{pmatrix} X^I(z) \\ F_J(z) \end{pmatrix}$ $(I, J, \ldots = 0, 1, \ldots, n_V)$ that depends holomorphically on the scalar fields z^m and transforms as

$$\mathcal{V}(z) \mapsto S \cdot \mathcal{V}(z), \qquad S \in \mathrm{Sp}(2n_V + 2, \mathbb{R}) \quad (8.80)$$

under electric–magnetic duality transformations. An obvious difference to the rigid case here is that $\mathcal{V}$ has two more components to match the number of components of the symplectic field strength vector $\mathbb{F}^+_{\mu\nu} = \begin{pmatrix} F^{I+}_{\mu\nu} \\ G_{\mu\nu J+} \end{pmatrix}$. The existence of this symplectic section ensures that the holomorphic reparameterization invariance of the scalars and the symplectic covariance of the vector field sector can coexist and be consistent with the supersymmetry transformation law of the gaugini, similar to the situation in rigid supersymmetry. This is explained in more detail in Appendix 8.A.1, where we will also explain the origin of the holomorphicity of $\mathcal{V}$.

2. Under a general Kähler transformation, $K(z, \bar{z}) \to K(z, \bar{z}) + h(z) + \overline{h(z)}$, the holomorphic symplectic section $\mathcal{V}$ transforms non-trivially as

$$\mathcal{V}(z) \mapsto e^{-h(z)} \mathcal{V}(z). \tag{8.81}$$

In rigid special Kähler geometry, by contrast, at most an additive constant vector to $\mathcal{V}(z)$ could generate a very specific type of Kähler transformation. This non-trivial Kähler transformation of the symplectic section is needed to reconcile the various Kähler transformation properties of the fermions with one another, as is explained in detail in Appendix 8.A.1.

3. The Kähler potential of the scalar manifold parameterized by the z^m is given by

$$K(z, \bar{z}) = -\log i \left(\overline{X^I}(\bar{z}) F_I(X(z)) - \overline{F_I}(\overline{X}(\bar{z})) X^I(z) \right) = -\log i \langle \overline{\mathcal{V}}, \mathcal{V} \rangle. \tag{8.82}$$

Similar to rigid supersymmetry, this is manifestly symplectically invariant, but it differs from the rigid expression (8.51) by the logarithm. Just as in $\mathcal{N} = 1$ supergravity, the transformation of the fermions under Kähler transformations leads to a global restriction and requires the Kähler manifolds to be Hodge-Kähler manifolds. The logarithmic form of the Kähler potential is due to the Kähler transformation property (8.81), as is also explained in Appendix 8.A.2.

4. In contrast to rigid special Kähler geometry, the lower components F_J are not necessarily the derivatives of a holomorphic prepotential $F(X)$ with respect to X^J. However, for $n_V > 1$, there is always a symplectic matrix $S \in \mathrm{Sp}(2n_V + 2, \mathbb{R})$ such that $\widetilde{\mathcal{V}} = S \cdot \mathcal{V}$ does have this property, i.e., such that $\widetilde{\mathcal{V}} = \begin{pmatrix} \widetilde{X}^I \\ \widetilde{F}_J \end{pmatrix}$ with $\widetilde{F}_J = \partial \widetilde{F}(\widetilde{X}) / \partial \widetilde{X}^J$.[3] More precisely, for $n_V > 1$, one has to impose an additional constraint on the symplectic section $\mathcal{V}(z)$ in order to ensure the symmetry and uniqueness of the gauge kinetic matrix $\mathcal{N}_{IJ}$. This condition is

$$\langle \mathcal{D}_m \mathcal{V}, \mathcal{D}_n \mathcal{V} \rangle = 0, \tag{8.83}$$

---

[3] For symplectic frames with $F_J \neq \partial F / \partial X^J$, one often says that this is a section "without prepotential" or "the prepotential does not exist." A more precise statement would be that (for $n_V > 1$) there is always a symplectic frame where such a prepotential does exist, but that this is just not the symplectic frame under consideration. We will nevertheless also often use the above less precise terminology. In fact, in the case of gauge interactions, the electric–magnetic duality is at least partially broken by the presence of "naked" vector fields A_μ^I without spacetime derivatives, so that the symplectic frames with the standard prepotential form might not be accessible in the usual way.

where $\mathcal{D}_m \mathcal{V} = (\partial_m + (\partial_m K))\mathcal{V}$ is the Kähler covariant derivative of $\mathcal{V}$. With this condition, one can prove the existence of a symplectic frame with prepotential for $n_V > 1$. For the case $n_V = 1$, however, the condition (8.83) is empty as it is antisymmetric in m and n. And in fact, for $n_V = 1$, counterexamples without a prepotential in any symplectic frame have been constructed [4]. More details on this issue are given in Appendix 8.A.3.

In symplectic frames where the identification $F_J = \partial F/\partial X^J$ is possible, there is also an analogue of the special coordinates we encountered in rigid special Kähler geometry, where one could (locally) always choose the z^m such that they can be identified with the $X^I(z)$. In *local* special Kähler geometry, this cannot work exactly this way because there is now one more X^I than z^m. However, due to the complex rescalings (8.81) induced by Kähler transformations, one can interpret the X^I as homogeneous coordinates of a projective space and then identify the z^m with the corresponding inhomogeneous coordinates, e.g., $z^m = X^m/X^0$. This is the reason for the alternative name "projective special Kähler geometry."

5. Another difference to the rigid case is that such a holomorphic prepotential $F(X)$ must be *homogeneous of degree two*, i.e., under a rescaling $X^I \to \kappa X^I$, one has $F(\kappa X) = \kappa^2 F(X)$, whereas in the rigid case, $F(X)$ can be an arbitrary holomorphic function (as long as the resulting metric $g_{m\bar{n}}$ is positive definite). This can, e.g., be understood from the consistency of the non-trivial Kähler transformation (8.81) with $F_I = \partial_I F$, but it is also necessary for ensuring the right properties of the gauge kinetic matrix, $\mathcal{N}_{IJ}$. Details on this can be found in the Appendices 8.A.3 and 8.A.4.

6. In a symplectic frame with a holomorphic prepotential $F(X)$, the gauge kinetic matrix $\mathcal{N}_{IJ}(z, \bar{z})$ is given by

$$\mathcal{N}_{IJ} = \overline{F_{IJ}(X)} + 2i \frac{\mathrm{Im}F_{IK}(X)\mathrm{Im}F_{JL}(X)X^K X^L}{\mathrm{Im}F_{KL}(X)X^K X^L}, \tag{8.84}$$

where $F_{IJ} \equiv \partial_I \partial_J F$. This expression differs from the corresponding expression (8.58) by the second, non-antiholomorphic piece. The reason for this extra term is that, in connection with the degree 2 of the prepotential, it makes the supersymmetry transformation law of the gravitini symplectically invariant without destroying the symplectic invariance of the supersymmetry transformation law of the gaugini. This is further explained in Appendix 8.A.4.

Altogether, we can therefore define the local special Kähler geometry as

---

Definition. A *local special Kähler manifold* is a n-dimensional Hodge–Kähler manifold of restricted type:

- it is equipped with a tensor bundle, $\mathscr{H}$, given by the product of a flat holomorphic vector bundle with a symplectic structure group and of a holomorphic line bundle, so that, on each patch, U_A, of a good cover, a section of $\mathscr{H}$ is described by a projective symplectic vector

$$\mathscr{V} := \begin{pmatrix} X^I \\ F_J \end{pmatrix}, \qquad I = 0, \dots, n, \tag{8.85}$$

  such that the transition functions between two different local trivializations of $\mathscr{H}$ on U_A and on U_B have the form

$$\mathscr{V}_A = e^{-h_{AB}} S_{AB} \mathscr{V}_B, \tag{8.86}$$

  where S_{AB} is a constant matrix in $\mathrm{Sp}(2n+2, \mathbb{R})$, h_{AB} is a holomorphic function;
- the Kähler potential is given by

$$K = -\log\left(i\langle \mathscr{V}, \overline{\mathscr{V}}\rangle = i\mathscr{V}^T \Omega \overline{\mathscr{V}}\right), \tag{8.87}$$

  where $\langle \cdot, \cdot \rangle$ is a Hermitian and symplectic metric on $\mathscr{H}$;
- the sections satisfy

$$\langle \mathscr{D}_m \mathscr{V}, \mathscr{D}_n \mathscr{V}\rangle = 0. \tag{8.88}$$

For $n > 1$, this last equation guarantees the existence of a prepotential in some symplectic frame and the symmetry of the kinetic vector matrix $\mathcal{N}_{IJ}$.

The transition functions are subject to the usual consistency conditions on triple overlaps:

$$S_{AB} S_{BC} S_{CA} = 1 = e^{h_{AB} + h_{BC} + h_{CA}}. \tag{8.89}$$

8.2.2 $\mathcal{N} = 2$ Hypermultiplets and Hyper-Kähler vs. Quaternionic Kähler Geometry

In this subsection, we discuss the scalar field geometry of $\mathcal{N} = 2$ hypermultiplets in rigid supersymmetry and supergravity. Hypermultiplets are $\mathcal{N} = 2$ field multiplets with spins/helicities not exceeding $1/2$. They can thus be viewed as $\mathcal{N} = 2$ generalizations of $\mathcal{N} = 1$ chiral multiplets. As we will now describe, one

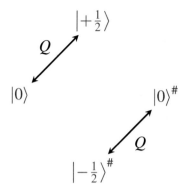

Fig. 8.1 The particle states corresponding to a CPT-completed chiral field multiplet in $\mathcal{N} = 1$ supersymmetry

hypermultiplet has the same field content as two chiral multiplets, but the proper embedding of $\mathcal{N} = 2$ supersymmetry and the relation to particle multiplets are a bit non-trivial. As we will see, this non-trivial structure straightforwardly leads to the peculiar scalar field geometries of hypermultiplets:hyper-Kähler manifolds for rigid supersymmetry and quaternionic Kähler manifolds for supergravity [9].

To begin with, we recall that, at the level of massless particle states, a massless $\mathcal{N} = 1$ chiral multiplet, $(\chi(x), \phi(x))$, corresponds to the direct sum of two unitary irreducible representations of the $\mathcal{N} = 1$ Poincaré superalgebra, as shown schematically in Fig. 8.1, where we label the states by their helicities, h, and indicate which states are superpartners with respect to the $\mathcal{N} = 1$ supersymmetry generator Q. The above two particle multiplets are CPT conjugates of one another, and the presence of both is required to make the field theory CPT-invariant. The transformation laws

$$\delta\chi_L = \frac{1}{2}\partial\!\!\!/\phi\epsilon_R, \qquad \delta\chi_R = \frac{1}{2}\partial\!\!\!/\phi^*\epsilon_L \tag{8.90}$$

suggest the associations

$$(\chi_L, \phi) \leftrightarrow (|+1/2\rangle, |0\rangle) \tag{8.91}$$

$$(\chi_R, \phi^*) \leftrightarrow (|-1/2\rangle^{\#}, |0\rangle^{\#}). \tag{8.92}$$

Let us now turn to analogous multiplets with $|h| \leq 1/2$ in $\mathcal{N} = 2$ supersymmetry. At the level of massless particle multiplets, there is only one type of $\mathcal{N} = 2$ supermultiplet with helicities between $+1/2$ and $-1/2$. It consists of two scalar particle states, one state with helicity $+1/2$ and one state with helicity $-1/2$. They are connected by the two supersymmetry generators $Q^{(1)}$ and $Q^{(2)}$ as in Fig. 8.2.

The two spinless states $|0\rangle, |0\rangle'$ form a doublet of the $SU(2)_R$ subgroup of the R-symmetry $U(2)_R$, which is possible because they are two linearly independent vectors in a *complex* vector space. By analogy with the $\mathcal{N} = 1$ chiral multiplet

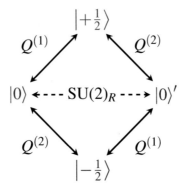

Fig. 8.2 The irreducible multiplet of particle states with $|h| \leq 1/2$ in $\mathcal{N} = 2$ supersymmetry. $Q^{(1)}$ and $Q^{(2)}$ denote the two independent supersymmetry generators, which are linked by the $SU(2)_R$ subgroup of the R-symmetry group $U(2)_R$. The two spinless states consequently form an $SU(2)_R$ doublet

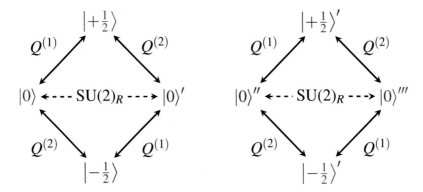

Fig. 8.3 An $\mathcal{N} = 2$ hypermultiplet in four dimensions corresponds to the direct sum of two irreducible $\mathcal{N} = 2$ particle multiplets, as shown. The red states can be viewed as one CPT-complete $\mathcal{N} = 1$ chiral multiplet of the type given in Fig. 8.1 and the blue states as another such multiplet

and Fig. 8.1, one might now be tempted to associate $|+1/2\rangle$ and $|-1/2\rangle$ with, respectively, the left- and right-handed component of a Majorana fermion field, χ, and $|0\rangle$ and $|0\rangle'$ with, respectively, the superpartners ϕ and ϕ^*, so that $\mathcal{N} = 2$ supersymmetry would ultimately be represented on a CPT-complete $\mathcal{N} = 1$ chiral multiplet. This is false, however, because the field space parameterized by the scalars ϕ and ϕ^* is $\mathbb{C} \cong \mathbb{R}^2$, which does not permit a non-trivial representation of $SU(2)$. Put differently, ϕ and ϕ^* describe a particle and its anti-particle, which cannot sit in one and the same $SU(2)_R$-doublet, contrary to what the above particle multiplet requires. We thus must not identify $|0\rangle$ as the anti-particle of $|0\rangle'$, but instead need to add another particle multiplet of the same type as in Fig. 8.3.

The anti-particles of $|+1/2\rangle$ and $|0\rangle$ can then be identified with $|-1/2\rangle'$ and $|0\rangle'''$ so that these four states can be viewed as one CPT-complete $\mathcal{N} = 1$ chiral multiplet, which is then accompanied by another, independent CPT-complete chiral multiplet consisting of the states $|+1/2\rangle', |0\rangle', |0\rangle'', |-1/2\rangle$. Putting everything together, a minimal $\mathcal{N} = 2$ *field* multiplet with $|h| \le 1/2$ thus must contain two independent Majorana fields χ^1, χ^2 and two independent complex scalars ϕ^1, ϕ^2, which we associate with the above states as follows:

$$(\chi_L^1, \phi^1) \leftrightarrow (|+1/2\rangle, |0\rangle) \tag{8.93}$$

$$(\chi_R^1, \phi^{1*}) \leftrightarrow (|-1/2\rangle', |0\rangle''') \tag{8.94}$$

$$(\chi_L^2, \phi^2) \leftrightarrow (|+1/2\rangle', |0\rangle'') \tag{8.95}$$

$$(\chi_R^2, \phi^{2*}) \leftrightarrow (|-1/2\rangle, |0\rangle'). \tag{8.96}$$

From (8.90), we then read off that (χ^1, ϕ^1) and (χ^2, ϕ^2) behave as two $\mathcal{N} = 1$ chiral multiplets with respect to $Q^{(1)}$, whereas (χ^1, ϕ^{2*}) and (χ^2, ϕ^{1*}) behave as two $\mathcal{N} = 1$ chiral multiplets with respect to $Q^{(2)}$. More precisely,

$$\delta\chi_L^1 = \frac{1}{2}\partial\!\!\!/\phi^1\epsilon_R^{(1)} + \frac{1}{2}\partial\!\!\!/\phi^{2*}\epsilon_R^{(2)} \tag{8.97}$$

$$\delta\chi_L^2 = \frac{1}{2}\partial\!\!\!/\phi^2\epsilon_R^{(1)} - \frac{1}{2}\partial\!\!\!/\phi^{1*}\epsilon_R^{(2)}, \tag{8.98}$$

where the minus sign in front of the ϕ^{1*} term is to ensure the desired invariance under the discrete R-symmetry (cf. (8.44)),

$$\begin{pmatrix} \epsilon^{(1)} \\ \epsilon^{(2)} \end{pmatrix} \longrightarrow \begin{pmatrix} \epsilon^{(2)} \\ -\epsilon^{(1)} \end{pmatrix}, \qquad \begin{pmatrix} \phi^1 \\ \phi^{2*} \end{pmatrix} \longrightarrow \begin{pmatrix} \phi^{2*} \\ -\phi^1 \end{pmatrix}. \tag{8.99}$$

Analogously, one has

$$\delta\phi^1 = \bar{\epsilon}_L^{(1)}\chi_L^1 - \bar{\epsilon}_R^{(2)}\chi_R^2 \tag{8.100}$$

$$\delta\phi^2 = \bar{\epsilon}_L^{(1)}\chi_L^2 + \bar{\epsilon}_R^{(2)}\chi_R^1. \tag{8.101}$$

Suppose now one has n_H hypermultiplets with analogous (linearized) transformation laws, i.e.,

$$\delta\chi_L^m = \frac{1}{2}\partial\!\!\!/\phi^m\epsilon_R^{(1)} + \frac{1}{2}\delta^{m\bar{l}}E_{\bar{l}\bar{n}}\partial\!\!\!/\phi^{\bar{n}}\epsilon_R^{(2)} \tag{8.102}$$

$$\delta\chi_R^{\bar{m}} = \frac{1}{2}\partial\!\!\!/\phi^{\bar{m}}\epsilon_L^{(1)} + \frac{1}{2}\delta^{\bar{m}l}E_{ln}\partial\!\!\!/\phi^n\epsilon_L^{(2)} \tag{8.103}$$

$$\delta\phi^m = \bar{\epsilon}_L^{(1)}\chi_L^m - \delta^{m\bar{n}}E_{\bar{n}\bar{r}}\bar{\epsilon}_R^{(2)}\chi_R^{\bar{r}} \tag{8.104}$$

$$\delta\phi^{\bar{m}} = \bar{\epsilon}_R^{(1)}\chi_R^{\bar{m}} - \delta^{\bar{m}n}E_{nr}\bar{\epsilon}_L^{(2)}\chi_L^r, \tag{8.105}$$

where $m, n, \ldots = 1, \ldots, 2n_H$, and

$$
(E_{mn}) := \begin{pmatrix} e & 0 & \cdots & 0 \\ 0 & e & & \vdots \\ \vdots & & \ddots & 0 \\ 0 & \cdots & 0 & e \end{pmatrix} = (E_{\overline{mn}}), \qquad e = \begin{pmatrix} 0 & 1 \\ -1 & 0 \end{pmatrix}. \tag{8.106}
$$

In Sect. 5.2.4, we showed that the above transformations for $\epsilon^{(2)} = 0$ maintain their structure if one acts on $\begin{pmatrix} \chi_L^m \\ \chi_R^{\overline{m}} \end{pmatrix}$ and $\begin{pmatrix} \partial \phi^m \\ \partial \phi^{\overline{m}} \end{pmatrix}$ with unitary $(4n_H \times 4n_H)$-matrices of the form

$$
U = \begin{pmatrix} A & 0 \\ 0 & A^* \end{pmatrix}, \qquad A \in U(2n_H). \tag{8.107}
$$

Preserving also the transformations with $\epsilon^{(2)} \neq 0$ then imposes the additional restriction $E \cdot A^* \overset{!}{=} A \cdot E$, which because of $A^\dagger = A^{-1}$ and the reality of E then implies $A^T \cdot E \cdot A = E$, showing that A also has to be symplectic,

$$
A \in U(2n_H) \cap Sp(2n_H, \mathbb{C}) \equiv Sp(n_H)(\equiv Usp(2n_H)). \tag{8.108}
$$

Together with the $SU(2)_R$ R-symmetry subgroup, which acts on the n_H doublets

$$
\begin{pmatrix} \phi^{2l-1} \\ (\phi^{2l})^* \end{pmatrix} \tag{8.109}
$$

for all $l = 1, \ldots, n_H$, we therefore have the invariance group $Sp(n_H) \times SU(2)$ of the linearized supersymmetry transformations. This structure should be preserved by the holonomy group of the scalar manifold.[4]

The supersymmetry parameters $\epsilon^{(i)}$ $(i = 1, 2)$ transform as an $SU(2)$ doublet. In $\mathcal{N} = 2$ supergravity, the supersymmetry parameters are no longer constant, and the $SU(2)$ part of the curvature has to be non-trivial. This is explained in more detail in Appendix 8.B.2. Manifolds of real dimension $4n_H$ with holonomy contained in $Sp(n_H) \times SU(2)$ and non-trivial $SU(2)$ holonomy are called *quaternionic Kähler manifolds*. They are in general not Kähler so that they could *not* be coupled to just

---

[4] A priori, one only has that the tangent space group of $\mathcal{M}_{\text{hyper}}$ should allow a restriction to $Sp(n_H) \times$ SU(2) or a subgroup thereof. In other words, $\mathcal{M}_{\text{hyper}}$ has a $Sp(n_H) \times$SU(2) structure. This implies that $\mathcal{M}_{\text{hyper}}$ admits a connection with holonomy group contained in $Sp(n_H) \times$ SU(2). A priori, this needs not be the torsion-free Levi–Civita connection, but supersymmetry invariance of the action requires (see Appendix 8.B.2) that this must be the case. Hence it is indeed the holonomy of the Levi–Civita connection that must be contained in $Sp(n_H) \times SU(2)$, and we can speak of Riemannian manifolds with holonomy in $Sp(n_H) \times SU(2)$.

$\mathcal{N} = 1$ supergravity at the full nonlinear level. In other words, constructing an $\mathcal{N} = 2$ supergravity theory with hypermultiplets by coupling an $\mathcal{N} = 1$ gravitino multiplet to $\mathcal{N} = 1$ supergravity with chiral multiplets requires (a) an even number of chiral multiplets (so that one has $4n$ real scalars) and (b) a deformation of the nonlinear couplings of the scalars among themselves, i.e., of the scalar field geometry such that the geometry is quaternionic Kähler and in general no longer Kähler.

In rigid $\mathcal{N} = 2$ supersymmetry, by contrast, the $\epsilon^{(i)}$ are constants, and the $SU(2)$ part of the curvature must be trivial. For rigid supersymmetry, the holonomy group of the scalar manifold is therefore contained in $Sp(n_H)$. Such manifolds are called hyper-Kähler. They form a subclass of Kähler manifolds that can be characterized by the presence of three independent complex structures that satisfy the algebra of the quaternions. The fact that they are still Kähler manifolds is consistent with the fact that a rigid $\mathcal{N} = 2$ theory is always also a particular rigid $\mathcal{N} = 1$ theory. Or, turning this around, one can infer from the fact that rigid $\mathcal{N} = 2$ supersymmetry must be a special case of rigid $\mathcal{N} = 1$ supersymmetry that the $SU(2)$ part of the holonomy group must be trivial for hypermultiplets in rigid supersymmetry, because otherwise the scalar manifold would be quaternionic Kähler instead of hyper-Kähler and hence in general no longer Kähler, in contradiction with the $\mathcal{N} = 1$ requirements.

8.3   Extended Supergravity with $\mathcal{N} \geq 3$

We have seen that for $\mathcal{N} = 2$ supersymmetry, the vector multiplet scalar geometry is constrained by an interplay of symplectic duality invariance and the local composite $U(1)$ symmetry of the gaugini and gravitini, which is closely related to Kähler transformations on the scalar manifold. This $U(1)$ may be associated with the $U(1)$-part of the R-symmetry group $U(2)$ of $\mathcal{N} = 2$ supersymmetry.

For the hypermultiplets, on the other hand, it is the non-trivial transformation of the scalars under the $SU(2)$ part of the R-symmetry group that leads to the reduced holonomy group $\text{Hol}(\mathcal{M}_{\text{hyper}}) \subset Sp(n_H) \times SU(2)$ of a quaternionic Kähler manifold, with the $SU(2)$ being non-trivial and of Planckian size (see (8.207)).

For higher $\mathcal{N}$ supersymmetry, the multiplets get bigger, and the scalar fields transform non-trivially under larger R-symmetry groups. By a reasoning completely analogous to our discussion of hypermultiplets, this then leads to drastic restrictions on the allowed holonomy groups of the scalar manifolds. In fact, for $\mathcal{N} \geq 3$, these restrictions are so strong that they fix the entire scalar manifold once the number of multiplets is specified, and these manifolds are certain symmetric spaces.

Let us look at this in more detail.

- $\mathcal{N} = 3$: For $\mathcal{N} = 3$ supersymmetry, the CPT-completed supergravity multiplet contains the graviton, three gravitini, three vector fields, and one spin-1/2 fermion. Apart from this multiplet, there is only one type of matter multiplet, namely, the $\mathcal{N} = 3$ vector multiplet. Its CPT-completed version has the same field content as an $\mathcal{N} = 4$ vector multiplet, namely, one vector field, four spin-1/2 fields, and six real scalars. In rigid supersymmetry, $\mathcal{N} = 3$ supersymmetry therefore also implies $\mathcal{N} = 4$ supersymmetry, but for local supersymmetry, this is no longer true, because the supergravity multiplets for $\mathcal{N} = 3$ and $\mathcal{N} = 4$ are different. The R-symmetry group of $\mathcal{N} = 3$ supersymmetry is $U(3)$. With respect to the $SU(3)$ subgroup thereof, the field content of a vector multiplet splits into (cf. Fig. 8.4) a singlet vector field, A_μ, a triplet of gaugini, $\lambda^{(i)}$, $(i, j, \ldots = 1, 2, 3)$, one $SU(3)$ singlet gaugino, $\lambda^{(4)}$, and three complex scalars, $\phi^{ij} = -\phi^{ji}$, which naturally transform in the antisymmetric tensor representation of $SU(3)$ (which is equivalent to the fundamental representation via the epsilon tensor). The linearized supersymmetry transformation laws are

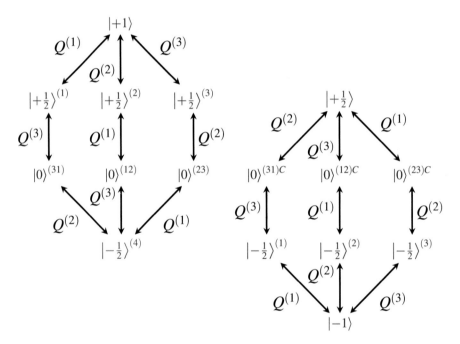

Fig. 8.4 A CPT-completed $\mathcal{N} = 3$ vector multiplet consists of the direct sum of two irreducible particle representations. The red states are $SU(3)_R$ singlets, whereas the blue states form obvious $SU(3)_R$ triplets. The states of the right multiplet are the CPT conjugates of the states of the left multiplet. The vector field A_μ is associated with $|\pm 1\rangle$, the gaugini triplet $\lambda_L^{(i)}$ corresponds to $|+\frac{1}{2}\rangle^{(i)}$, the gaugino singlet $\lambda_L^{(4)}$ to $|+\frac{1}{2}\rangle$, and the scalars ϕ^{ij} to $|0\rangle^{(ij)}$. The states can carry non-trivial $U(1)_R$ charges, which follow from the number and chirality of the supercharges that connect the states. This implies different $U(1)_R$ charge for the $SU(3)$ singlet and the $SU(3)$ triplet of fermions, as described in (8.112). For $\mathcal{N} = 4$ supersymmetry, there is a fourth supercharge $Q^{(4)}$ connecting, e.g., $1\rangle$ with $|+\frac{1}{2}\rangle^{(4)}$, which then implies that the four states with helicity $+\frac{1}{2}$ form an $SU(4)_R$ quartet and hence cannot carry different $U(1)_R$-charges anymore. The $U(1)_R$ must then be represented trivially on the vector multiplets, similar to the $U(1)_R$ for the $\mathcal{N} = 2$ hypermultiplets

(cf. Fig. 8.4 and the analogy with the hypermultiplet discussion)

$$\delta\lambda_L^{(i)} = \frac{1}{2}\partial\!\!\!/\phi^{ij}\epsilon_R^j + \frac{1}{8}\gamma^{\mu\nu}F_{\mu\nu}^-\epsilon_L^{(i)} \tag{8.110}$$

$$\delta\lambda_L^{(4)} = \frac{1}{4}\varepsilon^{ijk}\partial\!\!\!/\phi^{ij*}\epsilon_R^{(k)}. \tag{8.111}$$

The maximal internal symmetry group of these transformation laws is $SU(3) \times U(1)$ with the following $U(1)$ charge assignments:

$$A_\mu : 0, \quad \lambda_L^{(i)} : \frac{1}{2}q, \quad \phi^{ij} : +1q, \quad \lambda_L^{(4)} : -\frac{3}{2}q, \quad \epsilon_L^{(i)} : \frac{1}{2}q, \tag{8.112}$$

where q is some arbitrary real number, and charge conjugation changes the sign of the charge. This can obviously be identified with the $U(1)$ R-symmetry subgroup (cf Fig. 8.4), so that the $U(3)$ R-symmetry is actually the full invariance group of the transformation laws of one vector multiplet. For n such vector multiplets, there is thus just room for an additional $SU(n)$ that rotates the vector multiplets as a whole into each other. This is different from the $\mathcal{N} = 2$ hypermultiplets, where the smaller R-symmetry group $SU(2)$ was realized, which leaves an additional symmetry factor $Sp(1) \cong SU(2)$ even for one hypermultiplet and $Sp(n_H)$ instead of just $SU(n_H)$ for n_H hypermultiplets.

Correspondingly, the holonomy group of the scalar manifold of $\mathcal{N} = 3$ vector multiplets must have holonomy group contained in $U(3) \times SU(n)$, with the $U(3)$ holonomy group being non-trivial. According to a classification of special holonomy manifolds by Berger [10], a non-compact real $6n$-dimensional manifold with this property must be equivalent to

$$\mathcal{M}^{\mathcal{N}=3} = \frac{SU(3, n)}{S(U(3) \times U(n))}. \tag{8.113}$$

- $\mathcal{N} = 4$: For $\mathcal{N} = 4$ supersymmetry, the supergravity multiplet now also contains two real scalar fields in addition to the graviton, four gravitini, six vector fields, and four spin-1/2 fermions. The $\mathcal{N} = 4$ vector multiplet has the same field content as the $\mathcal{N} = 3$ vector multiplet just discussed. Now, however, the vector multiplet is CPT self-conjugate, i.e., it does not have to be completed by a CPT conjugate multiplet. This means that the six real scalars, which transform in the six-dimensional, real representation of the R-symmetry subgroup $SU(4) \cong SO(6)$ must be their own anti-particles and hence cannot carry a non-trivial charge with respect to the $U(1)$ part of the $U(4)$ R-symmetry group. Likewise, different charge assignments for $\lambda_L^{(4)}$ and the other gaugini as in (8.112) are no longer possible as that would break $SU(4)$. This can also be seen from the linearized supersymmetry transformations, which are obtained from the

$\mathcal{N} = 3$ transformations by obvious $SU(4)$ covariantization,

$$\delta\lambda_L^{(i)} = \frac{1}{2}\partial\!\!\!/\phi^{ij}\epsilon_R^{(j)} + \frac{1}{8}\gamma^{\mu\nu}F_{\mu\nu}^{-}\epsilon_L^{(i)} \tag{8.114}$$

with $i, j, \ldots, = 1, 2, 3, 4$, and the reality condition $(\phi^{ij})^* = \frac{1}{2}\varepsilon_{ijkl}\phi^{kl}$, which obviously does not allow any $U(1)$ charge assignments for the scalars anymore. This is similar to the $\mathcal{N} = 2$ hypermultiplet scalars being neutral with respect to the $U(1)$ subgroup of the R-symmetry group $U(2)$. The holonomy group of the real $6n$-dimensional scalar manifold of n vector multiplets in $\mathcal{N} = 4$ supergravity must therefore be contained in $SO(6) \times SO(n)$, with the $SO(6)$ factor being non-trivial and the $SO(n)$ arising here as the maximal subgroup of $SO(6n)$ that commutes with the action of $SO(6)$ on the scalars. $SO(n)$ simply rotates the different multiplets into each other. Berger's list then fixes the scalar manifolds of the vector multiplets to be of the form $SO(6, n)/(SO(6) \times SO(n))$. Together with the scalars from the supergravity multiplet, which parameterize $SU(1, 1)/U(1)$, we therefore have the scalar manifold

$$\mathcal{M}^{\mathcal{N}=4} = \frac{SU(1, 1)}{U(1)} \times \frac{SO(6, n)}{SO(6) \times SO(n)}. \tag{8.115}$$

- $\mathcal{N} = 5$: For $\mathcal{N} = 5$ supersymmetry, there are no longer matter multiplets with helicities between 1 and -1, and the only supermultiplet one could use is the supergravity multiplet. In its CPT-completed version, this multiplet contains one graviton, five gravitini, ten vector fields, eleven spin $1/2$ fields, and ten real scalar fields. The holonomy group must be $U(5)$, which then leaves the space

$$\mathcal{M}^{\mathcal{N}=5} = \frac{SU(5, 1)}{U(5)}. \tag{8.116}$$

- $\mathcal{N} = 6$: Here, the CPT-completed supergravity multiplet contains the graviton, 6 gravitini, 16 vector fields, 26 spin $1/2$ fermions, and 30 real scalar fields. The holonomy group of the scalar manifold must be the R-symmetry group $U(6)$, which leaves as scalar manifold

$$\mathcal{M}^{\mathcal{N}=6} = \frac{SO^*(12)}{U(6)}. \tag{8.117}$$

- $\mathcal{N} = 8$: As the CPT-completed $\mathcal{N} = 7$ supergravity multiplet is equivalent to the CPT-self-conjugate $\mathcal{N} = 8$ supergravity multiplet, there is no $\mathcal{N} = 7$ supergravity theory. The $\mathcal{N} = 8$ supergravity multiplet contains the graviton, 8 gravitini, 28 vector fields, 56 spin $1/2$ fermions, and 70 real scalars. As this multiplet is CPT-self-conjugate, the scalars must be neutral with respect to the $U(1)$ subgroup of the $U(8)$ R-symmetry group and only transform non-trivially with respect to the $SU(8)$ subgroup, which must therefore also be the holonomy group of the scalar manifold. This scalar manifold is given by

$$\mathcal{M}^{\mathcal{N}=8} = \frac{E_{7(7)}}{SU(8)}. \tag{8.118}$$

8.A Appendix: Details on and Origin of Local Special Kähler Geometry

In this appendix, we would like to understand the origin of the differences between rigid and local special Kähler geometry that are listed in Sect. 8.2.1.2. Our goal herein is not so much mathematical completeness or a full derivation of all terms and prefactors in the Lagrangian, but rather to emphasize the geometrical roots of these differences and to exhibit them as directly as possible and without introducing additional formalism [2, 4–8].

8.A.1 The Symplectic Section and Its Kähler Transformation

We begin with the necessity of the existence of a scalar field-dependent symplectic section, v. Just as for rigid special Kähler geometry, this necessity can be deduced from the requirement that the transformation laws of the gaugini should be symplectically invariant and covariant with respect to holomorphic scalar reparameterizations. More precisely, if we again consider the supersymmetry transformation with respect to the original $\mathcal{N} = 1$ supersymmetry parameter $\epsilon^{(1)}$, the analogy with the rigid transformations (8.63), (8.64) suggests that in supergravity one has

$$\delta^{(1)} \lambda_L^{m(1)} = \frac{1}{2} (\partial z^m) \epsilon_R^{(1)} \tag{8.119}$$

$$\delta^{(1)} \lambda_L^{m(2)} = -\frac{1}{4} g^{m\bar{n}} \overline{f_n^I} (\mathrm{Im} \mathcal{N}_{IJ}) F_{\mu\nu}^{J-} \gamma^{\mu\nu} \epsilon_L^{(1)}. \tag{8.120}$$

Here, $\mathcal{N}_{IJ}(z, \bar{z})$ is the scalar field-dependent gauge kinetic matrix of the vector field Lagrangian (8.39), and $f_n^I(z, \bar{z})$ is a scalar field-dependent object that transforms as a holomorphic covector, $f_n^I \mapsto (\partial z^m / \partial \tilde{z}^n) f_m^I$, under scalar field reparameterizations and forms the upper part of a symplectic vector

$$U_n(z, \bar{z}) = \begin{pmatrix} f_n^I(z, \bar{z}) \\ h_{Jn}(z, \bar{z}) \end{pmatrix}. \tag{8.121}$$

To be most general as possible, we allow for a possible non-holomorphic z^m-dependence, but, as in rigid supersymmetry, we do expect that also U_n will be related to a suitable derivative of another symplectic (and possibly non-holomorphic) section $V(z, \bar{z})$ with respect to z^n. In order to ensure the symplectic invariance of

(8.120), however, we only need to require

$$\overline{f_n^I} \mathcal{N}_{IJ} = \overline{h_{Jn}}, \tag{8.122}$$

because then

$$\overline{f_n^I} \mathrm{Im} \mathcal{N}_{IJ} F_{\mu\nu}^{J-} = \frac{1}{2i} \left(\overline{h_{Jn}} F_{\mu\nu}^{J-} - \overline{f_n^I} G_{\mu\nu I-} \right) = \frac{1}{2i} \langle \mathbb{F}_{\mu\nu}^-, \overline{U_n} \rangle, \tag{8.123}$$

which is manifestly symplectically invariant.

As mentioned above, we expect that U_n be related to a suitable derivative of a, possibly non-holomorphic, symplectic section, which we denote as

$$V(z, \bar{z}) = \begin{pmatrix} L^I(z, \bar{z}) \\ M_J(z, \bar{z}) \end{pmatrix}, \tag{8.124}$$

and which will be closely related to the advertised holomorphic section $\mathcal{V}(z)$ mentioned in item 1.

In order to understand this better, it is necessary to address also item 2 in Sect. 8.2.1.2, namely, the claim that in supergravity the scalar field-dependent symplectic sections have to transform non-trivially under Kähler transformations (see (8.81) for $\mathcal{V}(z)$). The root of this behavior is the particular non-trivial Kähler transformation property of the fermions we already discovered in $\mathcal{N} = 1$ supergravity. Repeating the arguments of Sect. 6.1.2 for each of the two supersymmetries, one now finds that the two gravitini $\psi_\mu^{(1)}, \psi_\mu^{(2)}$ have to transform under Kähler transformations as (cf. (6.18))

$$\psi_\mu^{(j)} \mapsto \exp\left[-\frac{i}{2} (\mathrm{Im}\, h) \gamma_5 \right] \psi_\mu^{(j)}, \qquad (j = 1, 2) \tag{8.125}$$

which immediately implies

$$\epsilon^{(j)} \mapsto \exp\left[-\frac{i}{2} (\mathrm{Im}\, h) \gamma_5 \right] \epsilon^{(j)}, \qquad (j = 1, 2) \tag{8.126}$$

via $\delta\psi_\mu^{(j)} = \mathscr{D}_\mu \epsilon^{(j)} + \ldots$, where $\epsilon^{(i)}$ are the two Majorana spinor parameters.

This behavior is consistent with the $SU(2)_R$ R-symmetry rotating the two $\psi_\mu^{(j)}$ and the two $\epsilon^{(j)}$ into each other and requires the introduction of Kähler covariant derivatives (cf. (6.16), (6.21)),

$$\mathscr{D}_{[\nu} \psi_{\rho]}^{(j)} \equiv D_{[\nu}(\omega) \psi_{\rho]}^{(j)} + \frac{i}{2} Q_{[\nu} \gamma_5 \psi_{\rho]}^{(j)} \tag{8.127}$$

$$\mathscr{D}_\nu \epsilon^{(j)} \equiv D_\nu(\omega) \epsilon^{(j)} + \frac{i}{2} Q_\nu \gamma_5 \epsilon^{(j)} \tag{8.128}$$

with $Q_\nu(z, \bar{z}) \equiv i/2[(\partial_{\bar{n}} K) \partial_\nu z^{\bar{n}} - (\partial_m K) \partial_\nu z^m]$.

While this is completely analogous to $\mathcal{N} = 1$ supergravity, an important difference occurs for the gaugini. If we again identify $\epsilon^{(1)}$ with the original $\mathcal{N} = 1$ supersymmetry parameter and $(\lambda^{m(1)}, \lambda^{m(2)}) = (\chi^m, \lambda^m/2)$, we expect a supersymmetry transformation rule with respect to $\epsilon^{(1)}$ of the form (cf. (8.119), (8.120))

$$\delta\lambda_L^{m(1)} = \frac{1}{2}\slashed{\partial}z^m \epsilon_R^{(1)} \tag{8.129}$$

$$\delta\lambda_L^{m(2)} = -\frac{1}{4}g^{m\bar{n}}\overline{f_n^I}\,\mathrm{Im}\,\mathcal{N}_{IJ}F_{\mu\nu}^{J-}\gamma^{\mu\nu}\epsilon_L^{(1)}. \tag{8.130}$$

From the first equation and (8.126), we deduce

$$\lambda_L^{m(1)} \mapsto \exp\left[+\frac{i}{2}(\mathrm{Im}\,h)\right]\lambda_L^{m(1)}, \tag{8.131}$$

which must then also hold for $\lambda_L^{m(2)}$ due to the assumed $SU(2)_R$ R-symmetry. This, however, seems to be in conflict with the second equation (8.130), which naively would imply a transformation with a minus sign in the exponent due to the appearance of $\epsilon_L^{(1)}$ instead of $\epsilon_R^{(1)}$. In $\mathcal{N} = 1$ supergravity, we encountered a similar problem for the supersymmetry transformations of the fermions in the presence of a non-trivial superpotential (see the discussion below Eq. (6.30)). This problem could be solved by assuming that the superpotential transforms non-trivially under Kähler transformations. To solve the analogous problem in (8.130), one likewise requires that the field-dependent term in front of $\gamma^{\mu\nu}$ transform non-trivially under Kähler transformations. In this term, the inverse scalar field metric $g^{m\bar{n}}$ is by construction Kähler invariant, and a Kähler transformation of $\mathcal{N}_{IJ}$ and/or $F_{\mu\nu}^{J-}$ would lead to Kähler non-invariances in the gauge field sectors. This leaves f_n^I as the only reasonable candidate for a non-trivial Kähler transformation, which then has to be of the form

$$f_n^I \mapsto \exp[-i(\mathrm{Im}h)]f_n^I \quad \Longleftrightarrow \quad U_n \mapsto \exp[-i(\mathrm{Im}h)]U_n \tag{8.132}$$

in order to yield the correct Kähler transformation of the right-hand side of (8.130).

Equation (8.132) now also implies that U_n cannot be just the partial derivative of a symplectic section, V, as that would also produce derivatives of h in (8.132). Instead, U_n can at most be a Kähler covariant derivative,

$$U_n = \mathcal{D}_n V \equiv \left(\partial_n + \frac{1}{2}(\partial_n K)\right)V, \tag{8.133}$$

where

$$V = V(z,\bar{z}) = \begin{pmatrix} L^I(z,\bar{z}) \\ M_J(z,\bar{z}) \end{pmatrix} \tag{8.134}$$

is a symplectic section that transforms as

$$V \mapsto \exp\left[-i(\mathrm{Im}h)\right] V \tag{8.135}$$

under Kähler transformations. This transformation property of V depends on $h(z)$ and $\overline{h}(\overline{z})$, which implies that V cannot depend holomorphically on the coordinates z^m of the scalar manifold, as indicated in (8.134).

We can however define a modified symplectic section, v, that does transform holomorphically under Kähler transformations,

$$\mathscr{V}(z) = \begin{pmatrix} X^I(z) \\ F_J(z) \end{pmatrix} := e^{-\frac{K(z,\overline{z})}{2}} V(z,\overline{z}) \quad \Rightarrow \quad \mathscr{V}(z) \mapsto \exp\left[-h(z)\right] \mathscr{V}(z) \tag{8.136}$$

and hence may sensibly be restricted to have a holomorphic coordinate dependence, as indicated. We thus have

$$U_n = \mathscr{D}_n \left(e^{\frac{K}{2}} \mathscr{V}\right) \equiv \left(\partial_n + \frac{1}{2}(\partial_n K)\right)\left(e^{\frac{K}{2}} v\right) = e^{\frac{K}{2}}\left(\partial_n + (\partial_n K)\right)v \equiv e^{\frac{K}{2}} \mathscr{D}_n \mathscr{V} \tag{8.137}$$

so that, in particular,

$$f_n^I = e^{\frac{K}{2}} \mathscr{D}_n X^I \equiv e^{\frac{K}{2}}\left(\partial_n + (\partial_n K)\right)X^I \tag{8.138}$$

and similarly for h_{Jn}. We finally note that the holomorphicity of v, $\partial_{\overline{n}} v = 0$, is equivalent to

$$\mathscr{D}_{\overline{n}} V \equiv \left(\partial_{\overline{n}} - \frac{1}{2}(\partial_{\overline{n}} K)\right) V = 0. \tag{8.139}$$

8.A.2 The Kähler Potential

After having discussed the necessity of a holomorphic symplectic section, $\mathscr{V}(z)$, we now come to the explicit form of the Kähler potential (8.82) mentioned in item 3 of Sect. 8.2.1.2. The analogy with Eq. (8.51) in rigid supersymmetry suggests that the Kähler potential should be related to the real symplectic invariant $i\langle \overline{\mathscr{V}}, \mathscr{V} \rangle = i(\overline{X^I} F_I - X^I \overline{F_I})$. As opposed to the rigid case, however, we now have the behavior (8.136) under Kähler transformations, which implies

$$i\langle \overline{\mathscr{V}}, \mathscr{V} \rangle \longrightarrow e^{-(h+\overline{h})} i\langle \overline{\mathscr{V}}, \mathscr{V} \rangle. \tag{8.140}$$

Hence, $i\langle\overline{\mathscr{V}},\mathscr{V}\rangle$ does not transform the right way to serve as the Kähler potential, but the logarithm of it would do the job. We are thus led to the logarithmic expression (8.82).

Let us summarize what we have shown so far. The scalar manifolds, $\mathscr{M}_{\text{vec}}$, of vector multiplets in $\mathscr{N} = 2$ supergravity are Kähler manifolds that allow for a symplectic section, $V(z,\bar{z}) = e^{K(z,\bar{z})/2}\mathscr{V}(z)$, that satisfies the following properties:

- $\partial_{\bar{m}}v = 0 \iff \mathscr{D}_{\bar{m}}V = 0$
- $K = -\log i\langle\overline{\mathscr{V}},\mathscr{V}\rangle \iff \langle V,\overline{V}\rangle = i$
- $\mathscr{V}(z) \mapsto e^{-h(z)}\mathscr{V}(z)$ under Kähler transformations,

where it is understood that $\mathscr{V}(z)$ must be such that the resulting metric $g_{m\bar{n}} = \partial_m\partial_{\bar{n}}K$ is positive definite for the z^m-domain of interest.

Geometrically, the above Kähler transformation property of $\mathscr{V}(z)$ means that $\mathscr{V}(z)$ is a section not only in an $\text{Sp}(2(n_V + 1), \mathbb{R})$ bundle but, similar to the superpotential in $\mathscr{N} = 1$ supergravity, also in a holomorphic line bundle. This line bundle obeys a topological restriction such that the (special) Kähler manifold must actually be a (special) Hodge-Kähler manifold, as we discussed for $\mathscr{N} = 1$ supergravity in Sect. 6.2.

8.A.3 The Existence of a Prepotential

The properties listed at the end of the previous subsection are still not sufficient to allow for a consistent embedding into $\mathscr{N} = 2$ supergravity and have to be supplemented by the additional requirement

$$\langle\mathscr{D}_m\mathscr{V}, \mathscr{D}_n\mathscr{V}\rangle = 0, \tag{8.141}$$

where $\mathscr{D}_n\mathscr{V} = (\partial_n + (\partial_n K))\mathscr{V}$. Due to the antisymmetry of the symplectic product, this condition is antisymmetric in m and n, and hence non-trivial only for $n_V > 1$, where (8.141) is necessary for the symmetry of the gauge kinetic matrix $\mathscr{N}_{IJ}$ and ensures its uniqueness, as we will discuss in Appendix 8.A.4. Moreover, for $n_V > 1$, (8.141) implies that there is always a symplectic transformation that brings $\mathscr{V}(z)$ into the familiar prepotential form

$$\mathscr{V} = \begin{pmatrix} X^I \\ F_J \end{pmatrix}, \quad \text{with} \quad F_J = \frac{\partial F(X)}{\partial X^J} \tag{8.142}$$

for a suitable holomorphic function $F(X)$ that exists at least locally and, unlike in rigid supersymmetry, is also restricted to be homogeneous of degree two. A simple way to understand the homogeneity requirement is provided by the behavior $X^I \mapsto e^{-h}X^I$ and $F_J \mapsto e^{-h}F_J$, which is consistent only if the $F_J = \partial F/\partial X^J$ are homogeneous functions of the X^I of degree one, implying homogeneity of degree two for $F(X)$.

The generic necessity of performing a suitable symplectic rotation in supergravity before the prepotential form (8.142) is achieved is an important difference to rigid special Kähler geometry, where the section is of a prepotential form in *any* symplectic duality frame. This difference is ultimately related to the fact that in supergravity the $(n_V + 1)$ components X^I cannot be swapped for the n_V independent coordinates z^m. This is best illustrated by a simple example [7]. Choosing, e.g., $n_V = 1$ with $F(X) = -i X^0 X^1$ and special coordinates $X^0 = 1$, $X^1 = z$, the section $\mathscr{V}(z)$ in the form (8.142) is

$$
\mathscr{V}(z) = \begin{pmatrix} 1 \\ z \\ -iz \\ -i \end{pmatrix}
\tag{8.143}
$$

which leads to the metric $g_{z\bar{z}} = (z + \bar{z})^{-2}$ corresponding to the manifold $\mathscr{M}_{\mathrm{vec}} = SU(1, 1)/U(1)$. Performing now a symplectic transformation of the form

$$
\mathscr{V} \mapsto S\mathscr{V} = \begin{pmatrix} 1 & 0 & 0 & 0 \\ 0 & 0 & 0 & -1 \\ 0 & 0 & 1 & 0 \\ 0 & 1 & 0 & 0 \end{pmatrix} \mathscr{V} = \begin{pmatrix} 1 \\ i \\ -iz \\ z \end{pmatrix},
\tag{8.144}
$$

we see that the transformed section can no longer be of the form (8.142), as the lower two components cannot be functions of the upper two components. The prepotential form thus in general only holds in some duality frames but not in all of them. If one does not impose (8.141), on the other hand, one could even write down sections $\mathscr{V}(z)$ that can never be rotated into a prepotential form (8.142), no matter what symplectic transformation one uses. The proof that (8.141) ensures the existence of a prepotential basis is a bit technical [2, 4] and will therefore not be repeated here. Instead we content ourselves with showing the opposite direction, namely, that in a basis (8.142) with second-degree prepotential, condition (8.141) follows. As (8.141) is automatically true for $n_V = 1$, we assume $n_V > 1$ unless stated otherwise. We start by showing that (8.141) is implied by[5]

$$
\langle \mathscr{V}, \partial_m \mathscr{V} \rangle = 0.
\tag{8.145}
$$

---

[5] For $n_V > 1$, (8.145) also follows from (8.141). To see this, one acts with $\mathscr{D}_{\bar{l}}$ on (8.141), where it just acts as $\partial_{\bar{l}}$ due to the Kähler weights of $\mathscr{D}_m \mathscr{V}$. This results in $g_{m\bar{l}} \langle \mathscr{V}, \partial_n \mathscr{V} \rangle + g_{n\bar{l}} \langle \partial_m \mathscr{V}, \mathscr{V} \rangle = 0$, which implies (8.145) upon contraction with $g^{\bar{l}p}$ and $p = m \neq n$.

To see this, we first note that $\langle \mathscr{V}, \mathscr{V} \rangle = 0$ due to antisymmetry of the symplectic product, so that (8.145) can be equivalently written as

$$\langle \mathscr{V}, \mathscr{D}_m \mathscr{V} \rangle \equiv \langle \mathscr{V}, (\partial_m + (\partial_m K)) \mathscr{V} \rangle = 0. \tag{8.146}$$

Acting with $\mathscr{D}_n$ and antisymmetrizing in n and m and using (8.145) and $\langle \mathscr{V}, \mathscr{V} \rangle = 0$ then imply (8.141). Using now the prepotential form (8.142), one finds

$$\langle \mathscr{V}, \partial_m \mathscr{V} \rangle = X^I (\partial_m F_I) - F_I (\partial_m X^I) = X^I \left(\frac{\partial X^J}{\partial z^m} \partial_J \partial_I F \right) - \partial_m F$$

$$= \frac{\partial X^J}{\partial z^m} X^I \partial_I (\partial_J F) - \partial_m F = \frac{\partial X^J}{\partial z^m} \partial_J F - \partial_m F = 0, \tag{8.147}$$

where we used that $\partial_J F = X^I \partial_I \partial_J F$ is homogeneous of degree one. This then implies (8.141).

Before we come to the discussion of the gauge kinetic matrix and its relation to (8.141), we comment on the case $n_V = 1$. In that case, (8.141) is empty and hence does not imply the existence of a prepotential. If one uses instead the (for $n_V = 1$ non-equivalent) condition (8.145), on the other hand, the existence of a prepotential basis can be proven in a similar way as for $n_V > 1$ [2]. However, as the condition (8.145) is no longer implied by the required symmetry of the gauge kinetic matrix $\mathscr{N}_{IJ}$ (see below), there seems to be no physical reason that enforces (8.145). And indeed, in [4], models without an action and one vector multiplet were constructed where (8.145) does not hold.

8.A.4 The Gauge Kinetic Matrix

We finally come to the discussion of the gauge kinetic matrix and its relation to condition (8.141). More precisely, we would like to show that (8.141) is necessary for a sensible gauge kinetic matrix, $\mathscr{N}_{IJ}$, for $n_V > 1$, and that, in a prepotential basis, this matrix can be written as announced in (8.84). We already have derived the constraint

$$\overline{f_m^I} \mathscr{N}_{IJ} = \overline{h_{Jm}}, \tag{8.148}$$

which follows from the requirement that the gaugino transformation law (8.120) be symplectically invariant and covariant with respect to scalar reparameterizations. As we will now show, the gravitino transformation law imposes the additional condition

$$L^I \mathscr{N}_{IJ} = M_J. \tag{8.149}$$

Note that there is no complex conjugation involved in this equation, as opposed to (8.148). In order to motivate this equation, we recall the rigid supersymmetry transformations of the $\mathcal{N} = 1$ gravitino multiplet discussed in Sect. 2.2.2,

$$\delta\psi_{\mu L} = -\gamma^\rho \epsilon_R \, F^-_{\mu\rho} \tag{8.150}$$

$$\delta A_\mu = \bar{\epsilon}\psi_\mu, \tag{8.151}$$

where we used $\gamma_{\mu\nu\rho} = i\epsilon_{\mu\nu\rho\sigma}\gamma^\sigma \gamma_5$ to write the gravitino transformation in terms of the anti-self-dual field strength.

These transformations leave invariant the Lagrangian

$$\mathcal{L} = -\frac{1}{2}\bar{\psi}_\mu \gamma^{\mu\nu\rho} \partial_\nu \psi_\rho - \frac{1}{4} F^{\mu\nu} F_{\mu\nu}. \tag{8.152}$$

This theory should be embeddable in pure $\mathcal{N} = 2$ supergravity by making the supersymmetry parameter x-dependent and by coupling it to the $\mathcal{N} = 1$ supergravity multiplet in the usual way. This will introduce a derivative of ϵ as an additional term into the gravitino transformation law and require spacetime covariantization, but as there are no scalar fields involved, there is not much room for a modification of the above existing terms from rigid $\mathcal{N} = 1$ supersymmetry. And indeed, one can write the above transformation laws and the Lagrangian formally in terms of local special Kähler geometry for the special case $n_V = 0$. To this end, we use the prepotential $F(X^0) = -\frac{i}{2}(X^0)^2$, which leads to

$$\mathcal{V} = \begin{pmatrix} X^0 \\ F_0 \end{pmatrix} = \begin{pmatrix} X^0 \\ -iX^0 \end{pmatrix}. \tag{8.153}$$

This implies via (8.82) and (8.84)

$$K = -\log|X^0|^2, \qquad \mathcal{N}_{00} = -i \tag{8.154}$$

so that, after choosing $X^0 = 1$, one has $V = (L^0, M_0)^T = (1, -i)^T$ and hence

$$\delta\psi_{\mu L} = L^0 \mathrm{Im}\mathcal{N}_{00} F^{0-}_{\mu\rho} \gamma^\rho \epsilon_R. \tag{8.155}$$

This expression is symplectically invariant, because

$$L^0 \mathcal{N}_{00} = -i = M_0 \tag{8.156}$$

implies $L^0 \mathrm{Im}\mathcal{N}_{00} F^{0-}_{\mu\rho} = (1/2i)(M_0 F^{0-}_{\mu\rho} - L^0 G_{\mu\rho 0-}) = (1/2i)\langle \mathbb{F}_{\mu\nu}, V \rangle$.

For $\mathcal{N} = 2$ supergravity coupled to n_V vector multiplets, the gravitino transformation law should then contain the analogous term

$$\delta\psi^{(1)}_{\mu L} \propto L^I \mathrm{Im}\mathcal{N}_{IJ} F^{J-}_{\mu\rho} \gamma^\rho \epsilon^{(2)}_R, \tag{8.157}$$

because this ensures that both sides transform with $e^{-\frac{i}{2}(\mathrm{Im}h)}$ and are symplectic invariant provided the analogue of equation (8.156) holds, which is just (8.149).

The gauge kinetic matrix $\mathcal{N}_{IJ}$ thus has to obey the two constraints (8.148) and (8.149). In fact, these conditions determine $\mathcal{N}_{IJ}$ uniquely. To see this, we observe from the gaugino transformation (8.130) and the gravitino transformation (8.157) with respect to $\epsilon^{(2)}$ that the n_V independent gaugini $\lambda_L^{m(1)} g_{m\bar{n}}$ transform into $f_n^I \mathrm{Im}\mathcal{N}_{IJ} F_{\mu\nu}^{J-}$, whereas the gravitino $\psi_\mu^{(1)}$ transforms into the linear combination $L^I \mathrm{Im}\mathcal{N}_{IJ} F_{\mu\nu}^{I-}$. For any fixed scalar field value z^m, these must therefore be $(n_V + 1)$ linearly independent combination of vector field strengths. This implies that the following $((n_V + 1) \times (n_V + 1))$-matrix, A, must be invertible:

$$A := (\overline{f_m^I}, L^I), \tag{8.158}$$

where I is the row index and m labels the first n_V columns, whereas L^I is the last column. Defining similarly

$$B := (\overline{h_{Im}}, M_I), \tag{8.159}$$

Eqs. (8.148) and (8.149) can be written as

$$A^T \cdot \mathcal{N} = B^T, \tag{8.160}$$

so that, with the invertibility of A, one has a unique expression for the gauge kinetic matrix

$$\mathcal{N} = A^{T-1} \cdot B^T. \tag{8.161}$$

This has to be symmetric, $\mathcal{N} - \mathcal{N}^T = 0$, which is then equivalent to

$$B^T A - A^T B = 0 \iff \begin{pmatrix} \langle \overline{U_m}, \overline{U_n} \rangle & \langle \overline{U_n}, V \rangle \\ \langle V, \overline{U_n} \rangle & \langle V, V \rangle \end{pmatrix} = 0. \tag{8.162}$$

This implies

$$\langle U_m, U_n \rangle = 0, \qquad \langle V, \overline{U_n} \rangle = 0. \tag{8.163}$$

The second of these equations simply follows from taking a covariant derivative of $\langle V, \overline{V} \rangle = i$, whereas the first one is equivalent to (8.141), which implies the existence of a basis with a prepotential for $n_V > 1$, as discussed earlier.

It remains to verify that in a basis with a prepotential $\mathcal{N}_{IJ}$ takes the form (8.84). To see this, we use

$$\langle \overline{\mathscr{D}_m V}, \mathscr{V} \rangle = 0 \tag{8.164}$$

$$(\partial_m X^I) F_{IK} = \partial_m F_K \tag{8.165}$$

$$X^I F_{IK} = F_K \tag{8.166}$$

$$\overline{f_n^I F_{IK}} = \overline{h_{Km}}. \tag{8.167}$$

The first of these equations is valid in any symplectic frame and equivalent to $\langle \overline{\mathscr{D}_m V}, V \rangle = 0$, which follows from $\langle \overline{V}, V \rangle = -i$ and $\mathscr{D}_{\overline{m}} V = 0$. The second uses the prepotential relations $F_{IK} = \partial_J \partial_K F$ and $F_K = \partial_K F$, which also imply the third equation due to the homogeneity of F of degree two. The fourth equation finally follows from the second and third upon using $f_n^I = e^{K/2}(\partial_m + (\partial_m K)) X^K$ and $h_{Km} = e^{K/2}(\partial_m + (\partial_m K)) F_K$.

We know that $\mathcal{N}_{IJ}$ is uniquely specified by (8.148) and (8.149), so it suffices to show that (8.84) satisfies these two equations in a prepotential basis. We begin with (8.148) and first show that

$$2i \overline{f_m^I}(\mathrm{Im}F_{IK}) X^K \quad = \quad \overline{f_n^I F_{IK}} X^K - \overline{f_n^I \overline{F}_{IK}} X^K \tag{8.168}$$

$$\stackrel{(8.166),(8.167)}{=} \overline{f_m^I F_I} - \overline{h_{Km}} X^K \stackrel{(8.164)}{=} 0$$
$$\tag{8.169}$$

The second term on the right-hand side of (8.84) thus does not contribute to (8.148), which then follows from $\overline{f_m^I \mathcal{N}_{IK}} = \overline{f_m^I F_{IK}} = \overline{h_{Km}}$.

Equation (8.149), finally, follows from $(L^I, M_K) = e^{K/2}(X^I, F_K)$ and (8.166), with the non-antiholomorphic piece in $\mathcal{N}_{IJ}$ now playing a crucial role. This term is thus the unique extension of $\overline{F_{IJ}}$ that ensures (8.149) (i.e., ultimately the symplectic invariance of the gravitino transformation law) without destroying (8.148) (i.e., the symplectic invariance of the gaugino transformation law).

8.B   Appendix: Quaternionic-Kähler vs. Hyper-Kähler Manifolds of Hypermultiplets

In this appendix we show in more detail how the invariance of the action implies a non-trivial $SU(2)$ curvature for hypermultiplets in $\mathcal{N} = 2$ supergravity and hence leads to the difference between hyper-Kähler geometry and quaternionic-Kähler geometry [9].

8.B.1 $Sp(n_H) \times SU(2)$-Adapted Vielbein

To begin with, we consider the real $4n_H$-dimensional scalar manifold, $\mathcal{M}_{\text{hyper}}$, of n_H $\mathcal{N} = 2$ hypermultiplets and parameterize it locally by $4n_H$ real scalar fields q^X ($X, Y, \ldots = 1, \ldots, 4n_H$). Denoting an ordinary vielbein on $\mathcal{M}_{\text{hyper}}$ by $\mathfrak{f}_X^{\Gamma}(q)$ with flat $SO(4n_H)$-indices $\Gamma, \Delta, \ldots = 1, \ldots 4n_H$, the metric components, $h_{XY}(q)$, can be expressed as

$$h_{XY}(q) = \mathfrak{f}_X^{\Gamma}(q)\delta_{\Gamma\Delta}\mathfrak{f}_Y^{\Delta}(q). \tag{8.170}$$

We now fix a particular point $p \in \mathcal{M}_{\text{hyper}}$ and switch to a new basis of the vielbein at that point p. To this end, we split the vielbein components into two sets,

$$(\mathfrak{f}_X^A) = \begin{pmatrix} \mathfrak{f}_X^{1A} \\ \mathfrak{f}_X^{2A} \end{pmatrix}, \qquad (A, B, \ldots = 1, \ldots, 2n_H). \tag{8.171}$$

We then make a basis change with the non-orthogonal matrix (cf. (8.106) for the definition of E)

$$M := \frac{1}{\sqrt{2}} \begin{pmatrix} i\mathbf{1}_{2n_H} & -iE \\ \mathbf{1}_{2n_H} & E \end{pmatrix} \Leftrightarrow M^{-1} = \frac{1}{\sqrt{2}} \begin{pmatrix} -i\mathbf{1}_{2n_H} & \mathbf{1}_{2n_H} \\ -iE & -E \end{pmatrix}, \tag{8.172}$$

which satisfies

$$M^T M = e \otimes E = \begin{pmatrix} 0 & E \\ -E & 0 \end{pmatrix}, \qquad (M^{-1})^* = (e \otimes E)M^{-1}. \tag{8.173}$$

More precisely, we define a new vielbein at p,

$$f_X = \begin{pmatrix} f_X^{1A} \\ f_X^{2A} \end{pmatrix}, \tag{8.174}$$

by

$$f_X = M^{-1}\mathfrak{f}_X \Leftrightarrow \begin{array}{l} f_X^1 = \frac{1}{\sqrt{2}}(-i\mathfrak{f}_X^1 + \mathfrak{f}_X^2) \\ f_X^2 = \frac{1}{\sqrt{2}}(-iE\mathfrak{f}_X^1 - E\mathfrak{f}_X^2) \end{array} \tag{8.175}$$

In terms of this new vielbein, we then have, with Eq. (8.173),

$$h_{XY} = \mathfrak{f}_X^T\mathfrak{f}_Y = f_X^T M^T M f_Y = f_X^T(e \otimes E)f_Y = f_X^{iA} e_{ij} E_{AB} f_Y^{jB}, \tag{8.176}$$

as well as the reality condition

$$
f_X^* = \begin{pmatrix} f_X^{1A} \\ f_X^{2A} \end{pmatrix}^* = M^{-1*} \mathfrak{f}_X = (e \otimes E) M^{-1} \mathfrak{f}_X = (e \otimes E) f_X \tag{8.177}
$$

i.e.,

$$
(f_X^{iA})^* = f_X^{jB} e_{ji} E_{BA} =: f_{XiA}. \tag{8.178}
$$

Equation (8.176) tells us that the new vielbein f_X^{iA} is adapted to an $Sp(n_H) \times SU(2)$ split of the tangent space at the point p. In Sect. 8.2.2, we used the linearized supersymmetry transformations (8.102), (8.103) to argue that the tangent space group of the scalar manifold, $\mathcal{M}_{\text{hyper}}$, of n_H $\mathcal{N} = 2$ hypermultiplets should be contained in $Sp(n_H) \times SU(2) \in SO(4n_H)$. In other words, the above $Sp(n_H) \times SU(2)$ covariant form of the vielbein at p can actually be extended smoothly across all of $\mathcal{M}_{\text{hyper}}$. We can thus consistently use a vielbein of the adapted form f_X^{iA} for all points.

The reality condition (8.178), on the other hand, indicates that it is natural to denote complex conjugation by lowering the $SU(2)$ indices, $i, j, \ldots$, with e_{ij} and the $Sp(n_H)$ indices, $A, B, \ldots$, by E_{AB} according to the convention

$$
V^i = e^{ij} V_j, \qquad V_i = V^j e_{ji}; \qquad V^A = E^{AB} V_B, \qquad V_A = V^B E_{BA} \tag{8.179}
$$

where $e^{ij} e_{ik} = \delta_k^j$ and $E^{AB} E_{AC} = \delta_C^B$.

Before we come to the supersymmetry transformation laws, we note two additional contraction identities,

$$
f_X^{iA} f_{YjA} + f_Y^{iA} f_{XjA} = h_{XY} \delta_j^i \tag{8.180}
$$

$$
h^{XY} f_X^{iA} f_{YjB} = \delta_j^i \delta_B^A \tag{8.181}
$$

The first of these is equivalent to $f_X^{iT} f_Y^{j*} + f_Y^{iT} f_X^{j*} = h_{XY} \delta_j^i$, which may be verified by using (8.175), the reality of $\mathfrak{f}_X$, and $\mathfrak{f}_X^{1T} \mathfrak{f}_Y^1 + \mathfrak{f}_X^{2T} \mathfrak{f}_Y^2 = \mathfrak{f}_X^T \mathfrak{f}_Y = h_{XY}$. The second identity is equivalent to $h^{XY} \mathfrak{f}_X \mathfrak{f}_Y^\dagger = \mathbf{1}_{4n_H} = \mathbf{1}_2 \otimes \mathbf{1}_{2n_H}$, which follows from (8.175), the reality of $\mathfrak{f}_X$, and the relations $h^{XY} \mathfrak{f}_X \mathfrak{f}_Y^T = \mathbf{1}_{4n_H}$ and $M^{-1} M^{-1\dagger} = \mathbf{1}_{4n_H}$.

Remark: In the literature, one often also finds the statement that the vielbein f_X^{iA} satisfies the additional relation

$$
f_X^{iA} f_{YiB} + f_Y^{iA} f_{XiB} = \frac{1}{n_H} h_{XY} \delta_B^A. \qquad \text{(False!)} \tag{8.182}
$$

This equation, however, is *false*, except for the special case $n_H = 1$. This can be verified in many ways, e.g., contracting (8.182) with $f_{kC}^X f^{YlD}$, which, using (8.181), results in

$$\delta_C^A \delta_B^D + E^{AD} E_{BC} = \frac{1}{n_H} \delta_C^D \delta_B^A.$$ (8.183)

Choosing $A = C \neq D = B$ such that $E^{AD} = 0$ (e.g., $C = 1, D = B = 4$ for $E = \mathbb{1}_{n_H} \otimes e$), one finds a contradiction. Alternatively, one can contract (8.182) with f_Z^{kB} and use (8.180), (8.182), and again (8.180) to arrive at

$$\left(1 - \frac{1}{n_H}\right) h_{YZ} f_X^{kA} + \left(1 - \frac{1}{n_H}\right) h_{XZ} f_Y^{kA} + \left(1 - \frac{1}{n_H}\right) h_{XY} f_Z^{kA} = 0$$ (8.184)

At the origin of Riemannian normal coordinates with $h_{XY} = \delta_{XY}$, one can then take $X = Y \neq Z$ to infer that, for $n_H > 1$, $f_Z^{kA} = 0$ for all $Z \neq X = Y$. Contracting this with $f_{Z'kA}$ and setting $Z = Z'$ then would give the contradiction $h_{ZZ} = \delta_{ZZ} = 0$. Yet another way to show that (8.182) cannot hold for $n_H > 1$ is to go to the origin of Riemannian normal coordinates with $h_{XY} = \delta_{XY}$ and $f_X^\Gamma = \delta_X^\Gamma$ and to use directly (8.175) for, e.g., $X = Y = 1$. This would lead (using $E = \mathbb{1}_{n_H} \otimes e$) to $\delta_1^A \delta_1^B - \delta^{AC} E_{C1} E_{1D} \delta^{DB} = \delta_1^A \delta_1^B + \delta_2^A \delta_2^B = (1/n_H)\delta_B^A$, which is not a valid identity for $n_H > 1$.

8.B.2 Holonomy and Curvature

Up to now, we have shown that the $4n_H$-dimensional scalar manifold, $\mathcal{M}_{\text{hyper}}$, of n_H hypermultiplets in $\mathcal{N} = 2$ supersymmetry admits a vielbein, $f_X^{iA}(q)$, that is adapted to the $Sp(n_H) \times SU(2)$ structure of $\mathcal{M}_{\text{Hyper}}$ and that satisfies (8.176), (8.178), (8.180), and (8.181). We also saw that the reality condition (8.178) motivates the convention that complex conjugation raises or lowers the $SU(2)$ index i and the $Sp(n_H)$-index A. According to our discussion around (8.102) and (8.103), the left-handed chiral fermions of hypermultiplets and the left-handed supersymmetry parameter transform in the fundamental representations of, respectively, $Sp(n_H)$ and $SU(2)$, so that we likewise use the analogous convention that charge conjugation lowers the corresponding indices, i.e., we set

$$\zeta^A := \chi_L^A, \qquad \zeta_A := \chi_R^{\overline{A}}$$ (8.185)

$$\epsilon^i := \epsilon_L^{(i)}, \qquad \epsilon_i := \epsilon_R^{(i)},$$ (8.186)

where, in contrast to the vielbein convention (8.178), the lower indices should not be thought of as arising from lowering the indices with e_{ij} or E_{AB}, as that cannot change the chirality.

The proper generalization of the linearized transformation laws (8.102)–(8.105) can then be written in the form

$$\delta \zeta^A = \frac{i}{2} f_X^{iA}(q) \not{\partial} q^X \epsilon_i \tag{8.187}$$

$$\delta \zeta_A = -\frac{i}{2} f_{XiA}(q) \not{\partial} q^X \epsilon^i \tag{8.188}$$

$$\delta q^X = -i f_{iA}^X(q) \bar{\epsilon}^i \zeta^A + i f^{XiA}(q) \bar{\epsilon}_i \zeta_A. \tag{8.189}$$

In fact, by setting

$$(q^X) = \begin{pmatrix} q^{\tilde{X}} \\ q^{X'} \end{pmatrix} = \begin{pmatrix} \sqrt{2} \delta_m^{\tilde{X}} \mathrm{Re}\,\phi^m \\ \sqrt{2} \delta_m^{X'} \mathrm{Im}\,\phi^m \end{pmatrix} \tag{8.190}$$

$$\Longleftrightarrow \phi^m = \frac{1}{\sqrt{2}} (\delta_{\tilde{X}}^m q^{\tilde{X}} + i \delta_{X'}^m q^{X'}), \tag{8.191}$$

where $\tilde{X}, \tilde{Y}, \ldots = 1, \ldots, 2n_H$, and $X', Y', \ldots = 1, \ldots, 2n_H$, and by going to the origin of Riemann normal coordinates with the choice $f_X^\Gamma = \delta_X^\Gamma$, i.e., $f_{\tilde{X}}^{1A} = \delta_{\tilde{X}}^A$, $f_{X'}^{2A} = \delta_{X'}^A$, and $f_{\tilde{X}}^{2A} = f_{X'}^{1A} = 0$, the relation (8.175) brings (8.187)–(8.189) to the form (8.102)–(8.105).

Using (8.189), the commutator of two supersymmetry transformations of q^X is, to lowest order in fermion fields,

$$[\delta_\eta, \delta_\epsilon] q^X = \frac{1}{2} (f_{iA}^X f_Y^{jA} + f^{XjA} f_{YiA})(\bar{\epsilon}^i \gamma^\mu \eta_j) \partial_\mu q^Y + c.c., \tag{8.192}$$

which is equal to the desired result $\frac{1}{2}(\bar{\epsilon}^i \gamma^\mu \eta_i) \partial_\mu q^X + c.c.$ precisely when (8.180) holds. For the fermions ζ^A, on the other hand, one uses the Fierz identities

$$\epsilon_i \bar{\eta}^j = -\frac{1}{2} \bar{\eta}^j \gamma^\nu \epsilon_i \gamma_\nu P_L \tag{8.193}$$

$$\epsilon_i \bar{\eta}_j = -\frac{1}{2} \bar{\eta}_j \epsilon_i P_R + \frac{1}{4} \bar{\eta}_j \gamma_{\nu\rho} \epsilon_i \gamma^{\nu\rho} P_R \tag{8.194}$$

to obtain for the terms involving $\partial_\mu \zeta^A$,

$$[\delta_\eta, \delta_\epsilon] \zeta^A = -\frac{1}{4} f_X^{iA} f_{jB}^X (\bar{\eta}^j \gamma^\nu \epsilon_i) \gamma^\mu \gamma_\nu \partial_\mu \zeta^B$$

$$+ \frac{1}{4} f_X^{iA} f^{XjB} \left(\bar{\eta}_j \epsilon_i \gamma^\mu \partial_\mu \zeta_B - \frac{1}{2} \bar{\eta}_j \gamma_{\nu\rho} \epsilon_i \gamma^\mu \gamma^{\nu\rho} \partial_\mu \zeta_B \right) - (\epsilon \leftrightarrow \eta)$$

Using the linearized field equation $\not{\partial}\zeta^A = 0$ and $\gamma^\mu\gamma_\nu = -\gamma_\nu\gamma^\mu + 2\delta^\mu_\nu$, this becomes

$$[\delta_\eta, \delta_\epsilon]\zeta^A = -\frac{1}{2}f^{iA}_X f^X_{jB}(\overline{\eta}^j\gamma^\mu\epsilon_i - \overline{\epsilon}^j\gamma^\mu\eta_i)\partial_\mu\zeta^B$$

$$-\frac{1}{8}f^{iA}_X f^{XjB}\left(\overline{\eta}_j\gamma_{\nu\rho}\epsilon_i - \overline{\epsilon}_j\gamma_{\nu\rho}\eta_i\right)\gamma^\mu\gamma^{\nu\rho}\partial_\mu\zeta_B.$$

The spinor bilinear in the second line is symmetric under exchange of i and j due to (1.44) and hence vanishes upon contraction with the vielbein terms due to (8.181), which also yields the desired result $\frac{1}{2}\overline{\epsilon}^i\gamma^\mu\eta_i\partial_\mu\zeta^A + c.c.$ for the first line. This shows that (8.180), (8.181) essentially ensure the closure of the supersymmetry algebra and that an equation of the (incorrect) form (8.182) is indeed not required.

We now come to the final part and discuss the invariance of the action. This will tell us that the connection compatible with the $Sp(n_H) \times SU(2)$ structure of $\mathcal{M}_{\text{Hyper}}$ is actually the torsion-free Levi–Civita connection and that the $SU(2)$ curvature must be non-trivial in supergravity.

Starting point is the kinetic term of the hypermultiplet fields,

$$e^{-1}\mathcal{L}_{\text{hyper,kin}} = e^{-1}(\mathcal{L}_{\text{scalar}} + \mathcal{L}_{\text{fermion}}) = -\frac{1}{2}h_{XY}(q)\partial_\mu q^X\partial^\mu q^Y - 2\overline{\zeta}^A\not{\mathcal{D}}\zeta_A,$$

$$(8.195)$$

where $\mathcal{D}_\mu$ is the spacetime and $Sp(n_H)$ covariant derivative with respect to an $Sp(n_H)$ connection compatible with the $Sp(n_H) \times SU(2)$ structure on $\mathcal{M}_{\text{hyper}}$. In the following, we will in general denote by $\mathcal{D}_\mu$ the covariant derivative with respect to spacetime coordinate transformations, local Lorentz transformations, local composite $Sp(n_H) \times SU(2)$ transformations, and general field reparameterizations $q^X \to \widetilde{q}^X(q)$.

We first consider the supersymmetry variation of the scalar kinetic term with respect to the scalar fields,

$$\delta\left[e^{-1}\mathcal{L}_{\text{scalar}}\right] = -\frac{1}{2}h_{XY,Z}\delta q^Z\partial_\mu q^X\partial^\mu q^Y - \frac{1}{2}h_{XY}\partial_\mu(\delta q^X)\partial^\mu q^Y$$

$$-\frac{1}{2}h_{XY}\partial_\mu q^X\partial^\mu(\delta q^Y)$$

$$= \frac{1}{2}\underbrace{\left(-h_{XY,Z} + h_{ZY,X} + h_{XZ,Y}\right)}_{\Gamma^W_{XY}h_{WZ}}\delta q^Z\partial_\mu q^X\partial^\mu q^Y$$

$$+h_{XY}\delta q^X\nabla_\mu\partial^\mu q^Y$$

$$= h_{XY}\delta q^X\mathcal{D}_\mu\partial^\mu q^Y,$$

$$(8.196)$$

where we used partial integration between the first and second line and $\mathcal{D}_\mu \partial^\mu q^Y = \nabla_\mu \partial^\mu q^Y + \Gamma^Y_{XZ} \partial_\mu q^X \partial^\mu q^Z = \partial_\mu q^Y + \Gamma^\mu_{\mu\nu} \partial^\nu q^Y + \Gamma^Y_{XZ} \partial_\mu q^X \partial^\mu q^Z$.

Now consider the variation of the fermionic term due to the variation of the fermions,

$$\delta[e^{-1}\mathscr{L}_{\text{fermion}}] = -2\overline{\zeta}^A \mathcal{D}(\delta\zeta_A) - 2\overline{\delta\zeta}^A \mathcal{D}\zeta_A = -2\overline{\zeta}^A \mathcal{D}(\delta\zeta_A) - 2\overline{\zeta}_A \mathcal{D}(\delta\zeta^A)$$

$$= -2\overline{\zeta}^A \mathcal{D}\left(-\frac{i}{2} f_{XiA} \slashed{\partial} q^X \epsilon^i\right) + c.c$$

$$= \left[i\overline{\zeta}^A \gamma^\mu (\mathcal{D}_\mu f_{XiA})(\slashed{\partial} q^X)\epsilon^i + c.c.\right] + \left[i\overline{\zeta}^A f_{XiA}(\mathcal{D}\slashed{\partial} q^X)\epsilon^i + c.c.\right]$$

$$+ \left[i\overline{\zeta}^A \gamma^\mu f_{XiA}(\slashed{\partial} q^X)\mathcal{D}_\mu \epsilon^i + c.c.\right], \tag{8.197}$$

where we used partial integration and (1.44) in the first line. Using $\gamma^\mu \gamma^\nu = \gamma^{\mu\nu} + g^{\mu\nu}$ and that $\mathcal{D}_{[\mu}\partial_{\nu]}q^X = 0$ due to the symmetry of the Christoffel symbols $\Gamma^\rho_{\mu\nu}$ and Γ^Z_{XY}, the second term becomes

$$if_{XiA}\overline{\zeta}^A \epsilon^i \mathcal{D}_\mu \partial^\mu q^X + c.c. = -h_{XZ}\delta q^Z \mathcal{D}_\mu \partial^\mu q^X = -\delta[e^{-1}\mathscr{L}_{\text{scalar}}], \tag{8.198}$$

and hence

$$\delta[e^{-1}\mathscr{L}_{\text{hyper,kin}}] = \left[i\overline{\zeta}^A \gamma^\mu (\mathcal{D}_\mu f_{XiA})(\slashed{\partial} q^X)\epsilon^i + c.c\right] + [\mathscr{J}^\mu_i \mathcal{D}_\mu \epsilon^i + c.c.] \tag{8.199}$$

with the supercurrent

$$\mathscr{J}^\mu_i := if_{XiA}\overline{\zeta}^A \gamma^\mu \slashed{\partial} q^X. \tag{8.200}$$

The vanishing of the first term in (8.199) requires

$$\mathcal{D}_X f^{iA}_Y \equiv \partial_X f^{iA}_Y - \Gamma^Z_{XY} f^{iA}_Z + f^{jA}_Y \omega_{Xj}{}^i + f^{iB}_Y \omega_{XB}{}^A = 0, \tag{8.201}$$

where $\mathcal{D}_X$ contains the $Sp(n_H)$ and $SU(2)$ connections, $\omega_{XA}{}^B(q)$ and $\omega_{Xi}{}^j(q)$, and the Christoffel connection, $\Gamma^Z_{XY}(q)$, on $\mathcal{M}_{\text{hyper}}$, as indicated. Just as for the spin connection (cf. (3.18) in Sect. 3.2), this equation is simply the statement that the $Sp(n_H) \times SU(2)$-compatible connection is equivalent to the torsion-free Levi–Civita connection, and we have $Sp(n_H) \times SU(2)$ holonomy with respect to the Levi–Civita connection, as announced earlier.

To eliminate the second term in (8.199), we add the usual Noether term to the Lagrangian,

$$\mathscr{L}_{\text{Noether}} = -\mathscr{J}^\mu_i \psi^i_\mu + c.c. \tag{8.202}$$

for which the gravitino variation $\delta\psi_\mu^i = \mathcal{D}_\mu\epsilon^i + \ldots$ just leads to the negative of the last term in (8.199). Just as in our discussion for $\mathcal{N} = 1$ supergravity, the variation of the matter fields in $\mathcal{J}_i^\mu$ itself leads to terms involving the energy momentum tensor and is cancelled by variations of the metric but also to a term involving an antisymmetrized product of three gamma matrices that cannot be so absorbed. It is instead cancelled by the non-trivial composite connection terms (here the composite $SU(2)$ connection) in the gravitino variation of the kinetic term of the gravitini. To see how this works in detail, we use again $\gamma^\mu\gamma^\nu = \gamma^{\mu\nu} + g^{\mu\nu}$ and (1.44) to write

$$e^{-1}\mathscr{L}_{\text{Noether}} = -if_{XiA}(\partial_\nu q^X)\overline{\psi}_\mu^i(-\gamma^{\mu\nu} + g^{\mu\nu})\zeta^A + c.c. \qquad (8.203)$$

The variation of ζ^A introduces one additional gamma matrix, so that a term involving $\gamma^{\mu\nu\rho}$ can only arise from the $\gamma^{\mu\nu}$-term in (8.203). We thus have

$$\delta_\zeta[e^{-1}\mathscr{L}_{\text{Noether}}]|_{\gamma^{\mu\nu\rho}} = if_{XiA}(\partial_\nu q^X)\overline{\psi}_\mu^i\gamma^{\mu\nu}\left(\frac{i}{2}f_Y^{jA}\gamma^\rho\partial_\rho q^Y\epsilon_j\right)|_{\gamma^{\mu\nu\rho}} + c.c.$$

$$= -\frac{1}{2}f_{XiA}f_Y^{jA}(\partial_\nu q^X)(\partial_\rho q^Y)\overline{\psi}_\mu^i\gamma^{\mu\nu\rho}\epsilon_j + c.c. \qquad (8.204)$$

This is cancelled by the gravitino variation $\delta\psi_\mu^i = \mathcal{D}_\mu\epsilon^i + \ldots$ in the kinetic term $\mathscr{L}_{\text{gravitino}} = -\overline{\psi}_\mu^i\gamma^{\mu\nu\rho}\mathcal{D}_\nu\psi_{\rho i}$,

$$\delta\mathscr{L}_{\text{gravitino}} = -\overline{\psi}_\mu^i\gamma^{\mu\nu\rho}\mathcal{D}_\nu\mathcal{D}_\rho\epsilon_i + c.c. \qquad (8.205)$$

$$= -\frac{1}{2}R_{XYi}{}^j(\partial_\nu q^X)(\partial_\rho q^Y)\overline{\psi}_\mu^i\gamma^{\mu\nu\rho}\epsilon_j, \qquad (8.206)$$

where $R_{XYi}{}^j$ is the composite $SU(2)$ curvature. Comparing with (8.204), we conclude that

$$R_{XYi}{}^j = M_P^{-2}f_{[XiA}f_{Y]}^{jA}, \qquad (8.207)$$

where we have reinstalled the Planck mass, which originates from $\delta\psi_\mu^i \propto M_P$ and $\mathscr{L}_{\text{Noether}} \propto M_P^{-1}$. Using the explicit relation (8.175) to the standard $SO(4n_H)$ vielbein f_X^Γ, one can verify that the right-hand side of the above equation (8.207) is not identically zero and hence that the $SU(2)$ curvature in $\mathcal{N} = 2$ supergravity must be non-trivial. Obviously, the above reasoning becomes empty in the case of rigid supersymmetry, and the $SU(2)$ curvature becomes flat, as also follows from (8.207) by formally sending $M_P \to \infty$.

Exercises

8.1. Construct the Kähler potential, the metric, and the Killing vectors and compute the action of the isometries on the symplectic sections for the STU model, i.e., for the prepotential

$$F(X) = \frac{X^1 X^2 X^3}{X^0}.$$

References

1. M.K. Gaillard, B. Zumino, Duality rotations for interacting fields. Nucl. Phys. **B193**, 221 (1981). Dedicated to Andrei D. Sakharov on occasion of his 60th birthday
2. B. Craps, F. Roose, W. Troost, A. Van Proeyen, What is special Kahler geometry? Nucl. Phys. B **503**, 565–613 (1997). [arXiv:hep-th/9703082 [hep-th]]
3. S. Weinberg, *The Quantum Theory of Fields. Vol. 3: Supersymmetry* (Cambridge University Press, Cambridge, 2000), 419 p.
4. P. Claus, K. Van Hoof, A. Van Proeyen, A symplectic covariant formulation of special Kahler geometry in superconformal calculus. Class. Quant. Grav. **16**, 2625–2649 (1999). [arXiv:hep-th/9904066 [hep-th]]
5. B. de Wit, P.G. Lauwers, A. Van Proeyen, Lagrangians of N=2 Supergravity - Matter Systems. Nucl. Phys. B **255**, 569–608 (1985)
6. A. Strominger, Special geometry. Commun. Math. Phys. **133**, 163 (1990)
7. A. Ceresole, R. D'Auria, S. Ferrara, A. Van Proeyen, Duality transformations in supersymmetric Yang-Mills theories coupled to supergravity. Nucl. Phys. B **444**, 92–124 (1995). [arXiv:hep-th/9502072 [hep-th]]
8. A. Ceresole, R. D'Auria, S. Ferrara, The symplectic structure of N=2 supergravity and its central extension. Nucl. Phys. Proc. Suppl. **46** , 67–74 (1996). [hep-th/9509160]
9. J. Bagger, E. Witten, Matter couplings in N=2 Supergravity. Nucl. Phys. B **222**, 1–10 (1983)
10. M. Berger, Sur les groupes d'holonomie homogènes des variétés a connexion affines et des variétés riemanniennes. Bull. Soc. Math. France **83**, 279–330 (1953)

Gauged Supergravity

9

In the previous chapter, we discussed the consequences of extended supersymmetry for the geometries of the scalar manifolds in supergravity. Some parts of these geometrical structures are caused by the larger R-symmetry groups, whereas others can be traced back to the electric–magnetic duality of the vector field sector. In the present chapter, we go one step further and turn on gauge interactions in the form of minimal couplings between the vector fields and the other matter fields in extended supergravity. Due to the extended supersymmetry, the possible gauge interactions are more restricted, and they are more tightly connected to the interactions of the other fields, in particular they completely determine the scalar potential for $\mathcal{N} \geq 2$. For this reason, one gives extended supergravity theories with gauge interactions a special name and calls them *gauged supergravity theories*. Another important aspect of the gauge interactions is that their presence breaks the electric–magnetic duality of the ungauged theories. As this duality underlies many geometrical structures, it is useful to discuss gauged supergravity also in a way that formally maintains the symplectic covariance. This formalism is called the *embedding tensor formalism* and will be used in this book. To arrive there, we first describe, in Sect. 9.1, some general features of gauged supergravity and how their connection with the scalar potential arises. In Sects. 9.2 and 9.3, we discuss some important subgroups of the electric–magnetic duality group as well as the relevance of the symplectic duality frames for gauge interactions. In Sect. 9.4, we introduce some terminology on coset manifolds, whereas Sect. 9.5 puts all these things together and presents the embedding tensor formalism for gauged supergravity, with the $\mathcal{N} = 8$ theory as an explicit illustration. In Sect. 9.6, we finally discuss how inequivalent gaugings can be classified and show, as an application, how this formalism can be used to uncover some gaugings in the $\mathcal{N} = 8$ theory that were unknown until a few years ago.

© Springer-Verlag GmbH Germany, part of Springer Nature 2021
G. Dall'Agata, M. Zagermann, *Supergravity*, Lecture Notes in Physics 991,
https://doi.org/10.1007/978-3-662-63980-1_9

9.1   Supergravities and Scalar Potentials

Generic matter multiplets in supergravity theories contain scalar fields, whose
expectation values determine the masses and the couplings of the other fields and
eventually the physics we would like to describe by these models. Such expectation
values are arbitrary as long as there is no scalar potential. On the other hand, the
existence of a scalar potential constrains their values and hence provides crucial
information on the resulting physics.

In $\mathcal{N} = 1$ supergravity, the scalar potential receives contributions from both
the superpotential, W, and the D-terms. There is a large amount of freedom
involved in the choice of a superpotential. For one thing, any non-vanishing
superpotential can be transformed to any other non-vanishing superpotential by a
suitable Kähler transformation such that the Kähler invariant combination of Kähler
and superpotential,

$$\mathcal{G} = K + M_P^2 \, \log \frac{|W|^2}{M_P^6}, \tag{9.1}$$

stays the same. But even if one fixes a particular Kähler potential, one usually has
a large amount of freedom in the choice of W, as long as one just respects any
gauge symmetries that might be present. It should also be emphasized that in $\mathcal{N} =$
1 supergravity a superpotential and the resulting F-term potential can always be
chosen at will if there is no gauge symmetry whatsoever.

The other contribution to the $\mathcal{N} = 1$ scalar potential, the D-term potential, by
contrast, is directly related to a *gauging*, i.e., to the process of turning some of the
global internal symmetries into local symmetries. These global internal symmetries
are essentially coincident (with few exceptions) with the isometries of the scalar
manifold, described by its Killing vectors, ξ_I^i,

$$\delta\varphi^i = \alpha^I \, \xi_I^i(\varphi), \quad \text{where} \quad \nabla_{(i}\xi_{I\, j)} = 0, \tag{9.2}$$

with infinitesimal symmetry parameters, α^I. An important difference to the scalar
potential that arises from a superpotential is that the D-term potential is *fixed* as soon
as the action of the gauge symmetries on the fields is fixed.[1]

In this chapter, we will mainly deal with this gauging procedure. The reason
is that extended supergravities have no scalar potential without gauging! In other
words, in extended supergravity there is no analogue of an F-term potential due
to a superpotential that can exist independently of any gauging. This means that if
we want to stabilize the scalar fields that inevitably appear in almost all extended
supergravity theories and that we know cannot be massless because of the limits
from fifth force experiments, we need to gauge the theory. This is also the case

---

[1] As described in Chap. 6, in $\mathcal{N} = 1$ supergravity this is also true for any Fayet–Iliopoulos
constants, as they entail gaugings of the $U(1)$ R-symmetry group that act on the fermions.

if we want to break supersymmetry in the vacuum, because if there is no scalar potential, the only maximally symmetric vacuum allowed is a fully supersymmetric Minkowski spacetime.

One may use this as an argument to focus exclusively on $\mathcal{N} = 1$ theories, which anyway can lead more easily to realistic phenomenology due to their exclusive compatibility with chiral gauge interactions. In many string theory compactifications, however, the closed string sector can be described by effective field theories with extended supersymmetry, which is reduced to $\mathcal{N} = 1$ only due to their coupling to open strings living on branes used to describe supersymmetric extensions of the Standard Model as well as orientifold projections.

Moreover, in many branches of string theory, it is only due to extended supersymmetry that certain computations can be carried out with sufficient technical control. In the context of the gauge/gravity correspondence, this is in particular true for the gauge theories and their renormalization group flows at the boundary, whose holographic description then also necessitates extended supersymmetry in the gravity dual in the bulk. These extended supergravity theories must have a nontrivial scalar potential to allow for AdS vacua or domain wall solutions, and hence they are also gauged supergravity theories with extended supersymmetry.

In this chapter, we will give a brief introduction to gauged supergravity, explaining how the gauging procedure works and what kind of generic features and constraints they involve.

Before entering into the details of such constructions, let us give a brief description of what the necessary steps are and what we should expect. As discussed in Sect. 5.3.5, isometries of the scalar manifold of a supergravity theory are generically global symmetries of the Lagrangian (or at least of the equations of motion). In fact, transformations of the form (9.2) obviously leave the kinetic couplings of the scalar fields,

$$g_{ij}(\varphi)\, \partial_\mu \varphi^i\, \partial^\mu \varphi^j, \tag{9.3}$$

invariant (for the sake of simplicity, we do not discuss now the other matter couplings).

As explained in Sect. 5.3.5, gauging such isometries means allowing for a non-constant symmetry parameter, $\alpha^I(x)$, which then implies that the variation of (9.3) gives uncancelled terms like

$$\partial_\mu \alpha^I(x) \partial^\mu \varphi^j\, \xi_{I\,j}. \tag{9.4}$$

As detailed in Sect. 5.3.5, these terms can be cancelled by introducing minimal couplings to the vector fields of the theory,

$$\partial_\mu \varphi^i \longrightarrow \widehat{\partial}_\mu \varphi^i \equiv \partial_\mu \varphi^i - A_\mu^I \xi_I^i, \tag{9.5}$$

and by requiring that the Maxwell-type transformation, $\delta A_\mu^I = \partial_\mu \Lambda^I$, of the ungauged theory is changed to

$$\delta A_\mu^I = \partial_\mu \alpha^I + A_\mu^J \alpha^K f_{JK}{}^I. \tag{9.6}$$

This procedure is in general not possible for all scalar field isometries. Instead, the gaugeable isometries usually only form subgroups, $G \subset \text{Iso}(\mathcal{M}_{\text{scalar}})$, that have to obey certain restrictions. The most obvious restrictions for G are that its dimension be at most equal to the number, n_V, of vector fields, $\dim G \leq n_V$, and that the vector fields transform in the adjoint representation of the subgroup G in the sense of a global symmetry prior to the gauging.

In order to also make the kinetic terms of the vector fields gauge invariant, one has to replace the Abelian field strengths, $F_{\mu\nu}^I$, by their non-Abelian counterparts, $\mathcal{F}_{\mu\nu}^I = 2\partial_{[\mu}A_{\nu]}^I + f_{JK}{}^I A_\mu^J A_\nu^K$.

Although we are not going to do this in later parts of this chapter, let us temporarily follow a common practice and rescale the vector fields by a coupling constant, $A \to g A$, so that our modification (9.5) of the derivative of the scalar fields is seen to be a modification of order g, and hence the scalar kinetic term corresponds to a modification of order g^2. Likewise, the field strengths of the vector fields have been modified at order g relative to the Abelian part, so that also their kinetic terms are modified at order g^2 relative to the ungauged case.

These modifications obviously break supersymmetry, because the scalars and vector fields have fermionic superpartners whose kinetic terms do not yet contain such gauge covariantizations. Covariantizing also these kinetic terms of the fermions as well as the supersymmetry transformation rules, many supersymmetry variations will in fact cancel in a way that is very similar to the ungauged theory. A small number of uncancelled variations, however, remains and requires further modifications of the action and the supersymmetry transformation laws at order g or g^2. More precisely, in order to compensate the new terms, we must further modify the supersymmetry transformation rules of the fermions,

$$\delta_{\text{SUSY}} \text{ (Fermions)} = \delta_{\text{SUSY,old}} \text{ (Fermions)} + \delta_{\text{SUSY,new}} \text{ (Fermions)}, \tag{9.7}$$

where

$$\delta_{\text{SUSY,new}} \text{ (Fermions)} = \mathcal{O}(g) \tag{9.8}$$

are "fermionic shifts" involving scalar field-dependent terms and one power of g. This, in turn, also enforces the introduction of Yukawa-like terms for the same fermions,

$$\mathscr{L}_{\text{Yuk}} = \mathcal{O}(g). \tag{9.9}$$

The variation of $\mathscr{L}_{\text{Yuk}}$ due to $\delta_{\text{SUSY,new}}$ will then lead to an uncancelled variation of order g^2, which can be cancelled by adding a scalar potential of order g^2,

$$V(\varphi) = \mathscr{O}(g^2). \tag{9.10}$$

For consistent gaugings, no $\mathscr{O}(g^3)$ terms are needed. These steps match precisely the ones we described in Chap. 4, when we added a cosmological constant to minimal pure supergravity, except that there was no gauge covariantization involved.

Although the procedure just described looks straightforward, it can be technically very challenging, and its application leads to the discovery of many interesting general properties of supergravity theories. For this reason, in what follows, we are going to consider a simplified instance, namely, the case where the scalar σ-model is described by a coset manifold,

$$\mathscr{M}_{\text{scalar}} = \frac{G}{H}. \tag{9.11}$$

Actually, for theories with $\mathscr{N} \geq 3$, this is always true and also for $\mathscr{N} = 2$ theories, one is often interested in the analysis of scalar manifolds that have this form. This case, as we will see, simplifies considerably many technical aspects, leaving the substance of the gauging procedure unchanged.

Before we can enter this discussion, however, we first have to make more precise what kind of global symmetries we could actually encounter when we talk about the process of gauging.

9.2 Duality

As we are going to see in the following, the gauging procedure is intimately related to the existence and action of duality symmetries. Actually, duality relations are at the basis of our understanding of the vast majority of the most interesting physical effects in supergravity and string theory. In the supergravity context, the subject was opened by the seminal paper by Gaillard and Zumino [1], which discusses the duality group of four-dimensional theories containing vector fields. We already introduced the electric–magnetic duality in Chap. 8. This, however, does not coincide with the U-duality group, which is the one playing the most prominent role in the gauging procedure. For this reason, we now present the details that are needed to reduce the electric–magnetic duality group to the *U-duality group* and how this group is at the basis of all possible deformations of extended supergravities via the gauging procedure.

9.2.1 From Electric–Magnetic Duality to U-Duality

In Chap. 8, we saw that the presence of n_V vector fields in a four-dimensional Lagrangian comes with a generalized electric–magnetic duality leaving invariant the set of Bianchi identities and equations of motion of the vector fields under $\mathrm{Sp}(2n_V, \mathbb{R})$ rotations. On the other hand, under an infinitesimal $\mathrm{Sp}(2n_V, \mathbb{R})$ transformation, the Lagrangian does in general not remain invariant but changes according to (8.28),

$$\delta \mathscr{L} = \frac{1}{4} C_{IJ} F^I_{\mu\nu} \tilde{F}^{J\,\mu\nu} + \frac{1}{4} B^{IJ} G_{I\,\mu\nu} \tilde{G}^{\mu\nu}_J + \delta \mathscr{L}_{\text{rest}}. \tag{9.12}$$

If we want the electric–magnetic duality to leave all the equations of motion unaffected, we should understand if and how it acts on the other fields present in the Lagrangian. This, as we will see, will imply further restrictions so that the resulting set of transformations is going to be encoded in the *U-duality group* G_U,

$$\mathrm{G}_U \subset \mathrm{Sp}\,(2n_V, \mathbb{R}). \tag{9.13}$$

In order to simplify the following discussion, we will focus on scalar fields, but the proof can be extended to fermion fields as well.

If we call φ^i the set of scalar fields on which the couplings and $\mathscr{L}_{\text{rest}}$ depend, we define their equation of motion operator, E_i, as

$$E_i[\mathscr{L}] \equiv \left(\frac{\partial}{\partial \varphi^i} - \partial_\mu \frac{\partial}{\partial \partial_\mu \varphi^i} \right) \mathscr{L} = 0. \tag{9.14}$$

If the duality transformations act non-trivially on the scalar fields, they are going to transform as

$$\delta \varphi^i = k^i(\varphi), \tag{9.15}$$

where $k^i(\varphi)$ are some as yet unspecified functions. Once again, we would like that the full set of Bianchi identities and equations of motion for all fields in $\mathscr{L}$ are invariant under the action of the duality transformations. This is easily achieved if the equations of motion of the scalar fields transform covariantly with respect to the action of the duality symmetry[2]

$$\delta E_i[\mathscr{L}] = -\frac{\partial k^j}{\partial \varphi^i} E_j[\mathscr{L}]. \tag{9.16}$$

---

[2] This follows straightforwardly from (9.15), recalling that for a contravariant vector $V^{i\prime} = \frac{\partial \varphi^{i\prime}}{\partial \varphi^j} V^j \simeq \left(\delta^i_j + \epsilon \frac{\partial k^i}{\partial \varphi^j} + O(\epsilon^2) \right) V^j$ and that a covariant vector transform with the inverse Jacobian.

This requirement produces an interesting result, once we apply the equation of motion operator to the variation (8.28). After some algebraic simplifications, we obtain that

$$E_i[\delta\mathscr{L}_{\text{rest}}] = 0, \tag{9.17}$$

and the only way to consistently satisfy this identity is that

$$\delta\mathscr{L}_{\text{rest}} = 0. \tag{9.18}$$

We therefore conclude that the subgroup of electric–magnetic duality transformations that leave the full set of equations of motion and Bianchi identities invariant is the set of transformations that also leave invariant $\mathscr{L}_{\text{rest}}$. If, as usual, we interpret the scalar fields as coordinates on a manifold, $k^i(\varphi)$ are going to be identified with the Killing vectors, $\xi^i(\varphi)$, associated with the isometries of this manifold, and we therefore see that the U-duality group is going to coincide with the symmetry group under which the reparameterization of the scalar fields leaves the Lagrangian invariant.[3]

As we said, this proof can be extended to more general Lagrangians containing also fermion fields, and therefore we will always have a reduction of the electric–magnetic group $\mathrm{Sp}(2n_V, \mathbb{R})$ to a U-duality subgroup that is the subgroup of transformations leaving invariant $\mathscr{L}_{\text{rest}}$. In general, the global symmetry group leaving invariant the set of Bianchi identities and equations of motions is $\mathrm{G}_{\text{global}} = \mathrm{G}_U \times \mathrm{G}_{\text{inert}}$, where $\mathrm{G}_{\text{inert}}$ is the global symmetry of the inert matter fields under duality transformations, i.e., those fields that do not have direct couplings to the vector fields, as, e.g., the hypermultiplets in $\mathscr{N} = 2$ supergravity.

Finally, one can also see that when gravity is present, Einstein's equation contains a stress tensor that is invariant under the U-duality transformations, as a consequence of its conservation law following from the equations of motion of the other fields [1]. This is a very important fact that implies that one can map different solutions of Einstein's equation into each other by carefully acting on the original set of fields defining the solution with the U-duality group.

9.3   Gauging and Symplectic Frames

By gauging we usually mean the procedure of making local a global symmetry, $\mathrm{G}_{\mathscr{L}}$, of a given Lagrangian, $\mathscr{L}$. However, when dealing with the duality group, G_U, we see that this is a symmetry group of the Bianchi identities and equations of motion

---

[3] We should note that in some instances there may be additional global symmetries that act trivially on the scalar manifolds and that there are also examples where there is still a restricted non-trivial duality group in models without scalar fields, like pure $\mathscr{N} = 2$ supergravity. In most of these cases, the additional factors come from the R-symmetry acting non-trivially on the fermion fields of the theory.

and not necessarily of the Lagrangian. Moreover, we should also notice that for a given theory with a fixed set of multiplets, we may have different Lagrangians with different $G_{\mathscr{L}}$. Actually, we are often allowed to have entirely different groups for different Lagrangian realizations. This is clear if we recall that the group of electric–magnetic duality transformations, $Sp(2n_V, \mathbb{R})$, leaves invariant the equations of motion and Bianchi identities of the vector fields, but only a subgroup, G_U, can be reabsorbed in a field redefinition of the other fields. In particular, only electric vectors appear in the Lagrangian, and therefore $G_{\mathscr{L}}$ transformations may only be represented by block triangular matrices acting on the symplectically covariant field strengths, $\mathbb{F}$, introduced in (8.10). A generic $Sp(2n_V, \mathbb{R})$ transformation is going to modify the Lagrangian, instead. Hence for the same theory, we can consider different Lagrangians having different realizations of $G_{\mathscr{L}}$. It is therefore important to understand which realizations of $G_{\mathscr{L}}$ are available before the gauging procedure is carried out.

The set of Lagrangians that cannot be mapped to each other by local field redefinitions is identified with the double quotient space

$$GL(n_V, \mathbb{R}) \setminus Sp(2n_V, \mathbb{R}) / G_U. \tag{9.19}$$

This follows from the fact that we can always perform local field redefinitions of the n_V vector fields of the theory, which corresponds to the $GL(n_V, \mathbb{R})$ quotient, as well as use the U-duality group G_U corresponding to the isometry group of the scalar manifold $\mathscr{M}_{\text{scalar}}$ to redefine the other fields. Each different Lagrangian corresponds then to a distinct symplectic frame and is invariant under a particular electric subgroup of the G_U duality group acting locally on the physical fields. Hence we can use this choice of frame to obtain different gauge groups by introducing minimal electric couplings. The choice of the symplectic frame is therefore important to find a purely electric realization of G_{gauge}. On the other hand, one should notice once more that the resulting equations of motion and Bianchi identities are equivalent for any Lagrangian defined by (9.19) before the gauging. In fact, even if the Lagrangians differ in different symplectic frames, the full set of Bianchi identities and equations of motion can still be mapped into each other by the symplectic field redefinition. So, classically we have different Lagrangian descriptions of the same physics. As we will see explicitly later, this is broken by the gauging, and hence this choice of a symplectic frame is relevant for the gauging procedure.

Alternatively, as we will see shortly, keeping the symplectic frame fixed, we can obtain $G_{\text{gauge}} \not\subset G_{\mathscr{L}}$ (while obviously $G_{\text{gauge}} \subset G_U$) by introducing magnetic charges and couplings for the non-perturbative symmetries that would not leave invariant $\mathscr{L}$. This requires writing the Lagrangian with the gauge interactions in a duality invariant way, and this can be done by introducing the embedding tensor formalism. As we will see, this formalism is especially suited for a general analysis of the gaugings of a given supergravity theory. The reason is that once we fix a given symplectic frame the embedding tensor describes precisely how G_{gauge} is embedded in the duality group G_U and eventually in the symplectic group $Sp(2n_V, \mathbb{R})$. This allows for a general formulation of the gauging procedure that is also transparent to the differences between theories that have the same G_{gauge}, but inequivalent symplectic embeddings.

9.4   Coset Manifolds and Gauging

Before entering the discussion of the embedding tensor formalism, we recall a few basic facts about coset manifolds, which play a prominent role as σ-model geometries of supergravity theories. We illustrate this terminology with a simple example that we are going to use throughout the chapter. As mentioned above, for the sake of simplicity, we will limit our discussion and examples of the embedding tensor applications to theories where the scalar manifold is homogeneous, so that we can use the coset structure to obtain linear realizations of the symmetries involved in the gauging procedure.

A *homogeneous space* is a manifold with a metric whose isometry group, G, has a transitive action on the space, meaning that any point on the space can be reached from any other by the group action. The subgroup, H, of G that leaves a chosen point, x, of the manifold fixed, is called the *isotropy subgroup* of x. Because of the transitive action of G, any other point, $x' = gx$ (obtained by the action of $g \in$ G, $g \notin$ H), is invariant under the subgroup gHg^{-1} of G, which is isomorphic to H. This implies that any homogeneous space can be described as the *coset space* G/H, defined as the set of equivalence classes of elements of G with respect to the (right) action of H elements:

$$g \sim g', \quad \text{if} \quad g = g'h \qquad \text{for} \qquad g, g' \in \text{G}, \; h \in \text{H}. \tag{9.20}$$

The dimension of such a coset space is then $d = \dim[\text{G}] - \dim[\text{H}]$.

Given this identification, we can now describe homogeneous spaces directly by their coset structure. The Lie algebra of the group G, $\mathfrak{g}$, can be split as

$$\mathfrak{g} = \mathfrak{h} \oplus \mathfrak{k}, \tag{9.21}$$

where $\mathfrak{h}$ is the Lie algebra of H and $\mathfrak{k}$ contains the remaining generators. We can then classify coset manifolds according to the structure of their algebra

$$[t_i, t_j] = f_{ij}{}^k t_k, \tag{9.22}$$

$$[t_i, t_a] = f_{ia}{}^k t_k + f_{ia}{}^b t_b, \tag{9.23}$$

$$[t_a, t_b] = f_{ab}{}^k t_k + f_{ab}{}^c t_c, \tag{9.24}$$

where $t_A \in \mathfrak{g}$, $t_i \in \mathfrak{h}$, and $t_a \in \mathfrak{k}$. A very useful concept is that of the Cartan–Killing metric

$$\eta_{AB} = f_{AD}{}^C f_{BC}{}^D, \tag{9.25}$$

which is non-degenerate for semi-simple groups. This implies that, for semi-simple groups, we can always map the generators to a basis so that the Cartan–Killing metric on $\mathfrak{g}$ is diagonal. With respect to that basis, any coset manifold G/H is *reductive*, i.e., the decomposition above satisfies

$$f_{ia}{}^{j} = 0 \quad \Leftrightarrow \quad \text{reductive,} \tag{9.26}$$

so that $[\mathfrak{h}, \mathfrak{k}] \subseteq \mathfrak{k}$. Actually, one can prove the existence of a reductive decomposition also for the sum of a semi-simple and an Abelian Lie algebra [2]. In the following, we will therefore focus on reductive coset spaces. Another useful definition appears when the commutator of generators in $\mathfrak{k}$ closes on $\mathfrak{h}$. This is called a *symmetric coset manifold*:

$$f_{ab}{}^{i} = 0 \quad \Leftrightarrow \quad \text{symmetric,} \tag{9.27}$$

or $[\mathfrak{k}, \mathfrak{k}] \subseteq \mathfrak{h}$. This typically happens when G is simple and H is maximal, i.e., when there is no other proper subgroup of G that contains H strictly. Moreover, this is always the case when we have a non-compact reductive G/H and maximally compact H. In fact, a reductive non-compact coset space can be obtained from its compact counterpart by multiplying the generators in $\mathfrak{k}$ by the imaginary unit i. This, however, forces $[\mathfrak{k}, \mathfrak{k}] \subseteq \mathfrak{h}$, because if, in the compact case, the commutator between elements of $\mathfrak{k}$ also involved terms in $\mathfrak{k}$, the closure of the algebra could in general no longer work with real structure constants when one multiplies such generators by i.

Another useful concept, when dealing with non-compact coset manifolds, is given by the solvable or *Iwasawa decomposition* [2]. This decomposition assures that for any Euclidean non-compact maximal homogeneous manifold G/H there is a *solvable subalgebra* of $\mathfrak{g}$, $\mathfrak{solv} \subset \mathfrak{g}$, acting transitively on G/H, such that

$$\mathfrak{g} = \mathfrak{h} + \mathfrak{solv}, \qquad \dim \mathfrak{solv} = \dim \text{G/H}, \tag{9.28}$$

where a solvable algebra is an algebra whose derivative algebras vanish at some finite order, i.e., repeated commutators of the generators vanish at a finite order. Such a solvable algebra can be constructed as follows. First one considers the maximal Abelian subspace of $\mathfrak{k}$, $\mathscr{C}_{\mathfrak{k}}$, which can be proven to be given by all the non-compact elements of the Cartan subalgebra of $\mathfrak{g}$, named $\mathscr{C}$. Hence, $\mathscr{C}_{\mathfrak{k}} = \mathscr{C} \cap \mathfrak{k}$. Then one fixes a set of positive roots $\Delta_{+} \subset \mathfrak{g}$ with respect to $\mathscr{C}_{\mathfrak{k}}$, for instance, by taking a hyperplane passing through the origin of the algebra root diagram that does not contain any generator in addition to those appearing at the origin and then considering all generators on one side of such hyperplane (Fig. 9.1). A solvable algebra is then given by

$$\mathfrak{solv} = \mathscr{C}_{\mathfrak{k}} + \Delta_{+}. \tag{9.29}$$

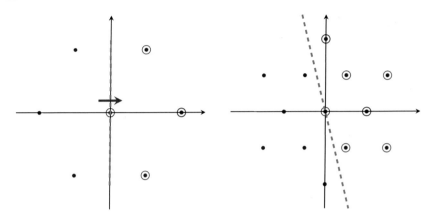

Fig. 9.1 Examples of the Iwasawa decomposition for the isometries of the SU(2,1)/[SU(2) × U(1)] manifold (root diagram on the left) and $G_{2(2)}$/[SO(4)] manifold (root diagram on the right). For $\mathfrak{su}(2,1)$ only one generator of the Cartan subalgebra is non-compact, the one whose weights are plotted following the red arrow. The set of positive roots (in red) plus the non-compact Cartan give the four generators of the corresponding manifold. For the $\mathfrak{g}_{2(2)}$ algebra, we have that $\mathscr{C} = \mathscr{C}_{\mathfrak{k}}$; hence the set of positive roots plus the two Cartan generators give a total of eight generators

Let us now give an example, by considering the coset manifold

$$\mathscr{M} = \mathrm{SU}(1,1)/\mathrm{U}(1). \tag{9.30}$$

The generators of the corresponding algebras can be taken as

$$t_A = \left\{\frac{1}{2}\sigma_1, \frac{i}{2}\sigma_2, \frac{1}{2}\sigma_3\right\} \in \mathfrak{su}(1,1), \quad t_h = t_2 = \frac{i}{2}\sigma_2 \in \mathfrak{u}(1), \tag{9.31}$$

satisfying

$$[t_1, t_2] = -t_3, \; [t_3, t_1] = t_2, \; [t_2, t_3] = -t_1, \tag{9.32}$$

so that the Cartan–Killing metric is

$$\eta = \mathrm{diag}\{1, -1, 1\}. \tag{9.33}$$

The root diagram can be described by placing the non-compact Cartan generator $t_3 \in \mathscr{C}_{\mathfrak{k}}$ at the origin and taking the positive and negative roots as $t_\pm = \frac{1}{2}(t_1 \pm t_2)$, which then have weight one,

$$[t_3, t_\pm] = \pm t_\pm. \tag{9.34}$$

Fig. 9.2 Root diagram and decomposition for $\mathfrak{su}(1, 1)$

A generic coset element can be described by the exponential of the generators in the quotient subspace, $\mathfrak{k} = \text{span}\{t_1, t_3\}$,

$$L(\phi, \psi) = \exp(\psi\, t_1) \exp(\phi\, t_3) = \begin{pmatrix} e^{\frac{\phi}{2}} \cosh \frac{\psi}{2} & e^{-\frac{\phi}{2}} \sinh \frac{\psi}{2} \\ e^{\frac{\phi}{2}} \sinh \frac{\psi}{2} & e^{-\frac{\phi}{2}} \cosh \frac{\psi}{2} \end{pmatrix}, \tag{9.35}$$

or, using the Iwasawa decomposition, by the exponential of the generators of the solvable algebra, $\mathfrak{solv} = \text{span}\{t_3, t_+\}$,

$$L(\phi, C) = \exp(C\, t_+) \exp(\phi\, t_3) = \begin{pmatrix} e^{\frac{\phi}{2}} & e^{-\frac{\phi}{2}} C \\ 0 & e^{-\frac{\phi}{2}} \end{pmatrix}. \tag{9.36}$$

In this case, as expected, the result is an exponential for the coordinates corresponding to the generators in the Cartan subalgebra and a polynomial in the coordinates associated with the positive roots generators (Fig. 9.2).

As $L(x)$ is just a coset representative, the action with a constant group element, g, from the left in general gives the coset representative at the transformed coordinate, x', only modulo a local H transformation from the right,

$$gL(x) = L(x')h(x). \tag{9.37}$$

For instance, if we choose the finite $O(1,1)$ transformation

$$g = \exp(\log 2\, t_1) = \frac{1}{2\sqrt{2}} \begin{pmatrix} 3 & 1 \\ 1 & 3 \end{pmatrix}, \tag{9.38}$$

then

$$gL(\phi, C) = L(\tilde{\phi}, \tilde{C})h(\phi, C), \tag{9.39}$$

where

$$h(\phi, C) = \frac{1}{\sqrt{(C+3)^2 + e^{2\phi}}} \begin{pmatrix} C+3 & -e^{\phi} \\ e^{\phi} & C+3 \end{pmatrix} \tag{9.40}$$

is an element of U(1) and the new representative coordinates are specified by

$$\tilde{\phi} = 2 \log \left[\frac{2\sqrt{2}\,e^{\phi/2}}{\sqrt{(C+3)^2 + e^{2\phi}}} \right], \tag{9.41}$$

$$\tilde{C} = 3 - \frac{8(3+C)}{(C+3)^2 + e^{2\phi}}. \tag{9.42}$$

9.4.1 Vielbein, Metric, and Isometries of G/H

From the coset manifold representative, L, we can construct the left-invariant one-form

$$\Omega(x) = L^{-1}(x)dL(x). \tag{9.43}$$

This is called the *Maurer–Cartan form*, and it is Lie algebra valued and may be expanded in terms of the $\mathfrak{g}$ generators

$$\Omega(x) = e^a(x)\,t_a + \omega^i(x)\,t_i, \tag{9.44}$$

where e^a is a covariant vielbein on G/H, and ω is called the H-connection. Recalling (9.37), under left multiplication by a constant element of G, the Maurer–Cartan form is not invariant, but transforms as

$$\Omega(x') = h\Omega h^{-1} + h dh^{-1}. \tag{9.45}$$

This is a very interesting property, because, once projected onto the subalgebras $\mathfrak{k}$ and $\mathfrak{h}$, this means that the vielbein rotates under the action of the H subgroup, and ω^i transforms as a connection:

$$e^a(x') = e^b(x) D_b{}^a(h^{-1}), \tag{9.46}$$

$$\omega^i(x') = \omega^j D_j{}^i(h^{-1}) + (h dh^{-1})^i, \tag{9.47}$$

where $D_A{}^B(g)$ is the adjoint representation of G acting on the algebra generators.

Once we construct a (left-invariant) metric from the vielbein obtained in (9.44), we can ask ourselves how to derive the isometries, which should naturally respect the free action of the group G on the manifold. Let us consider once again (9.37) at the infinitesimal level. Expanding the constant g transformation as $g = \mathbb{1} + \alpha^A t_A$ and the local transformation $h(x) = \mathbb{1} + \alpha^A w_A^i(x)t_i$ and interpreting the result as the action of the corresponding Killing vector

$$L(x') = L(x) + \alpha^A \xi_A{}^\beta(x)\partial_\beta L(x), \tag{9.48}$$

we obtain

$$t_A L(x) = \xi_A(x)[L(x)] - L(x) t_i w_A^i(x), \tag{9.49}$$

where $\xi_A(x)[L(x)] \equiv \xi_A{}^\beta(x)\partial_\beta L(x)$, and $w_A^i(x)$ is often called the H-compensator. It can be readily verified that

$$[\xi_A, \xi_B] = -f_{AB}{}^C \xi_C, \tag{9.50}$$

and multiplying (9.49) by L^{-1} from the left and projecting on the generators, one gets an explicit expression for the Killing vectors and for the H-compensators

$$\xi_A{}^\alpha(x) = D_A{}^a(L(x))e_a^\alpha(x), \tag{9.51}$$

$$w_A^i(x) = \omega_\alpha{}^i(x)\xi_A^\alpha(x) - D_A{}^i(L(x)), \tag{9.52}$$

In order to make all this concrete, let us explicitly compute the metric and isometries of the SU(1,1)/U(1) manifold. We start from the representative (9.36). The Maurer–Cartan form is then

$$\Omega = \begin{pmatrix} \frac{d\phi}{2} & e^{-\phi}dC \\ 0 & -\frac{d\phi}{2} \end{pmatrix} = d\phi\, t_3 + e^{-\phi}\, dC\, t_+ = e^1\, t_1 + e^2\, t_3 + \omega_{U(1)}\, t_2. \tag{9.53}$$

This means that we can identify the vielbein with

$$e^a = \{d\phi, e^{-\phi}dC\} \tag{9.54}$$

and the U(1) connection with

$$\omega_{U(1)} = e^{-\phi}dC, \tag{9.55}$$

whose curvature is

$$R_{U(1)} = d\omega_{U(1)} = e^{-\phi}dC \wedge d\phi. \tag{9.56}$$

This implies that the metric is proportional to $e^a e^b \delta_{ab}$ with a coefficient of proportionality fixing the curvature

$$ds^2 = \ell^2 \left(d\phi^2 + e^{-2\phi}dC^2 \right). \tag{9.57}$$

The space is indeed an Einstein space with negative Riemannian curvature:

$$R_{\alpha\beta} = -\frac{1}{\ell^2} g_{\alpha\beta}. \tag{9.58}$$

For what concerns the isometries, we can then use (9.49) to obtain

$$t_1 L = \frac{1}{2} \begin{pmatrix} 0 & e^{-\phi/2} \\ e^{\phi/2} & e^{-\phi/2}C \end{pmatrix} = \frac{1}{2}(1 + e^{2\phi} - C^2)\partial_C L - C\partial_\phi L - Lt_2\, e^{-\phi}, \quad (9.59)$$

$$t_2 L = \frac{1}{2} \begin{pmatrix} 0 & -e^{-\phi/2} \\ e^{\phi/2} & e^{-\phi/2}C \end{pmatrix} = \frac{1}{2}(1 - e^{2\phi} + C^2)\partial_C L + C\partial_\phi L + Lt_2\, e^{-\phi}, \quad (9.60)$$

$$t_3 L = \frac{1}{2} \begin{pmatrix} e^{\phi/2} & e^{-\phi/2}C \\ 0 & -e^{-\phi/2} \end{pmatrix} = C\partial_C L + \partial_\phi L. \quad (9.61)$$

From these we then recognize the Killing vectors

$$\xi_1 = \frac{1}{2}(1 + e^{2\phi} - C^2)\partial_C - C\partial_\phi, \quad (9.62)$$

$$\xi_2 = \frac{1}{2}(1 - e^{2\phi} + C^2)\partial_C + C\partial_\phi, \quad (9.63)$$

$$\xi_3 = C\partial_C + \partial_\phi. \quad (9.64)$$

Notice that we can associate prepotentials, $\mathscr{P}_I$, to these isometries, through the U(1) curvature,

$$d\mathscr{P}_I = \iota_I R_{U(1)}, \quad (9.65)$$

so that

$$\mathscr{P}_1 = \frac{1}{2}(e^\phi - e^{-\phi} + e^{-\phi}C^2), \quad (9.66)$$

$$\mathscr{P}_2 = \frac{1}{2}(e^\phi + e^{-\phi} + e^{-\phi}C^2), \quad (9.67)$$

$$\mathscr{P}_3 = C\, e^{-\phi}. \quad (9.68)$$

In fact, one can prove that SU(1,1)/U(1) is a Kähler manifold and that $R_{U(1)}$ can be identified with the Kähler form J.

9.4.2 The Special-Kähler Manifold SU(1,1)/U(1) and Inequivalent Symplectic Frames

The homogeneous manifold $SU(1,1)/U(1)$ is also an instance of a special Kähler manifold. One way to see this is by introducing the prepotential

$$F(X) = -i\, X^0 X^1, \tag{9.69}$$

where X^I are the projective coordinates defined in Chap. 8. Choosing the independent coordinate z according to $X^0 = 1$, $X^1 = z$, we have $F_0 = -iz$ and $F_1 = -i$, so that under $Sp(4,\mathbb{R})$, the combination $\{X^I, F_I\}$ forms a vector. This leads to the Kähler potential

$$K = -\log[i\,(\overline{X}^I F_I - X^I \overline{F}_I)] = -\log[2(z + \overline{z})], \tag{9.70}$$

and metric

$$ds^2 = g_{z\overline{z}}\, dz \otimes d\overline{z} = \frac{dz \otimes d\overline{z}}{(z + \overline{z})^2}. \tag{9.71}$$

Notice that we recover the metric of the previous section by setting

$$z = e^{\phi} + i\, C, \tag{9.72}$$

with $\ell = 1/2$. Notice also that the real part of the complex coordinate z has to be positive definite, because of the Kähler potential (9.70). Special geometry fixes also the gauge kinetic matrix to be diagonal

$$\mathcal{N}_{00} = -i\, z, \qquad \mathcal{N}_{11} = -\frac{i}{z}. \tag{9.73}$$

Finally, the isometries we discussed in the previous section can be described by holomorphic Killing vectors:

$$\xi_1 = \frac{i}{2}\left(1 + z^2\right) \partial_z, \tag{9.74}$$

$$\xi_2 = \frac{i}{2}\left(-1 + z^2\right) \partial_z, \tag{9.75}$$

$$\xi_3 = z\, \partial_z. \tag{9.76}$$

We can now see explicitly the existence of different symplectic frames for the $\mathcal{N} = 2$ supergravity theory constructed by supergravity coupled to a single vector multiplet with prepotential (9.69). Let us focus on the part of the Lagrangian quadratic in the vector field strengths, i.e., the first two terms in (8.5). In the frame defined by the prepotential above, we have

$$
e^{-1}\mathscr{L}_{F^2} = -\frac{e^{\phi}}{4} F^{0}_{\mu\nu} F^{0\,\mu\nu} - \frac{e^{\phi}}{4|z|^2} F^{1}_{\mu\nu} F^{1\,\mu\nu} + \frac{C}{4} F^{0}_{\mu\nu} \widetilde{F}^{0\,\mu\nu} - \frac{C}{4|z|^2} F^{1}_{\mu\nu} \widetilde{F}^{1\,\mu\nu} .
\tag{9.77}
$$

Now we can construct the full symplectic vector of field strengths on which one acts with symplectic and duality transformations to find the inequivalent symplectic frames. Focusing again on $\mathscr{L}_{F^2}$, we have

$$
G_{0\mu\nu} = C\, F^{0}_{\mu\nu} + e^{\phi} \widetilde{F}^{0\rho\sigma},
\tag{9.78}
$$

$$
G_{1\mu\nu} = -\frac{C}{|z|^2} F^{0}_{\mu\nu} + \frac{e^{\phi}}{|z|^2} \widetilde{F}^{0\rho\sigma} .
\tag{9.79}
$$

Altogether, $\mathbb{F}$ is subject to $\mathrm{Sp}(4,\mathbb{R})$ transformations, whose generators can be split into those of $G_U = \mathrm{SU}(1,1)$,

$$
\mathfrak{su}(1,1) = \frac{1}{2} \left\{ \sigma_1 \otimes \sigma_3, -i\,\sigma_2 \otimes \mathbb{1}, -\sigma_3 \otimes \sigma_3 \right\},
\tag{9.80}
$$

the vector field redefinitions $\mathrm{GL}(2,\mathbb{R})$,

$$
\mathfrak{gl}(2,\mathbb{R}) = \frac{1}{2} \left\{ -\sigma_3 \otimes \sigma_3, \sigma_3 \otimes \mathbb{1}, \sigma_3 \otimes \sigma_1, \mathbb{1} \otimes i\,\sigma_2, \right\},
\tag{9.81}
$$

where the first generator is in common between $\mathfrak{gl}(2,\mathbb{R})$ and $\mathfrak{su}(1,1)$, and finally the remaining generators of the quotient

$$
\mathrm{GL}(2,\mathbb{R}) \backslash \mathrm{Sp}(4,\mathbb{R}) / \mathrm{SU}(1,1),
\tag{9.82}
$$

namely,

$$
\frac{1}{2} \left\{ \sigma_1 \otimes \mathbb{1}, i\,\sigma_2 \otimes \sigma_3, \sigma_1 \otimes \sigma_1, i\sigma_2 \otimes \sigma_1 \right\}.
\tag{9.83}
$$

The identification (9.80) can be explicitly confirmed by computing its action on the holomorphic sections, $\mathscr{V} = (X^I, F_I)$, and the resulting action on the independent coordinate z:

$$
\delta_I z = \delta_I \left(\frac{X^1}{X^0} \right) = \frac{\delta_I X^1}{X^0} - \frac{X^1}{(X^0)^2} \delta_I X^0 = \xi_I(z),
\tag{9.84}
$$

where $\delta_I X^J$ can be read from the direct action $\delta_I \mathscr{V} = t_I \mathscr{V}$.

Let us now see the effect of an equivalent frame, obtained by a duality transformation, and that of an inequivalent one, obtained by the action of an element of $GL(2, \mathbb{R})\backslash Sp(4, \mathbb{R})/SU(1, 1)$. The simplest example corresponds to the shift symmetry $C \rightarrow C + \alpha$, which is generated by $t_+ = \frac{1}{2}(\sigma_1 \otimes \sigma_3 - i\sigma_2 \otimes \mathbb{1})$. The application of the symplectic matrix

$$S_d = \exp(\alpha\, t_+) \tag{9.85}$$

on $\mathbb{F}$ produces the Lagrangian

$$e^{-1}\mathscr{L}'_{F^2} = -\frac{e^{\phi}}{4}\,F^0_{\mu\nu}F^{0\,\mu\nu} - \frac{e^{\phi}}{4(e^{2\phi} + (C+\alpha)^2)}\,F^1_{\mu\nu}F^{1\,\mu\nu}$$

$$+ \frac{C+\alpha}{4}\,F^0_{\mu\nu}\widetilde{F}^{0\,\mu\nu} - \frac{C+\alpha}{4(e^{2\phi} + (C+\alpha)^2)}\,F^1_{\mu\nu}\widetilde{F}^{1\,\mu\nu}, \tag{9.86}$$

which straightforwardly goes back to the previous Lagrangian by the obvious redefinition of $C' = C+\alpha$, which is also an isometry of the scalar manifold. On the other hand, if we now use the transformation generated by $t_{sp} = \frac{1}{2}(\sigma_1 \otimes \sigma_1 - i\sigma_2 \otimes \sigma_3)$, which belongs to $GL(2, \mathbb{R})\backslash Sp(4, \mathbb{R})/SU(1, 1)$, we obtain

$$e^{-1}\mathscr{L}'_{F^2} = -\frac{e^{\phi}}{4}\,F^0_{\mu\nu}F^{0\,\mu\nu} - \frac{e^{\phi}}{4(e^{2\phi} + (C-\alpha)^2)}\,F^1_{\mu\nu}F^{1\,\mu\nu}$$

$$+ \frac{C+\alpha}{4}\,F^0_{\mu\nu}\widetilde{F}^{0\,\mu\nu} - \frac{C+\alpha}{4(e^{2\phi} + (C-\alpha)^2)}\,F^1_{\mu\nu}\widetilde{F}^{1\,\mu\nu}, \tag{9.87}$$

which now cannot be reabsorbed by a field redefinition. Notice that while introducing $C' = C - \alpha$ produces just a total derivative difference, we explicitly see that the two Lagrangians are not identical, a fact that appears even more explicitly for other isometries. A particularly interesting case is that of the duality transformation (the missing entries are zero)

$$S = \begin{pmatrix} 1 & & & \\ & 0 & 1 & \\ & & 1 & \\ -1 & & & 0 \end{pmatrix} = \exp\left[\frac{\pi}{4}(i\sigma_2 \otimes \mathbb{1} + i\sigma_2 \otimes \sigma_3)\right], \tag{9.88}$$

which, as explained in Chap. 8, moves us to a frame where the prepotential does not exist. In this frame, the Lagrangian becomes

$$e^{-1}\mathscr{L}_{F^2} = -\frac{e^{\phi}}{4}\,F^0_{\mu\nu}F^{0\,\mu\nu} - \frac{e^{\phi}}{4}\,F^1_{\mu\nu}F^{1\,\mu\nu} + \frac{C}{4}\,F^0_{\mu\nu}\widetilde{F}^{0\,\mu\nu} + \frac{C}{4}\,F^1_{\mu\nu}\widetilde{F}^{1\,\mu\nu}, \tag{9.89}$$

and the transformations related to t_+ and t_{sp} are now lower-triangular

$$S^{-1}t_+S = \begin{pmatrix} 0 & & & \\ & 0 & & \\ & 1 & 0 & \\ & & 1 & 0 \end{pmatrix}, \qquad S^{-1}t_{\mathrm{sp}}S = \begin{pmatrix} 0 & & & \\ & 0 & & \\ & 1 & 0 & \\ & & -1 & 0 \end{pmatrix}, \tag{9.90}$$

hence showing again the fact that the first can be reabsorbed by a field redefinition and the second cannot. In fact, in the first case, the term in front of $F^0_{\mu\nu}\widetilde{F}^{0\,\mu\nu}$ and $F^1_{\mu\nu}\widetilde{F}^{1\,\mu\nu}$ are corrected in the same way, sending $C \to C + \alpha$, while in the second case, we have that the term in front of $F^0_{\mu\nu}\widetilde{F}^{0\,\mu\nu}$ is shifted as $C \to C + \alpha$, while the term in front of $F^1_{\mu\nu}\widetilde{F}^{1\,\mu\nu}$ gets shifted as $C \to C - \alpha$.

9.5   Gauging and the Embedding Tensor

Let us now assume that we have fixed our starting Lagrangian by choosing a specific symplectic frame and that we want to perform the gauging of a generic $G_{\mathrm{gauge}} \subset G_U \subset \mathrm{Sp}(2n_V, \mathbb{R})$. As we mentioned in the previous sections, the best way to do so is by means of the embedding tensor formalism [3, 4, 12], and we will make the simplifying assumptions that the scalar manifold is homogeneous and that there is no inert matter under the duality group. We therefore review in the following this formalism by explaining the general idea and the constraints and discuss some of the Lagrangian terms that are crucially related to the non-vanishing of some of the components of the embedding tensor.

The Bianchi identities and equations of motion of a generic supergravity theory are invariant under global G_U transformations

$$\begin{cases} \delta L = \alpha^\beta \left(t_\beta L + L\, t_i\, w_\beta{}^i \right) \\ \delta A^M_\mu = -\alpha^\beta \left(t_\beta \right)_N{}^M A^N_\mu \end{cases} \tag{9.91}$$

and local $\mathrm{U}(1)^{2n_V}$ gauge transformations

$$\delta A^M_\mu = \partial_\mu \Lambda^M(x). \tag{9.92}$$

Only the subset that does not introduce magnetic vectors in the Lagrangian will have a Lagrangian description, and an even smaller subset will be a global symmetry of the Lagrangian, according to the discussion of Sect. 9.3.

To proceed with the gauging, we need to know which generators are going to constitute the gauge algebra. This is done by selecting some of the generators $t_\alpha \in \mathfrak{g}_U \subset \mathfrak{sp}(2n_V, \mathbb{R})$ and associating to these generators a linear combination of the vector fields (electric and magnetic), which is going to become the gauge vector. To do so we have to specify the *embedding tensor*

$$\Theta_M{}^\alpha, \tag{9.93}$$

a $2n_V \times \dim \mathfrak{g}_U$ matrix, which will give us the gauge algebra generators as combinations of the duality ones:

$$X_M \equiv \Theta_M{}^\alpha t_\alpha. \tag{9.94}$$

Once we know this, the formalism is so constrained that all the couplings and interactions of the corresponding gauged supergravity Lagrangian are fixed. Notice that, by construction, $\dim(G_{\text{gauge}}) = \text{rank}(\Theta_M{}^\alpha)$, and consistency forces

$$\dim(G_{\text{gauge}}) = \text{rank}(\Theta_M{}^\alpha) \le n_V. \tag{9.95}$$

We could think of this tensor as the charge matrix. In fact, we can construct covariant derivatives in terms of the full set of Lagrangian and dual vectors $A^M = \{A_\mu^I, A_{I\,\mu}\}$:

$$\widehat{\partial}_\mu \equiv \partial_\mu - A_\mu^M \, \Theta_M^\alpha \, t_\alpha = \partial_\mu - A_\mu^I \, \Theta_I{}^\alpha \, t_\alpha - A_{I\,\mu} \, \Theta^{I\,\alpha} \, t_\alpha. \tag{9.96}$$

These covariant derivatives appear in the local transformations of the vector fields under a G_{gauge} transformation

$$\begin{cases} \delta L = \alpha^M(x) \left(X_M L + L \, X_i \, w_M{}^i \right) \\ \delta A_\mu^M = \partial_\mu \alpha^M + A_\mu^N \, X_{NP}{}^M \, \alpha^P \equiv \widehat{\partial}_\mu \alpha^M, \end{cases} \tag{9.97}$$

and $X_{NP}{}^M \equiv \Theta_N{}^\alpha \, (t_\alpha)_P{}^M$. This covariantization procedure obviously cannot be consistent for an arbitrary choice of Θ. For instance, we would like the X generators to close into a Lie algebra (or at least be part of a free differential algebra [5], as we will see later). For this reason we have to impose constraints on the embedding tensor, which we will now analyze.

Before moving to the analysis of such constraints though, let us pause for some comments.

Before the gauging procedure, one can rotate the electric F^I and magnetic G_I vector field strengths among themselves, without changing the equations of motion nor the Lagrangian, provided one uses a transformation that is part of the U-duality group (see again Sect. 9.3). This means that we can choose which are the vectors we consider as fundamental degrees of freedom in the Lagrangian. Still, one can have many different Lagrangians for a given supergravity theory (once the number of supersymmetries and the matter content has been fixed), because one could change couplings by a symplectic transformation that is not in the U-duality group. Once the Lagrangian is fixed, the embedding tensor $\Theta_M{}^\alpha$ tells us which among the electric and magnetic vectors enter in the gauging procedure. This obviously breaks the invariance of the equations of motion and Bianchi identities under the U-duality group. Also, at most n_V vectors can enter with their potential in the Lagrangian, while the others should never appear naked.

The same gauge group may have different embeddings in the duality group for different choices of the embedding tensor, and the embedding of the duality group in the symplectic group may be chosen differently to generate different theories. Since Θ is telling us how we make local the isometries of the U-duality group, the resulting theory does not depend only on the gauge group but also on its embedding.

9.5.1 Constraints on the Embedding Tensor

Once the gauge group has been chosen and the embedding tensor fixed, the gauging can be carried out in the standard way, by introducing the covariant derivatives and possibly some topological terms, in order for the resulting Lagrangian to be invariant under local G_{gauge} transformations. This procedure, however, generically breaks supersymmetry, and therefore one has to restore it by further deforming the original Lagrangian in order to restore supersymmetry and preserve G_{gauge} invariance. This procedure can be completed successfully if we use embedding tensors that satisfy certain constraints. We now discuss them according to their origin.

- **Gauge invariance.** The embedding tensor must be a singlet of the gauge group, and hence it should be invariant under the action of the X_M generators:

$$\delta_M \Theta_N{}^\alpha = \Theta_M{}^\beta \delta_\beta \Theta_N{}^\alpha =$$
$$= \Theta_M{}^\beta (t_\beta)_N{}^P \Theta_P{}^\alpha + \Theta_M{}^\beta f_{\beta\gamma}{}^\alpha \Theta_N{}^\gamma = 0, \tag{9.98}$$

  where the first term is the action on the N index and the second term comes from the identification of the adjoint generators with the structure constants $(t_\beta)_\gamma{}^\alpha = -f_{\beta\gamma}{}^\alpha$. The contraction of this constraint with the duality algebra generators t_α gives

$$\Theta_M{}^\beta (t_\beta)_N{}^P \Theta_P{}^\alpha t_\alpha + \Theta_M{}^\alpha [t_\alpha, t_\beta] \Theta_N{}^\beta = 0, \tag{9.99}$$

  which implies the closure of the gauge algebra

$$[X_M, X_N] = -X_{MN}{}^P X_P, \tag{9.100}$$

  so that $X_{[MN]}{}^P$ play a role similar to that of structure constants. Note, however, that this constraint also contains $X_{(MN)}{}^P X_P = 0$, which is generically a non-trivial condition for magnetic gaugings.

- **Locality.** We obviously want that at most n_V mutually local vectors appear in the gauging process. This is expressed by

$$\Theta_M{}^\alpha \Theta_N{}^\beta \Omega^{MN} = 0 \qquad \Leftrightarrow \qquad \Theta^{I[\alpha} \Theta_I{}^{\beta]} = 0. \tag{9.101}$$

  This equation means that electric and magnetic charges should be mutually local, which in turn implies that there always is a choice of symplectic frame such that the gauging can be made purely electric.
- **Supersymmetry.** The embedding tensor $\Theta_M{}^\alpha$ is in a definite representation of the duality group, which is the product of the adjoint (α) and the fundamental (M, also in the fundamental of the symplectic group). Supersymmetry restricts the allowed representations with a linear projection,

$$X_{(MNP)} = X_{(MN}{}^Q \Omega_{P)Q} = 0, \tag{9.102}$$

which generically removes the highest weight, together with other representations. This projection appears as one requires the cancellation of the $O(g)$ terms originating from the supersymmetry variations of the minimal couplings by means of new terms depending on the fermion shifts we discussed in previous chapters. For instance, for $\mathcal{N} = 8$ supergravity the embedding tensor sits in the $\mathbf{56} \otimes \mathbf{133} = \mathbf{56} \oplus \mathbf{912} \oplus \mathbf{6480}$, while supersymmetry closes if Θ is in the representation $\mathbf{912}$. (This condition can be relaxed, so as to leave also the $\mathbf{56}$ representation if we keep the trombone symmetry, a scaling symmetry of the theory that can be gauged if one gives up a Lagrangian description.) We should stress that (9.102) may get corrected for $\mathcal{N} = 1$ theories whenever anomalies are present [6].

In theories where all scalar fields sit in the same multiplets as the vectors, the first two conditions become equivalent once the linear representation constraint is imposed; otherwise the locality and closure conditions should be imposed independently. In fact, the two conditions are independent whenever there are scalar isometries with no duality action, which means that the symplectic representation of the isometry algebra is not faithful, i.e.,

$$(t_\alpha)_M{}^N = 0, \qquad \text{for some } t_\alpha \in \mathfrak{g}_{Iso(\mathcal{M})}, \tag{9.103}$$

This is the case for $\mathcal{N} \leq 2$ theories, where we have matter multiplets that contain scalar fields, but no vectors, like chiral or hypermultiplets.

9.5.2   Couplings

Clearly, there are many different ways to solve the constraints on the embedding tensor, and we will discuss in the next section how, at least in principle, a full classification of all the possibilities that can be obtained. Let us, for now, provide some details on some important classes of gaugings and their physical properties.

Electric gaugings are gaugings where only the electric vectors are used:

$$\Theta^{I\alpha} = 0 \quad \Rightarrow \quad X^I = 0. \tag{9.104}$$

The other constraints further restrict the form of the remaining generators to

$$(X_I)_M{}^N = \begin{pmatrix} X_I{}^J{}_K & 0 \\ X_{IJK} & -X_{IJ}{}^K \end{pmatrix}, \tag{9.105}$$

with

$$X_I{}^J{}_K = X_{[IK]}{}^J, \quad \text{and} \quad X_{I[JK]} = 0, \quad X_{(IKJ)} = 0, \tag{9.106}$$

where the last expression may get modified in $\mathcal{N} = 1$ theories with quantum anomalies [6]. It is interesting to note that already in this case the structure constants are not always restricted only to the adjoint, but there is also a term mixing the electric and magnetic vectors. Actually, the $X_{I,JK}$ term generates Chern–Simons-like couplings of the form

$$\mathcal{L}_{CS} = -\frac{1}{3} \epsilon^{\mu\nu\rho\sigma} X_{I,JK} A_\mu^I A_\nu^K \left(\partial_\rho A_\sigma^J + \frac{3}{8} X_{LP}{}^J A_\mu^L A_\nu^P \right), \tag{9.107}$$

which are needed to cancel the gauge variation of the kinetic term.

On the other hand, the generic case may also have magnetic gaugings, and in this case

$$X_{MN}{}^P = X_{[MN]}{}^P + Z^P{}_{MN}, \tag{9.108}$$

with $Z^P{}_{MN} = Z^P{}_{NM}$, which, from (9.100), must satisfy

$$Z^P{}_{MN} X_P = 0. \tag{9.109}$$

Note that now Jacobi identities do not close in the usual form because of the Z terms:

$$X_{[MN]}{}^P X_{[QP]}{}^R + X_{[QM]}{}^P X_{[NP]}{}^R + X_{[NQ]}{}^P X_{[MP]}{}^R = -Z^R{}_{P[Q} X_{MN]}{}^P. \tag{9.110}$$

The important consequence of this fact is that ordinary gaugings, where one has an ordinary Lie algebra generating the gauge group, require $Z = 0$. On the other hand, whenever $Z \neq 0$, gauge invariance does not close trivially and often imposes the introduction of tensor fields so that the gauge transformations close in a free differential algebra.

The introduction of tensor fields is needed for two reasons. The first is that without tensor fields one would not have gauge invariance for a generic gauging. The second is that electric and magnetic vectors are related by non-local field redefinitions, and in order to have a local Lagrangian, the magnetic ones should be related to the gauge degrees of freedom of tensor fields. In fact, the full set of $2n_V$ gauge connections must transform as in (9.97), but this implies that the set of $2n_V$ curvatures

$$\mathbb{F}^M_{\mu\nu} = 2\partial_{[\mu} A^M_{\nu]} + X_{[NP]}{}^M A^N_\mu A^P_\nu \tag{9.111}$$

transform as

$$\delta \mathbb{F}^M = -\alpha^N X_{NP}{}^M \mathbb{F}^P + Z^M{}_{NP}(2\alpha^N \mathbb{F}^P - A^N \delta A^P). \tag{9.112}$$

so that the first term gives a covariant transformation, while the second is non-covariant and must be absorbed by new terms in the Lagrangian. The second problem is manifest when one looks at the Bianchi identities for such curvatures:

$$\widehat{d}\mathbb{F}^M = d\mathbb{F}^M + X_{NP}{}^M A^N \wedge \mathbb{F}^P = Z^M_{PQ} A^P \wedge \left(dA^Q + \frac{1}{3} X_{RS}{}^Q A^R \wedge A^S\right) \neq 0. \tag{9.113}$$

This is related to the failure of the Jacobi identity and in particular implies that $\widehat{d}F^I \neq 0$. This is not shocking if we think that the true gauge fields are defined as the combinations of the A^M connections singled out by the contraction with the embedding tensor $\Theta_M{}^\alpha$, and such combinations are well defined because

$$\widehat{d}\mathbb{F}^M X_M = 0. \tag{9.114}$$

The solution to restore gauge covariance (and invariance) is to introduce tensor fields, coupled to the vectors by means of the non-trivial Z coefficients

$$H^M_{\mu\nu} = \mathbb{F}^M_{\mu\nu} + Z^M{}_{NP} B^{NP}_{\mu\nu}. \tag{9.115}$$

Note that these new field strengths no longer satisfy the standard Bianchi identities, but rather

$$\widehat{\partial}_{[\mu} H^M_{\nu\rho]} = \frac{1}{3} Z^M{}_{NP} H^{NP}_{\mu\nu\rho}, \tag{9.116}$$

where $H_{\mu\nu\rho}^{NP}$ have to be interpreted as the field strengths of the two-forms. These newly defined curvatures indeed transform covariantly under the gauge group by means of the free differential algebra relations

$$\delta A_\mu^M = \widehat{\partial}_\mu \alpha^M - Z^M{}_{NP}\, \Sigma_\mu^{NP}, \tag{9.117}$$

$$\delta B_{\mu\nu}^{MN} = 2\widehat{\partial}_{[\mu} \Sigma_{\nu]}^{MN} - 2\alpha^{(M} H_{\mu\nu}^{N)} + 2A_{[\mu}^{(M} \delta A_{\nu]}^{N)}, \tag{9.118}$$

where Σ is the gauge transformation of B, and the magnetic A^M (selected by the Z coefficients) are pure gauge under this transformation, so that H^M transforms covariantly

$$\delta H_{\mu\nu}^M = -X_{PN}{}^M \alpha^P H_{\mu\nu}^N. \tag{9.119}$$

It is interesting to note that the coupling (9.115) creates an analogue of the Higgs mechanism, where the vector fields are used to give mass to a set of tensor fields (originally dual to scalar fields when massless), as the new kinetic Lagrangian is

$$e^{-1}\mathscr{L}_{kin} = \frac{1}{4}\, \mathscr{I}_{IJ}\, H_{\mu\nu}^I H^{J\mu\nu} + \frac{1}{4}\, \mathscr{R}_{IJ}\, H_{\mu\nu}^I \widetilde{H}^{J\mu\nu}. \tag{9.120}$$

We should also point out, though, that the H^M do not form the correct symplectic vector we would expect from our discussion on electric–magnetic duality. In fact the correct symplectic vector transforming linearly under electric–magnetic duality transformations is

$$\mathbb{G}^M = \begin{pmatrix} H^I \\ G_I \end{pmatrix}, \tag{9.121}$$

where now

$$\widetilde{G}_{I\mu\nu} = 2\frac{\partial \mathscr{L}}{\partial H^{I\mu\nu}}, \tag{9.122}$$

and H_I and $\mathbb{G}_I$ do not necessarily coincide. This also means that we would rather have $\mathbb{G}^M$ and not H^M to be the well-defined curvatures under gauge transformations. One therefore introduces further modifications to achieve this, but the beauty of the whole construction is that H^M and $\mathbb{G}^M$ as well as their gauge transformations do coincide on-shell. This happens if the tensor equations of motion give

$$(H^M - \mathbb{G}^M)\Theta_M{}^\alpha = 0, \tag{9.123}$$

which correctly produces the correct identification of the dual vectors. This happens if we introduce a topological term in the action of the form

$$\mathscr{L}_{\text{top}} \simeq \epsilon^{\mu\nu\rho\sigma} \left(\Theta^{I\alpha} \partial_\mu A_{I\,\nu} B_{\alpha\,\rho\sigma} + \frac{1}{8} \Theta^{I\alpha} \Theta_I{}^\beta B_{\alpha\,\mu\nu} B_{\beta\,\rho\sigma} + \dots \right), \qquad (9.124)$$

where we used the identification $B_\alpha = -t_{\alpha\,NP} B^{NP}$, so that the tensor equations of motion imply the covariant version of the duality equation between the massive vector and massive tensor degrees of freedom and hence allows us to remove one in favor of the other. Notice also that whenever we switch off the gauging, i.e., $\Theta = 0$, such relations disappear and the tensor fields decouple.

In fact, we should not expect that the tensor fields $B_{\mu\nu}^{MN}$ we just introduced add new degrees of freedom to the theory, but rather they should just be dual to some of the scalar fields of the original ungauged formulation. This is indeed the case once one consistently builds all the couplings according to this formalism. Moreover, the presence of tensor fields fits nicely with the derivation of some of these models from flux compactifications of string theory, where form fields are naturally present and where the reduction process may generically give rise to massive charged tensor fields.

The duality between tensors and scalars becomes manifest, once we build the full duality covariant Lagrangian and analyze the equations of motion of the vector fields:

$$\frac{1}{2} \epsilon^{\mu\nu\rho\sigma} D_\nu \mathbb{G}_{\rho\sigma}^M = \Omega^{MN} \frac{\partial \mathscr{L}}{\partial A_\mu^M}, \qquad (9.125)$$

where the right-hand side arises from the minimal couplings of the vectors to the matter fields, which clearly vanish in the ungauged theory. Originating from the minimal couplings, this "current" term must be proportional to the embedding tensor $\Theta_M{}^\alpha$, and thus we can write

$$\frac{1}{2} \epsilon^{\mu\nu\rho\sigma} D_\nu \mathbb{G}_{\rho\sigma}^M = \Omega^{MN} \Theta_N{}^\alpha J_\alpha^\mu. \qquad (9.126)$$

The locality constraint tells us then that

$$\frac{1}{2} \epsilon^{\mu\nu\rho\sigma} D_\nu \mathbb{G}_{\rho\sigma}^M \Theta_M{}^\alpha = 0, \qquad (9.127)$$

which are the Bianchi identities for the gauge fields. This is consistent with our previous discussion where we showed that the Bianchi identities for the gauge curvature have no magnetic source term, so that the gauge connections are well defined. For what concerns the remaining equations, the relation (9.126) becomes a dualization equation, relating the field strength of the two-form fields $B_{\mu\nu}^{MN}$ to the scalars, whose covariant derivative appears in the right-hand side current.

Substituting its solution in the action, we are eliminating the tensor fields by their equations of motion and effectively performing a rotation to the electric frame.

While this discussion shows that somehow one can always find a symplectic rotation so that the gauging becomes electric, generic string compactifications will naturally give rise to models with tensor fields, and therefore it definitely pays off to be general and write the Lagrangian and couplings in a symplectically invariant form with the help of tensor fields.

9.5.3 An Example: The Maximal Theory

As a working example of the procedure outlined in the previous sections, we now reconstruct the main points leading to the gauging of maximal supergravity in four dimensions [7], making explicit the steps mentioned in Sect. 9.1.

Maximal supergravity contains a single gravity multiplet, whose fields are the graviton $g_{\mu\nu}$; 8 gravitini, ψ_μ^i ($i = 1, \ldots, 8$); 28 vector fields, A_μ^I (conventionally $I = 0, \ldots, 27$); 56 spin 1/2 dilatini, $\chi_{ijk} = \chi_{[ijk]}$; and 70 real scalar fields, φ^u ($u = 1, \ldots, 70$).

The scalar fields describe a nonlinear σ-model given by a homogeneous manifold

$$\mathcal{M}_{\text{scalar}} = \frac{E_{7(7)}}{SU(8)}. \tag{9.128}$$

Hence, we see that the U-duality group of the theory is $G_U = E_{7(7)}$, which has **133** generators t_α. The vector fields and their duals transform in the **56**-dimensional fundamental representation of $E_{7(7)}$, which is a symplectic representation, defining an embedding of $E_{7(7)}$ in $Sp(56, \mathbb{R})$. The coset representative is customarily described by complex 56-dimensional vectors, $L_M^{ij} = -L_M^{ji}$, and their complex conjugates, $L_{M\,ij}$, which together build a matrix

$$L_M{}^{\underline{N}} = \left(L_M{}^{ij}, L_{M\,kl}\right). \tag{9.129}$$

This matrix transforms under rigid $E_{7(7)}$ transformations from the left and under local $SU(8)$ transformations from the right. We also note the following properties of $L_M{}^{\underline{N}}$, which follow from their definition,

$$L_M{}^{ij} L_{N\,ij} - L_N{}^{ij} L_{M\,ij} = i\,\Omega_{MN}, \tag{9.130}$$

$$\Omega^{MN} L_M{}^{ij} L_{N\,kl} = i\,\delta_{kl}^{ij}, \tag{9.131}$$

$$\Omega^{MN} L_M{}^{ij} L_N^{kl} = 0. \tag{9.132}$$

The gauging procedure is going to promote a subgroup of G_U to G_{gauge}. In order to do so, we introduce the embedding tensor that specifies which generators of $E_{7(7)}$

are going to be part of G_{gauge} and which linear combinations of the vector fields are going to be the associated gauge connections. This means introducing the covariant derivative (where we explicitly introduce the coupling g to keep track of the gauging terms)

$$\widehat{\mathscr{D}}_\mu L_M{}^{ij} = \partial_\mu L_M{}^{ij} - \widehat{Q}_{\mu\,kl}{}^{ij} L_M{}^{kl} - g\, A_\mu^P \Theta_P{}^\alpha (t_\alpha)_M{}^N L_N{}^{ij}, \tag{9.133}$$

where $\widehat{Q}$ represents the composite H-connection for the coset manifold, derived in the usual way from the Maurer–Cartan form (9.43), now including a gauging term,

$$\widehat{Q}_{\mu\,kl}{}^{ij} = \frac{2}{3}\, i\, \delta^{[i}_{[k} \left(L_{I\,l]m} \partial_\mu L^{I\,j]m} - L^I_{l]m} \partial_\mu L_I^{j]m} \right) - i\, g\, A_\mu^M \Omega^{NP} L_{N\,ij} X_{M P}{}^Q L_Q^{kl}. \tag{9.134}$$

The scalar kinetic terms therefore now are modified with $O(g)$ and $O(g^2)$ terms, following from the covariantization of the vielbein on the scalar manifold,

$$\widehat{P}_{\mu\,ijkl} = i\, \Omega^{MN} L_{M\,ij}\, \widehat{\mathscr{D}}_\mu L_{N\,kl}. \tag{9.135}$$

Having introduced covariant derivatives that make local some of the isometries of $E_{7(7)}$, we also have to modify the gauge field strengths, according to the rules mentioned in the previous sections. As we explained above, consistency requires that the embedding tensor satisfies a number of linear and quadratic constraints. How is this reflected in the construction of the Lagrangian, though? Having introduced $O(g)$ terms in the covariant derivatives and then in the Lagrangian, supersymmetry is broken and should be restored by the modification of the supersymmetry transformations and by the addition of further $O(g)$ and $O(g^2)$ terms in the Lagrangian, as mentioned in Sect. 9.1. Since all these modifications follow from the gauging procedure, they should in some way depend on the gauge structure constants and eventually on the embedding tensor. In fact they are actually encoded in the tensorial structure provided by the scalar-dressed structure constants, called T-tensor. The T-tensor is defined by

$$T_{\underline{MN}}{}^{\underline{P}}[\Theta, \varphi] = L_{\underline{M}}^{-1\,N} L_{\underline{N}}^{-1\,N} L_P{}^{\underline{P}} X_{MN}{}^P. \tag{9.136}$$

This is clearly a constrained tensor, whose constraints are induced by the constraints on the embedding tensor. For the maximal theory, we know from the previous discussion that such constraints force Θ to live in the representation **912** of $E_{7(7)}$. Once we decompose this in terms of SU(8), we get that $\mathbf{912} \to \mathbf{36} + \overline{\mathbf{36}} + \mathbf{420} + \overline{\mathbf{420}}$. This means that the T-tensor, which is a tensor with indices transforming under local SU(8) transformations, can be decomposed in terms of two simpler complex tensors, $A_1^{ij} = A_1^{ji}$ and $A_{2\,i}{}^{jkl} = A_{2\,i}{}^{[jkl]}$, $A_{2\,i}{}^{jki} = 0$, which live in the representations **36** and **420** of SU(8). This also implies that all modifications of the supersymmetry transformations and of the Lagrangian can be written in terms of these tensors.

Among the various constraints, we should single out a quadratic constraint called supersymmetric Ward identity

$$\frac{1}{24} A_{2\,m}{}^{jkl} A_{2\,jkl}^{n} - \frac{3}{4} A_{1}^{nj} A_{1\,mj} = \frac{1}{8} \delta_{m}^{n} \left(\frac{1}{24} A_{2i}{}^{jkl} A_{2\,jkl}^{i} - \frac{3}{4} A_{1}^{ij} A_{1\,ij} \right). \quad (9.137)$$

This identity is indeed necessary to close the supersymmetry algebra and, as we will see shortly, relates the fermion $O(g)$ shifts to the scalar potential.

As described in Sect. 9.1, we need to modify the supersymmetry transformations of the fermions by

$$\delta_{g} \psi_{\mu}^{i} = \sqrt{2}\, g\, A_{1}^{ij} \gamma_{\nu} \epsilon_{j},$$

$$\delta_{g} \chi^{ijk} = -2 g\, A_{2\,l}{}^{ijk} \epsilon^{l}. \quad (9.138)$$

This further requires the introduction of $O(g)$ Yukawa terms[4] for the fermions,

$$\mathcal{L}_{\text{Yuk}} = e.g. \left(\frac{1}{\sqrt{2}} A_{1\,ij} \overline{\psi}_{\mu}^{i} \gamma^{\mu\nu} \psi_{\nu}^{j} + \frac{1}{6} A_{2i}{}^{jkl} \overline{\psi}_{\mu}^{i} \gamma^{\mu} \chi_{jkl} \right.$$

$$\left. + \frac{\sqrt{2}}{144} \epsilon^{ijkpqrlm} A_{2\,pqr}^{n} \overline{\chi}_{ijk} \chi_{lmn} \right) + h.c., \quad (9.139)$$

whose coefficients are precisely given by the irreducible components of the T-tensor. Applying once more supersymmetry transformations to $\mathcal{L}_{\text{Yuk}}$, one now obtains $O(g^2)$ terms, which can be cancelled by the variation of a new addition to the Lagrangian in the form of a scalar potential,

$$V = g^{2} \left(\frac{1}{24} A_{2i}{}^{jkl} A_{2\,jkl}^{i} - \frac{3}{4} A_{1}^{ij} A_{1\,ij} \right). \quad (9.140)$$

This cancellation happens, thanks to the supersymmetric Ward identity (9.137), which assumes now a deeper meaning than just the consequence of a quadratic constraint, because it relates the scalar potential to the squares of the shift of the fermions, a general property we have seen already in Chap. 4, when dealing with the minimal theory. It is interesting to notice that this potential can also be written in terms of a real, symmetric and field-dependent (56×56)-matrix, $\mathcal{M}_{MN}$, defined from the coset representatives,

$$\mathcal{M}_{MN} \equiv L_{M}{}^{ij} L_{N\,ij} + L_{N}{}^{ij} L_{M\,ij}. \quad (9.141)$$

---

[4] Sometimes these terms are referred to as mass terms, because when the scalar fields pick a vacuum expectation value they indeed generate masses for the fermion fields.

This is positive definite, with inverse $\mathcal{M}^{MN} = \Omega^{MP}\Omega^{NQ}\mathcal{M}_{PQ}$. The scalar potential in terms of this matrix is especially interesting because it makes explicit the quadratic dependence of the scalar potential on the embedding tensor:

$$V = \frac{1}{672}g^2\left(X_{MN}{}^R X_{PQ}{}^S \mathcal{M}^{MP}\mathcal{M}^{NQ}\mathcal{M}_{RS} + 7 X_{MN}{}^Q X_{PQ}{}^N \mathcal{M}^{MP}\right).$$

$$(9.142)$$

Finally, to complete the construction, we have to replace the Abelian field strengths in the Lagrangian by the field strengths H^I, as described in the previous section, and include the topological and Chern–Simons-like terms. Supersymmetry is then restored by fixing an appropriate supersymmetry transformation for the tensor fields.

The final Lagrangian, up to 4-fermi interactions, is described by

$$e^{-1}\mathcal{L} = \frac{1}{2}R - \frac{1}{2}\epsilon^{\mu\nu\rho\sigma}\left(\overline{\psi}^i_\mu \gamma_\nu \widehat{\mathcal{D}}_\rho \psi_{i\sigma} - \overline{\psi}^i_\mu \overset{\leftarrow}{\widehat{\mathcal{D}}}_\rho \gamma_\nu \psi_{i\sigma}\right)$$

$$- \frac{1}{12}\left(\overline{\chi}^{ijk}\gamma^\mu \widehat{\mathcal{D}}_\mu \chi_{ijk} - \overline{\chi}^{ijk}\gamma^\mu \overset{\leftarrow}{\widehat{\mathcal{D}}}_\mu \chi_{ijk}\right) - \frac{1}{12}|\widehat{P}^{ijkl}_\mu|^2$$

$$- \frac{1}{6}\sqrt{2}\left(\overline{\chi}_{ijk}\gamma^\nu\gamma^\mu\psi_{\nu l}\,\widehat{P}^{ijkl}_\mu + \text{h.c.}\right) + \left(H^{\mu\nu+\Lambda}\mathcal{O}^+_{\mu\nu\Lambda} + \text{h.c.}\right)$$

$$- \frac{1}{4}i\left(\mathcal{N}_{IJ}H^{I+}_{\mu\nu}H^{J+\mu\nu} - \overline{\mathcal{N}}_{IJ}H^{I-}_{\mu\nu}H^{J-\mu\nu}\right)$$

$$(9.143)$$

$$+ \frac{1}{8}ig\,\epsilon^{\mu\nu\rho\sigma}\,\Theta^{I\alpha}\,B_{\mu\nu\alpha}\left(2\partial_\rho A_{\sigma I} + g\,X_{MNI}\,A_\rho{}^M A_\sigma{}^N - \frac{1}{4}g\,\Theta_I{}^\beta B_{\rho\sigma\beta}\right)$$

$$+ \frac{1}{3}ig\,\epsilon^{\mu\nu\rho\sigma}\,X_{MNI}\,A_\mu{}^M A_\nu{}^N\left(\partial_\rho A_\sigma{}^I + \frac{1}{4}g X_{PQ}{}^I A_\rho{}^P A_\sigma{}^Q\right)$$

$$+ \frac{1}{6}ig\,\epsilon^{\mu\nu\rho\sigma}\,X_{MN}{}^I A_\mu{}^M A_\nu{}^N\left(\partial_\rho A_{\sigma I} + \frac{1}{4}g X_{PQI} A_\rho{}^P A_\sigma{}^Q\right)$$

$$+ e^{-1}\mathcal{L}_{\text{Yuk}} - V,$$

where $L^{\Lambda ij}\mathcal{O}^+_{\mu\nu\Lambda} = \frac{i}{4}\mathcal{O}^{+ij}_{\mu\nu}$, and

$$\mathcal{O}^{+ij}_{\mu\nu} = \frac{1}{2}\sqrt{2}\,\overline{\psi}^i_\rho \gamma^{[\rho}\gamma_{\mu\nu}\gamma^{\sigma]}\psi^j_\sigma - \frac{1}{2}\overline{\psi}_{\rho k}\gamma_{\mu\nu}\gamma^\rho \chi^{ijk} - \frac{\sqrt{2}}{144}\epsilon^{ijklmnpq}\overline{\chi}_{klm}\gamma_{\mu\nu}\chi_{npq}.$$

$$(9.144)$$

Looking at the Lagrangian (9.143), we notice all the structures we described in our discussion so far. The first two lines contain the kinetic terms for the graviton, the gravitini, the dilatini, and the scalar fields, where gauge covariant derivatives have been introduced where necessary. The third line describes Noether couplings between the scalars and the fermions as well as Pauli-like couplings between the vectors and the fermions. The fourth and fifth lines describe the general structure of the vector kinetic terms and their couplings to the fermions, now improved by the couplings to tensor fields. The sixth line is the topological term required to be able to properly relate the tensor fields to their scalar duals. The seventh and eighth lines are the Chern–Simons-like couplings needed to ensure gauge invariance, and in the last line, we have the Yukawa couplings and the scalar potential.

9.6 Classifying Gaugings

Our discussion on the gauging procedure made clear that even when we fix a choice of gauge group, there is still the possibility that the linear and quadratic constraints admit more than one solution, leading to gauged supergravities that are potentially inequivalent even if they share the same set of gauge symmetry generators, $t_r \in \mathfrak{g}_{\text{gauge}} \subset \mathfrak{g}_U$, because they differ in the choice of the (electric and magnetic) vector fields that form the gauge connection. The aim of this section is to characterize group-theoretically the space of these inequivalent theories, showing the relation between the set of consistent choices of gauge connections for fixed $t_r \in \mathfrak{g}_{\text{gauge}} \subset \mathfrak{g}_U$ and symplectic transformations. In the following, we will argue heuristically for the various requirements, but the interested reader can find a detailed proof in [8].

So, let us assume that we completely fixed once and for all the gauge group and therefore fixed which among the various generators $t_\alpha \in \mathfrak{g}_U$ are selected as

$$t_r \in \mathfrak{g}_{\text{gauge}} \subset \mathfrak{g}_U, \qquad r = 1, \ldots, \dim \mathfrak{g}_{\text{gauge}} \leq n_V, \tag{9.145}$$

where the last constraint is related to the existence of enough vector fields to gauge G_{gauge}. To provide a consistent gauging, we therefore have to provide an embedding tensor $\Theta_M{}^\alpha$, such that

$$\Theta_M{}^\alpha t_\alpha = \theta_M{}^r t_r \tag{9.146}$$

and obviously satisfying the necessary linear and quadratic constraints, which, for instance, for $\mathcal{N} \geq 3$ supergravities are summarized by

$$[t_r, t_s] = f_{rs}{}^t t_t, \qquad f_{[rs}{}^u f_{t]u}{}^v = 0,$$
$$\theta_M{}^s f_{rs}{}^t = -t_{rM}{}^N \theta_N{}^t, \qquad \theta_{(M}{}^r t_{r\,NP)} = 0. \tag{9.147}$$

The question we want to answer therefore is if there are multiple solutions of (9.147) for $\theta_M{}^r$ which are inequivalent. One can prove that all such solutions are mapped onto each other by $\mathrm{Sp}(2n_V, \mathbb{R})$ transformations. The answer is then characterized group theoretically as the set of transformations of $\mathrm{Sp}(2n_V, \mathbb{R})$ that leave G_{gauge} invariant, which is the definition of the *normalizer* of G_{gauge} in $\mathrm{Sp}(2n_V, \mathbb{R})$: $\mathscr{N}_{\mathrm{Sp}(2n_V, \mathbb{R})}(G_{\mathrm{gauge}})$. In detail, if the original solution is specified by $\theta_M{}^r$, any other solution is specified by

$$\hat{\theta}_M{}^r = N_M{}^N \theta_N{}^s g_s^{-1r}, \qquad \text{for} \qquad N_M{}^N \, t_{r\,N}{}^P \left(N^{-1}\right)_P{}^Q = g_r{}^s t_{s\,M}{}^Q,$$

$$\text{with} \quad N \in \mathscr{N}_{\mathrm{Sp}(2n_V, \mathbb{R})}(G_{\mathrm{gauge}}), \qquad g \in \mathrm{GL}(\dim G_{\mathrm{gauge}}, \mathbb{R}).$$

$$(9.148)$$

In particular, we are interested only in the set of transformations that change the resulting gauge structure, which means the set of transformations that give rise to $X_{MN}{}^P$ differ from those generated by $\theta_M{}^r$. This means that we want to remove from the previous set all the transformations in the stabilizer, $\mathscr{S}$, of $X_{MN}{}^P = \theta_M{}^r t_{r\,N}{}^P$, which is represented by the quotient (here from the left)

$$S_0 = \mathscr{S}_{\mathrm{Sp}(2n_V, \mathbb{R})}(X) \backslash \mathscr{N}_{\mathrm{Sp}(2n_V, \mathbb{R})}(G_{\mathrm{gauge}}). \qquad (9.149)$$

The question that we are left to answer now is which among the symplectic transformations in S_0 give inequivalent theories. Before proceeding, though, we should mention that, depending on the context, what we regard as inequivalent can change. For instance, for our purposes it is more natural to regard as equivalent those theories that differ from each other only in the value of the gauge coupling constant, even if it is of course a physically relevant quantity. It is of course straightforward to include it back. More importantly, we can decide to distinguish between theories that have the same set of equations of motion and Bianchi identities but differ at the quantum level or regard them as equivalent if we are only interested in the classical regime. We will begin with the first option and therefore assume that we have fixed the choice of an electric frame, so that we can quotient $\mathscr{N}_{\mathrm{Sp}(2n_V, \mathbb{R})}(G_{\mathrm{gauge}})$ by the action of local redefinitions of the physical fields only. The resulting set is also a quotient space that we call $\mathfrak{S}$.

9.6.1 The Quotient Space $\mathfrak{S}$

As mentioned earlier, we assume for simplicity that the scalar manifold is homogeneous. Two ungauged Lagrangians of $D = 4$ supergravity are related by $\mathrm{Sp}(2n_V, \mathbb{R})$ transformations $S_M{}^N$ acting on the G_U/H coset as

$$L(\varphi)_M{}^{\underline{N}} \quad \rightarrow \quad S_M{}^P L(\varphi)_P{}^{\underline{N}}. \qquad (9.150)$$

In the gauged models, the change of symplectic frame also acts on the embedding tensor according to

$$X_{MN}{}^P \rightarrow S_M{}^Q S_N{}^R X_{QR}{}^S (S^{-1})_S{}^P. \tag{9.151}$$

This ensures that the T-tensor, defined as

$$T(\varphi)_{\underline{MN}}{}^{\underline{P}} = L^{-1}(\varphi)_{\underline{M}}{}^M L^{-1}(\varphi)_{\underline{N}}{}^N X_{MN}{}^P L(\varphi)_P{}^{\underline{P}}, \tag{9.152}$$

and hence the fermionic shifts in the supersymmetry transformations as well as the scalar potential are independent of the choice of symplectic frame. This in turn guarantees that the combination of equations of motion and Bianchi identities is invariant under symplectic transformations.

As it should be clear from our discussion above, any consistent embedding tensor, $\theta_M{}^r$, can be mapped to the standard electric one, $\delta_M{}^r$, which selects the first dim G_{gauge} vectors as gauge connections, by an element, N, of $\mathcal{N}_{Sp(2n_V,\mathbb{R})}(G_{gauge})$. In such reference symplectic frame, we therefore define

$$X^0_{MN}{}^P \equiv \delta_M{}^r t_{r\,N}{}^P \tag{9.153}$$

and call $T^0(\varphi)_{\underline{MN}}{}^{\underline{P}}$ the associated T-tensor. We then notice that if we apply an N^{-1} transformation *only* to the coset representatives, namely,

$$L(\varphi)_M{}^{\underline{N}} \rightarrow N^{-1}{}_M{}^P L(\varphi)_P{}^{\underline{N}}, \qquad X^0_{MN}{}^P \text{ unchanged}, \tag{9.154}$$

then the T-tensor transforms as

$$T^0_{\underline{MN}}{}^{\underline{P}} \equiv L^{-1}{}_{\underline{M}}{}^M L^{-1}{}_{\underline{N}}{}^N X^0_{MN}{}^P L_P{}^{\underline{P}}$$

$$\overset{N^{-1}}{\rightarrow} L^{-1}{}_{\underline{M}}{}^M L^{-1}{}_{\underline{N}}{}^N N_M{}^Q N_N{}^R X^0_{QR}{}^S N^{-1}{}_S{}^P L_P{}^{\underline{P}}$$

$$= T^\theta_{\underline{MN}}{}^{\underline{P}}. \tag{9.155}$$

As a result, the gauge kinetic functions and moment couplings transform accordingly with the N^{-1} symplectic transformation. Clearly the equations of motion and Bianchi identities are not invariant under (9.154), as is reflected by the fact that the T-tensor changes. We then interpret (9.154) as a *symplectic deformation*, namely, a map between two (potentially) inequivalent gauged models. The requirement $N \in \mathcal{N}_{Sp(2n_V,\mathbb{R})}(G_{gauge})$ ensures that t_r are a good choice of gauge generators also after the symplectic deformation, i.e., they belong to the $\mathfrak{g}_U$ algebra of both the old and the new symplectic frame.

Let us then see which transformations give inequivalent models. One obvious field redefinition we are allowed to make is a local redefinition of the vector fields (i.e., a redefinition involving only the vector fields appearing in the original Lagrangian). We will not change theory if by such a redefinition we rescale at most the structure constants by an overall factor, i.e., if such transformations are in the stabilizer of X^0, $\mathscr{S}_{GL(n_V,\mathbb{R})}(X^0)$. The other redefinitions that should not modify the theory are those U-duality transformations that leave the gauge group invariant,[5] namely, the transformations in the normalizer of G_{gauge} within G_U: $\mathscr{N}_{G_U}(G_{gauge})$. The combination of the two is indeed what we are looking for.

Take then two transformations N, $N' \in \mathscr{N}_{Sp(2n_V,\mathbb{R})}(G_{gauge})$, related by

$$N = u\,N's, \quad u \in \mathscr{N}_{G_U}(G_{gauge}), \quad s \in \mathscr{S}_{GL(n_V,\mathbb{R})}(X^0). \tag{9.156}$$

Substituting in (9.155), we get

$$T_{\underline{MN}}^{0}{}^{\underline{P}} \xrightarrow{N^{-1}} (L^{-1}u\,N's)_{\underline{M}}{}^{M}(L^{-1}u\,N's)_{\underline{N}}{}^{N}\,X_{MN}^{0}{}^{P}\,(s^{-1}N'^{-1}u^{-1}L)_{P}{}^{\underline{P}}, \tag{9.157}$$

and at the same time the vector kinetic terms and moment couplings transform with N^{-1}. The G_U transformation $u_M{}^N$ can be reabsorbed in the scalar fields, and therefore it does not affect the physics. Since we have required that the action of s on X^0 is trivial up to an overall rescaling, so that

$$s_M{}^Q s_N{}^R\,X_{QR}^{0}{}^{S}\,s_S^{-1P} \propto X_{RN}^{0}{}^{P}, \tag{9.158}$$

we can reabsorb the rescaling in the gauge coupling constant, and similarly $s_M{}^N$ can be reabsorbed in a local field redefinition of the electric vectors A_μ^Λ in the covariant derivatives and in the non-minimal couplings:

$$A_\mu{}^I \rightarrow A_\mu{}^J s_J{}^I. \tag{9.159}$$

We conclude that N and N' in (9.156) define the same gauged theory up to local field redefinitions and rescalings of the gauge coupling constant.

We therefore arrive at the result that symplectic deformations are classified by the space

$$\mathfrak{S} \equiv \mathscr{S}_{GL(n_V,\mathbb{R})}(X^0) \setminus \mathscr{N}_{Sp(2n_V,\mathbb{R})}(G_{gauge}) / \mathscr{N}_{G_U}(G_{gauge}), \tag{9.160}$$

---

[5] Actually one should enlarge the possible transformations to all automorphisms of the U-duality group. While this coincides with the U-duality group itself in most cases, there are instances where additional discrete factors become relevant. A particularly relevant example is maximal supergravity, where there is an additional $\mathbb{Z}_2$ parity symmetry that can be used.

where the quotients correspond to local field redefinitions. Notice that this definition carries a dependence on the initial choice of electric frame, to the extent that such choice affects the explicit form of $X^0_{MN}{}^P$ (for instance, it can affect the Chern–Simons-like couplings in the gauge generators). Therefore we must specify the explicit form of X^0 that we use to compute $\mathfrak{S}$, or equivalently the specific choice of electric frame in which we construct the gauged theory whose symplectic deformations we want to compute.

Once again, we can make this explicit in the example of an $\mathcal{N} = 2$ supergravity theory coupled to a vector multiplet, with homogeneous scalar manifold SU(1,1)/U(1), parametrized as in Sect. 9.4.2. Let us look once again at the shift symmetry of the imaginary part of the z field $C \rightarrow C + \alpha$. This is generated by $t_+ \in \mathfrak{su}(1, 1)$, which we have seen becomes a perturbative symmetry in the symplectic frame defined by the symplectic transformation S in (9.88). We therefore analyze the $\mathfrak{sp}(4, \mathbb{R})$ generators in this basis, which is related to the one in Sect. 9.4.2 by $t_I \rightarrow S^{-1} t_I S$. In this basis, the generators of SU(1,1) take the form

$$\frac{1}{2} \left\{ \sigma_1 \otimes \sigma_1, -i\, \sigma_2 \otimes \mathbb{1}, -\sigma_3 \otimes \mathbb{1} \right\} \qquad \text{(generators of } \mathfrak{su}(1, 1)), \qquad (9.161)$$

whereas the generators for the vector field redefinitions, GL(2,$\mathbb{R}$), read

$$\frac{1}{2} \left\{ -\sigma_3 \otimes \mathbb{1}, \sigma_3 \otimes \sigma_3, -\sigma_3 \otimes \sigma_1, -\mathbb{1} \otimes i\sigma_2 \right\} \qquad \text{(generators of } \mathfrak{gl}(2, \mathbb{R})). \qquad (9.162)$$

Obviously, the first generator is in common between $\mathfrak{gl}(2,\mathbb{R})$ and $\mathfrak{su}(1, 1)$. The remaining generators of the quotient,

$$\text{GL}(2, \mathbb{R}) \backslash \text{Sp}(4, \mathbb{R}) / \text{SU}(1, 1), \qquad (9.163)$$

are

$$\frac{1}{2} \left\{ \sigma_1 \otimes \sigma_3, i\, \sigma_2 \otimes \sigma_3, \sigma_1 \otimes \sigma_1, i\, \sigma_2 \otimes \sigma_1 \right\}. \qquad (9.164)$$

Clearly, there are no generators in $\mathfrak{su}(1, 1)$ that commute with t_+ and therefore $\mathcal{N}_{\text{SU}(1,1)}(\text{U}(1)_{t_+}) = \text{U}(1)_{t_+}$. On the other hand, there are generators in $\mathfrak{sp}(4, \mathbb{R})$ that commute with t_+, so that

$$\mathcal{N}_{\text{Sp}(4,\mathbb{R})}\left(\text{U}(1)_{t_+} \right) = \left\{ t_\bullet, t_\uparrow, t_\downarrow \right\}, \qquad (9.165)$$

where

$$t_\bullet = -\frac{1}{2} \mathbb{1} \otimes i\,\sigma_2, \tag{9.166}$$

$$t_\uparrow = \frac{1}{2} \left(\sigma_1 \otimes \sigma_1 - i\,\sigma_2 \otimes \sigma_1 \right), \tag{9.167}$$

$$t_\downarrow = \frac{1}{2} \left(\sigma_1 \otimes \sigma_3 - i\,\sigma_2 \otimes \sigma_3 \right). \tag{9.168}$$

These generators satisfy the algebra of the isometries of the Euclidean plane, $\mathbb{E}_2$,

$$[t_\bullet, t_\uparrow] = t_\downarrow, \quad [t_\bullet, t_\downarrow] = -t_\uparrow, \quad [t_\uparrow, t_\downarrow] = 0. \tag{9.169}$$

We see that in this symplectic frame $t_\bullet$ is in $\mathfrak{gl}(2, \mathbb{R})$ and hence as generators of inequivalent gaugings we are left with the generators $t_\uparrow$ and $t_\downarrow$. As already discussed at the end of Sect. 9.4.2, these are transformations that generate total derivative terms, and hence we really have inequivalent gauged Lagrangians only at the quantum level. There are also examples where the inequivalent gaugings have a more sizeable modification to the Lagrangian, introducing new couplings and modifying the scalar potential and its critical points. The most striking example is given by the maximal supergravity with gauge group SO(8), which we therefore discuss in the next section.

9.6.2 The $\mathfrak{S}$ Space of SO(8) Maximal Gauged Supergravity

In this section we want to repeat the discussion of the $\mathfrak{S}$ space for the SO(8) gauging of maximal supergravity, to show that there actually is an infinite family of inequivalent supergravity theories with the same gauge group [9].

Let us then consider the SO(8) gauged maximal supergravity, taken in its standard electric frame with SL(8, $\mathbb{R}$) as electric group. This is the standard form of this theory as it is obtained, for instance, by reduction of M-theory on the seven sphere S^7 [10]. The space of inequivalent theories is then

$$\mathfrak{S} = \mathscr{N}_{GL(28,\mathbb{R})}(SO(8)) \setminus \mathscr{N}_{Sp(56,\mathbb{R})}(SO(8)) \, / \, \mathscr{N}_{\mathbb{Z}_2 \ltimes E_{7(7)}}(SO(8)), \tag{9.170}$$

where in the last factor we used the full group of automorphisms of the U-duality group, which for the maximal theory includes an additional $\mathbb{Z}_2$ parity. Moreover, we replaced $\mathscr{S}_{GL(28,\mathbb{R})}(X^0)$ with $\mathscr{N}_{GL(28,\mathbb{R})}(SO(8))$ in this expression, because they coincide and therefore we can remove any explicit reference to the embedding tensor.

A detailed computation [8] shows that this reduces to the following quotient

$$\mathfrak{G} = \mathrm{GL}(1, \mathbb{R}) \times S_3 \setminus \mathrm{SL}(2, \mathbb{R}) \times S_3 / D_8, \qquad (9.171)$$

where S_3 is the triality discrete automorphism of SO(8), and the SL(2, $\mathbb{R}$) in $\mathcal{N}_{\mathrm{Sp}(56,\mathbb{R})}$ can be parameterized as follows:

$$G_\lambda \equiv \begin{pmatrix} \lambda & \\ & \lambda^{-1} \end{pmatrix} \otimes \mathbb{1}_{28}, \qquad \lambda \in \mathbb{R} \setminus \{0\}, \qquad (9.172)$$

$$W_\theta \equiv \begin{pmatrix} 1 & -g^2\theta/2\pi \\ & 1 \end{pmatrix} \otimes \mathbb{1}_{28}, \qquad \theta \in \mathbb{R}, \qquad (9.173)$$

$$U_\omega \equiv \begin{pmatrix} \cos\omega & -\sin\omega \\ \sin\omega & \cos\omega \end{pmatrix} \otimes \mathbb{1}_{28}, \qquad \omega \in [0, 2\pi], \qquad (9.174)$$

in the basis where the SO(8) generators are given by

$$\mathfrak{so}(8) \ni t_r = \begin{pmatrix} \Lambda_r & \\ & \Lambda_r \end{pmatrix}, \qquad (9.175)$$

with Λ_r the SO(8) generators in the adjoint representation.

In order to understand this, let us make contact with the embedding tensor formalism. The consistency constraints on the embedding tensor require that it is a singlet under SO(8). In fact, the **912** $E_{7(7)}$ representation in which $\Theta_M{}^\alpha$ sits contains two SO(8)-singlets in its manifestly triality-invariant decomposition:

$$\mathbf{912} \rightarrow \mathbf{36} + \mathbf{36'} + \mathbf{420} + \mathbf{420'} \qquad (9.176)$$

$$\rightarrow \mathbf{1}_\theta + \mathbf{1}_\xi + 2 \cdot (\mathbf{35}_\mathrm{v} + \mathbf{35}_\mathrm{s} + \mathbf{35}_\mathrm{c} + \mathbf{350}). \qquad (9.177)$$

The subscripts "θ" and "ξ" denote the relation to the symmetric tensors θ_{ij} and ξ^{ij} that (when positive-definite) define the SO(8) generators inside SL(8, $\mathbb{R}$) and that we will always assume to be in the standard form $\theta_{ij} \propto \xi^{ij} \propto \delta_{ij}$. The original SO(8) gauged maximal supergravity [10] corresponds to $\theta_{ij} \propto \delta_{ij}$, $\xi = 0$, and it is electrically gauged in the standard formulation. What we call X^0 corresponds to this particular embedding tensor. The "ω-deformed" SO(8) gaugings are then defined by turning on $\xi \neq 0$, and they are no longer electric in the original symplectic frame. This is clearly achieved in the above parametrization by acting on X^0 with the matrix U_ω. Following our analysis in Sect. 9.6.1, we prefer to regard the symplectic deformations as leaving X^0 unchanged, but acting on the coset representatives, thus yielding the deformed theories in their respective electric frames.

In practice, most of the transformations in $SL(2, \mathbb{R}) \times S_3$ either leave X^0 invariant up to a rescaling of the gauge coupling constant, $g \to \lambda g$, and their effect on the kinetic terms can be reabsorbed in a local redefinition of the vector fields, or have no effect on X^0 altogether, but with a non-trivial effect on the vector kinetic term of the form

$$\mathcal{N}(\varphi)_{\Lambda\Sigma} \to \mathcal{N}(\varphi)_{\Lambda\Sigma} + g^2 \frac{\theta}{2\pi} \delta_{\Lambda\Sigma}. \tag{9.178}$$

This transformation clearly represents a constant, $SO(8)$ invariant shift in the θ-angle of the gauge theory; hence, it has no effect on the (classical) equations of motion and supersymmetry variations. In fact, it is clear that we can always add a term $\propto \delta_{\Lambda\Sigma} F^\Lambda \wedge F^\Sigma$ to the gauged $SO(8)$ electric action, and the analysis above proves that there is no $E_{7(7)}$ transformation or local field redefinition that can remove it. The transformation that has the most striking effect is instead the ω-deformation of the $SO(8)$ gauged maximal supergravity [9]. If we keep $\theta = 0$, one indeed reproduces the known parameter space for the ω-deformation of the $SO(8)$ theory, namely, S^1/D_8, with identifications $\omega \simeq \pm\omega + k\pi/4$, $k \in \mathbb{Z}$ and fundamental domain $\omega \in [0, \pi/8]$. If we include θ, the $\mathfrak{S}$-space of symplectic deformations of $SO(8)$ gauged maximal supergravity, in its standard electric frame, is a quotient of a hyperboloid: $(dS_2/\mathbb{Z}_8)/\mathbb{Z}_2$. If we also impose periodicity in θ, the resulting space has the topology of a two-sphere.

This space of inequivalent theories becomes evident if we analyze their couplings and in particular their scalar potential. All such inequivalent models have an $\mathcal{N} = 8$ vacuum with a negative cosmological constant, and obviously the quadratic spectra around such vacua coincide. However, higher-order couplings change, as expected for inequivalent models. We will now show explicitly some of the couplings and compute their dependence on the parameter ω, with a special emphasis on the scalar potential, which now shows a different spectrum of vacua according to the parameter's choice.

For the sake of clarity, we restrict the analysis of the potential to the G_2-invariant sector of the theory. It is known that for $\omega = 0$ one finds one $\mathcal{N} = 8$ vacuum with $SO(8)$ symmetry, two parity conjugated vacua with $\mathcal{N} = 0$ and $SO(7)^-$ residual symmetry, another $\mathcal{N} = 0$ vacuum with $SO(7)^+$ symmetry, self-conjugate under parity, and two parity conjugated $\mathcal{N} = 1$ vacua with G_2 symmetry [11]. The G_2-invariant truncation contains two real scalar fields, $\vec{\varphi} = (\varphi_1, \varphi_2)$, and the potential can be written as the sum of three pieces,

$$V(\vec{\varphi}) = A(\vec{\varphi}) - \cos(2\omega) f(\varphi_1, \varphi_2) - \sin(2\omega) f(\varphi_2, \varphi_1), \tag{9.179}$$

where (in the following $x \equiv e^{|\vec{\varphi}|}$)

$$A(\vec{\varphi}) = \frac{(1+x^4)^3}{64|\vec{\varphi}|^4 x^{14}} \Big[4(1+x^4)^2(1-5x^4+x^8)(\varphi_1^4+\varphi_2^4)$$

$$+\varphi_1^2\varphi_2^2(1+4x^4-106x^8+4x^{12}+x^{16}) \Big], \qquad (9.180)$$

which is an even function of φ_1 and φ_2 and symmetric in their exchange, and

$$f(\varphi_1, \varphi_2) = \frac{(-1+x^4)^5 \varphi_1^3}{64|\vec{\varphi}|^7 x^{14}} \Big[4(1+5x^4+x^8)\varphi_1^4+ \qquad (9.181)$$

$$+ 7(1+6x^4+x^8)\varphi_1^2\varphi_2^2 + 7(1+x^4)^2\varphi_2^4 \Big], \qquad (9.182)$$

which is odd in the first argument and even in the second. Three symmetry operations leave the scalar potential invariant:

$$\begin{cases} \omega \leftrightarrow -\omega \\ \varphi_2 \leftrightarrow -\varphi_2 \end{cases} , \quad \begin{cases} \omega \leftrightarrow \omega + \frac{\pi}{2} \\ \vec{\varphi} \leftrightarrow -\vec{\varphi} \end{cases} , \quad \begin{cases} \omega \leftrightarrow \omega - \frac{\pi}{4} \\ \varphi_1 \rightarrow \varphi_2 \\ \varphi_2 \rightarrow -\varphi_1 \end{cases} . \qquad (9.183)$$

The first one results from a parity-related symmetry, while the last two result from $E_{7(7)}$-duality transformations. Altogether this implies that we get inequivalent potentials only in the expected range $\omega \in [0, \pi/8]$. In fact, depending on the parameter ω, the scalar potential exhibits a different number of vacua, as shown in Fig. 9.3. The $\omega = 0$ case corresponds to the usual truncation of the scalar potential that keeps the SO(8) vacuum (although seemingly unstable, all the masses satisfy

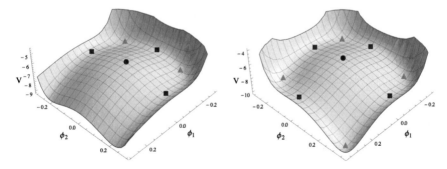

Fig. 9.3 Scalar potential of the G_2 truncation for $\omega = 0$ (left) and for $\omega = \pi/8$ (right). The red dot is the SO(8) vacuum, the blue squares are vacua with SO(7) symmetry, and orange triangles represent vacua with G_2 residual gauge symmetry. New SO(7) and G_2 vacua appear with respect to the $\omega = 0$ case

the Breitenlohner–Freedman bound), the $SO(7)^{\pm}$ vacua, and the G_2 ones. When $\omega \neq 0$, a new $SO(7)$ vacuum and new G_2 vacua appear. In fact, not only the number of vacua changes when $\omega \neq 0$, but also the value of their cosmological constant, as can be seen by looking at Fig. 9.3. In particular, the ratio of the value of the cosmological constant of the various vacua in the two potentials with respect to that of the $\mathcal{N} = 8$ vacuum in the center is an ω-dependent function, different for each one of the vacua.

References

1. M.K. Gaillard, B. Zumino, Duality rotations for interacting fields. Nucl. Phys. **B193**, 221 (1981). Dedicated to Andrei D. Sakharov on occasion of his 60th birthday
2. S. Helgason, *Differential Geometry and Symmetric Spaces* (Academic Press, New York, 1962)
3. H. Nicolai, H. Samtleben, Maximal gauged supergravity in three-dimensions. Phys. Rev. Lett. **86**, 1686–1689 (2001) [arXiv:hep-th/0010076 [hep-th]]
4. H. Samtleben, Lectures on gauged supergravity and flux compactifications. Class. Quant. Grav. **25**, 214002 (2008) [arXiv:0808.4076 [hep-th]]
5. G. Dall'Agata, R. D'Auria, S. Ferrara, Compactifications on twisted tori with fluxes and free differential algebras. Phys. Lett. B **619**, 149–154 (2005) [arXiv:hep-th/0503122 [hep-th]]
6. J. De Rydt, T.T. Schmidt, M. Trigiante, A. Van Proeyen, M. Zagermann, Electric/magnetic duality for chiral gauge theories with anomaly cancellation. JHEP **12**, 105 (2008) [arXiv:0808.2130 [hep-th]]
7. B. de Wit, H. Samtleben, M. Trigiante, The maximal D = 4 supergravities. JHEP **06**, 049 (2007) [arXiv:0705.2101 [hep-th]]
8. G. Dall'Agata, G. Inverso, A. Marrani, Symplectic deformations of gauged maximal supergravity. JHEP **07**, 133 (2014) [arXiv:1405.2437 [hep-th]]
9. G. Dall'Agata, G. Inverso, M. Trigiante, Evidence for a family of SO(8) gauged supergravity theories. Phys. Rev. Lett. **109**, 201301 (2012) [arXiv:1209.0760 [hep-th]]
10. B. de Wit, H. Nicolai, Phys. Lett. **B108**, 285 (1982); B. de Wit, H. Nicolai, Nucl. Phys. **B208**, 323 (1982)
11. N.P. Warner, Some new extrema of the scalar potential of gauged N = 8 supergravity. Phys. Lett. B **128**, 169 (1983)
12. M. Trigiante, Gauged supergravities. Phys. Rept. **680**, 1–175 (2017) [arXiv:1609.09745 [hep-th]]

Supergravity in Arbitrary Dimensions   **10**

While in the first nine chapters of this book, we focused on four-dimensional theories, we use this final chapter to give a brief outlook on supergravity theories in a number of dimensions other than four. Many details of their structure differ from the four-dimensional case, but the main terms and relations remain valid. We begin by analyzing the eleven- and ten-dimensional theories, as they are directly related to the low-energy limit of string theory and then discuss the five-dimensional models, which are very useful in the context of the gauge/gravity correspondence.

10.1   Higher-Dimensional Theories

$\mathcal{N} = 1$ supersymmetry in 4D admits a short representation containing only the graviton, $g_{\mu\nu}$, and the gravitino, ψ_μ. On-shell, both fields have two degrees of freedom, and we can use them to describe a pure supergravity theory. As we have seen in Chap. 4, the presence of the gravitino provides a lot of new interesting features and constraints to the standard gravity theory described by the Einstein–Hilbert action. This model can then be further coupled to matter fields resulting in even richer and more interesting theories. However, the minimal model can contain just the two basic fields $g_{\mu\nu}$ and ψ_μ.

For supergravity models in arbitrary dimensions higher than $D = 4$, the situation changes. The Lorentz group becomes $SO(1, D - 1)$, and hence massless physical states are classified by $SO(D - 2)$ representations. This means that while in four dimensions we can classify states by their helicities, in higher dimensions one has generically many more states, and the representations of supersymmetry will no longer admit short representations with only the graviton and gravitino fields. For a generic D, the graviton field $g_{\mu\nu}$ describes a total of

$$\frac{(D - 2)(D - 2 + 1)}{2} - 1 \tag{10.1}$$

© Springer-Verlag GmbH Germany, part of Springer Nature 2021
G. Dall'Agata, M. Zagermann, *Supergravity*, Lecture Notes in Physics 991,
https://doi.org/10.1007/978-3-662-63980-1_10

Table 10.1   The possible values for η and ϵ together with the resulting minimal spinor types, the minimal number of real supercharges, and the general form of the R-symmetry groups (M = Majorana, SM = Symplectic Majorana, W = Weyl, MW = Majorana–Weyl, SMW = Symplectic Majorana–Weyl)

D	η	ϵ	Min. spinor type	Min. # of real supercharges	R-symmetry group
2	+1	+1	MW	1	$SO(N_L) \times SO(N_R)$
	−1	+1			
3	+1	+1	M	2	$SO(N)$
4	+1	+1	M or W	4	$U(N)$
	−1	−1			
5	−1	−1	SM	8	$Usp(2N)$
6	+1	−1	SMW	8	$Usp(2N_L) \times Usp(2N_R)$
	−1	−1			
7	+1	−1	SM	16	$Usp(2N)$
8	+1	−1		16	$U(N)$
	−1	+1	M or W		
9	−1	+1	M	16	$SO(N)$
10	+1	+1	MW	16	$SO(N_L) \times SO(N_R)$
	−1	+1			
11	+1	+1	M	32	$SO(N)$
12	+1	+1	M or W	64	$U(N)$
	−1	−1			
...	...	...	...	...	...

on-shell degrees of freedom (the metric fluctuations are symmetric traceless matrix states). At the same time, the number of degrees of freedom of a spinor representation grows even faster with D, being

$$\# \, \text{dof} = k \, 2^{[D/2]-1}, \tag{10.2}$$

with $k = 2$ for Dirac spinors, $k = 1$ for Majorana spinors and for Weyl spinors, and $k = 1/2$ for Majorana–Weyl spinors (cf. Table 10.1 and Appendix 10.A). This implies that the gravitino ψ_μ has

$$k \, 2^{[D/2]-1}(D - 2 - 1) \tag{10.3}$$

states, where the $(D - 2 - 1)$ factor comes from the vector index and the fact that the Rarita–Schwinger action is invariant under $\delta \psi_\mu = \partial_\mu \lambda$ and one has to remove the auxiliary spinor $\gamma^\mu \psi_\mu$. By a simple comparison of (10.1) with (10.3), we immediately see that only in four dimensions, one can simply match bosonic and fermionic degrees of freedom in a multiplet by using only the graviton and gravitino fields. As soon as one moves to higher dimensions, one needs more fields, both bosonic and fermionic ones. For instance, in five dimensions the graviton has five degrees of freedom and the gravitino has eight (one has to take a Dirac spinor; no Weyl or Majorana spinors are allowed.[1]) To complete the graviton multiplet, the matching of bosonic and fermionic states thus requires three additional bosonic degrees of freedom represented by a massless vector field, A_μ, the graviphoton. This is just a special example of what is needed to construct a full supergravity multiplet in higher dimensions: antisymmetric tensor fields. These are rank n fields $B_{\mu_1...\mu_n}$ with complete antisymmetry of their indices and a tensor gauge invariance,

$$\delta B_{\mu_1...\mu_n} = n \, \partial_{[\mu_1} \Lambda_{\mu_2...\mu_n]}. \tag{10.4}$$

The vector field is a special instance where $n = 1$, and we usually don't see higher-rank tensor fields in four-dimensional theories because for $n = 2$ they are equivalent to scalar fields (as long as they are massless) and for $n = 3, 4$ they have no physical states. However, in D dimensions the number of physical states of a rank-n tensor is

$$\frac{(D-2)!}{(D-2-n)!n!} \tag{10.5}$$

and they are further reducible into the self- and anti-self-dual parts when $n = D/2$. This means they can play a fundamental role to provide the necessary bosonic degrees of freedom needed to complete a supergravity multiplet.

10.2 Example: $D = 11$ Supergravity

In four and in any other dimension, the maximal number of real supercharges allowed in constructing a theory with fields of spin ≤ 2 is 32. A Majorana spinor in 11 dimensions has $2^5 = 32$ components, and hence 11D supergravity is the highest-dimensional supergravity model that can be constructed without introducing higher-spin fields. The 11D supersymmetry algebra naturally contains a central charge,

$$\{Q_\alpha, Q_\beta\} = (C\Gamma)^m_{\alpha\beta} P_m + (C\Gamma)^{mn}_{\alpha\beta} Z_{mn}, \tag{10.6}$$

which is associated with the existence of membrane-like objects in the theory.

---

[1] In five dimensions, one can combine two Dirac spinors and impose a symplectic Majorana condition, but the resulting number of degrees of freedom is as for a Dirac spinor. For more details, see Appendix 10.A.

The massless 11D graviton has 44 degrees of freedom, while the gravitino has 128 physical states. This implies the need of additional higher-rank tensor fields. It is actually easy to see that a three-form, $C_{\mu\nu\rho}$, with a gauge transformation

$$\delta C_{\mu\nu\rho} = 3\, \partial_{[\mu} \Lambda_{\nu\rho]}$$

has exactly the 84 missing physical states needed to complete a supermultiplet

$$\{e_\mu^a, \ \psi_\mu, \ C_{\mu\nu\rho}\}. \tag{10.7}$$

This theory clearly has no scalar potentials (there are no scalars at all), and it has a unique parameter given by the 11D gravitational constant, κ_{11}.

The action of 11-dimensional supergravity was constructed first by Cremmer, Julia, and Scherk [1] and consists of very few terms:

$$
\begin{aligned}
S \ = \ & \frac{1}{2\kappa_{11}^2} \int d^{11}x\, e \left[R(\omega) - \overline{\psi}_\mu \Gamma^{\mu\nu\rho} D_\nu \left(\frac{\omega + \widehat{\omega}}{2} \right) \psi_\rho - \frac{1}{24} G_{\mu\nu\rho\sigma} G^{\mu\nu\rho\sigma} \right. \\
& - \frac{2\sqrt{2}}{(144)^2} \epsilon^{\mu_1 \ldots \mu_{11}} G_{\mu_1 \ldots \mu_4} G_{\mu_5 \ldots \mu_8} C_{\mu_9 \mu_{10} \mu_{11}} \\
& \left. - \frac{\sqrt{2}}{192} \left(\overline{\psi}_\mu \Gamma^{\mu\nu\rho\sigma\tau\eta} \psi_\nu + 12\, \overline{\psi}^\rho \Gamma^{\sigma\tau} \psi^\eta \right) \left(2\, G_{\rho\sigma\tau\eta} - \frac{3}{2}\sqrt{2}\kappa\, \overline{\psi}_{[\rho} \Gamma_{\sigma\tau} \psi_{\eta]} \right) \right],
\end{aligned}
\tag{10.8}
$$

where

$$G_{\mu\nu\rho\sigma} = 4\, \partial_{[\mu} C_{\nu\rho\sigma]}, \tag{10.9}$$

$$
\omega_{\mu ab} = \omega_{\mu ab}(e) - \frac{1}{8} \left[\overline{\psi}_\alpha \Gamma_{\mu ab}{}^{\alpha\beta} \psi_\beta + 2\overline{\psi}_\mu \Gamma_b \psi_a \right.
$$

$$
\left. -2\overline{\psi}_a \Gamma_\mu \psi_b + 2\overline{\psi}_b \Gamma_a \psi_\mu \right], \tag{10.10}
$$

$$
\widehat{\omega}_{\mu ab} = \omega_{\mu ab} - \frac{1}{4} \left(\overline{\psi}_\mu \Gamma_b \psi_a - \overline{\psi}_a \Gamma_\mu \psi_b + \overline{\psi}_b \Gamma_a \psi_\mu \right). \tag{10.11}
$$

This action obviously includes the kinetic terms for the graviton, the gravitino, and the three-form field C in the first line. The modified spin connection $\widehat{\omega}$ has been introduced to take into account the four-Fermi interactions between the gravitino fields. The second line is a special Chern–Simons-like term, which does not depend on the metric. The last line, finally, contains further interaction terms between the three-form and the gravitino. The invariance under supersymmetry follows from the application of

$$\delta e_\mu^a = \frac{1}{2}\bar{\epsilon}\Gamma^a\psi_\mu, \tag{10.12}$$

$$\delta\psi_\mu = D_\mu(\widehat{\omega})\epsilon + \frac{\sqrt{2}}{288}\left(\Gamma_\mu{}^{\nu\rho\sigma\tau} - 8\delta_\mu^\nu\Gamma^{\rho\sigma\tau}\right)\left(G_{\nu\rho\sigma\tau}\right.$$

$$\left. + \frac{3}{2}\sqrt{2}\,\overline{\psi}_{[\nu}\Gamma_{\rho\sigma}\psi_{\tau]}\right)\epsilon, \tag{10.13}$$

$$\delta C_{\mu\nu\rho} = -\frac{3}{4}\sqrt{2}\,\bar{\epsilon}\Gamma_{[\mu\nu}\psi_{\rho]}. \tag{10.14}$$

We don't discuss here the proof of invariance under supersymmetry of this action nor the peculiar new features appearing, thanks to the presence of a three-form field. On the other hand, we are going to use this theory as a starting point for a qualitative discussion of the features of the models that can be obtained by dimensional reduction.

10.3 Dimensional Reduction and Ten-Dimensional Supergravities

The 11-dimensional supergravity action we just presented should describe the low-energy limit of M-theory, a supposedly consistent quantum theory of membranes in 11 dimensions, which arises as the strong coupling limit of ten-dimensional type IIA string theory. It is therefore conceivable that one could show a detailed relation between supergravity in 11 dimensions and the low-energy limits of ten-dimensional string theory models. Hence, we will now discuss the Kaluza–Klein reduction of 11-dimensional supergravity to ten dimensions and less.

A simple way to dimensionally reduce a theory is to consider a spacetime metric where one of the coordinates runs on a circle of fixed radius. In this way, the fluctuations of the fields over this coordinate will be constrained by the geometry and will result in effective masses and couplings for the fluctuations in the rest of the spacetime. For instance, if we reduce 11D supergravity over a circle and keep only the massless modes in the resulting effective 10D theory, we obtain a much richer spectrum than the one we discussed in the previous section. If we split the 11D coordinates, x^M, into the non-compact ones, x^μ, and the circle coordinate, $\theta \equiv x^{10}$, we see that the reduction of the 11D metric gives rise to three different fields, with different transformation properties with respect to the residual 10D diffeomorphism invariance:

$$g_{MN} \rightarrow \begin{cases} g_{\mu\nu} & \text{ten-dimensional metric,} \\ g_{\mu 10} \sim A_\mu & \text{ten-dimensional vector,} \\ g_{10\,10} \sim \phi & \text{ten-dimensional scalar.} \end{cases} \qquad (10.15)$$

From the 11-dimensional line element,

$$ds_{11}^2 = e^{2\phi(x)}(d\theta + dx^\mu A_\mu(x))^2 + ds_{10}^2, \qquad (10.16)$$

we see that the ten-dimensional photon field, A_μ, describes the fibration of the circle on the ten-dimensional base spacetime and the dilaton field, ϕ, is associated with the radius of the circle of the internal direction. In the same fashion, we can reduce the 11-dimensional rank-3 tensor,

$$C_{MNP} \rightarrow \begin{cases} C_{\mu\nu\rho} & \text{ten-dimensional three-form,} \\ C_{\mu\nu 10} \equiv B_{\mu\nu} & \text{ten-dimensional two-form,} \\ C_{\mu 10\,10} = 0 & \text{vanishing because of antisymmetry,} \end{cases} \qquad (10.17)$$

and the 11-dimensional gravitino,

$$\Psi_M \rightarrow \begin{matrix} \psi_\mu^+, \quad \psi_\mu^- & \text{Two ten-dimensional gravitini} \\ \\ \psi_{10}^+ \equiv \lambda^+, \ \psi_{10}^- \equiv \lambda^- & \text{Two ten-dimensional spin } 1/2 \text{ fields.} \end{matrix} \qquad (10.18)$$

The plus and minus signs on the fermions refer to the chirality of the resulting ten-dimensional spinor fields. In fact, the 32-component Majorana spinor in 11D reduces to a 32-component Majorana spinor in 10D, which, however, can be further decomposed into two 10D Majorana–Weyl spinors of opposite chiralities, each having 16 independent real components. The same is true for the supercharges, $Q^{11} \rightarrow \{Q^+, Q^-\}$, which split into two Majorana–Weyl representations of opposite chirality in ten dimensions. The resulting model is therefore a *non-chiral* ten-dimensional supergravity theory with two supersymmetry generators: *type IIA supergravity*.

The splitting of the 11-dimensional supercharge implies that one can construct other supergravity models with only one supercharge (type I models) or with both supercharges of the same chirality (type IIB supergravity). It is actually useful to summarize the resulting spectrum of type IIA supergravity as follows (the numbers indicate the independent on-shell degrees of freedom of each field):

$$\{ \phi \ , \ \lambda^- \ , \ B_{\mu\nu} \ , \ \psi_\mu^+ \ , \ g_{\mu\nu} \} \quad \text{Common Sector}$$

$$1 \quad 8 \quad 28 \quad 56 \quad 35 \quad = 64_B + 64_F$$

$$\{ \lambda^+ \ , \ C_{\mu\nu\rho} \ , \ A_\mu \ , \ \psi_\mu^- \} \quad \text{IIA RR sector}$$

$$8 \quad 56 \quad 8 \quad 56 \quad = 64_B + 64_F. \tag{10.19}$$

The first line is a consistent 10D supermultiplet by itself when one restricts oneself to a single supersymmetry, while the second line collects the fields that complete the multiplet to the type IIA multiplet. The type IIA supergravity theory based on the above fields arises as the low energy limit of type IIA string theory. In the context of this string theory, the fields in the first line are referred to as the Neveu–Schwarz–Neveu–Schwarz (NSNS) sector, whereas the fields in the second line are called the Ramond–Ramond (RR) sector.

As the fields in the first line of (10.19) form a consistent supermultiplet of the 10D superalgebra with one supersymmetry generator, it is natural to suspect that there is also a consistent supergravity action with this reduced field content. This action indeed exists and is referred to as type I supergravity. Type I supergravity by itself, however, has quantum anomalies, and in order to cancel these anomalies, one has to add suitable super-Yang–Mills sectors based on 10D vector multiplets. We will come back to this point and its relation to string theory at the end of this section.

The Lagrangian of type IIA supergravity is obtained by dimensional reduction of the one of the 11-dimensional theory presented in the previous section [2]. For instance, from the kinetic term of the four-form, G, we obtain the kinetic terms of the ten-dimensional four- and three-forms, $F_4 = dC_3 + A \wedge H_3$ and $H_3 = dB$, respectively:

$$\frac{1}{48} G_{MNPQ} G^{MNPQ} \rightarrow \frac{1}{48} F_{\mu\nu\rho\sigma} F^{\mu\nu\rho\sigma} + \mathrm{e}^{-2\phi} \frac{1}{12} H_{\mu\nu\rho} H^{\mu\nu\rho}. \tag{10.20}$$

The 11D Ricci scalar reduces to the 10D Ricci scalar, plus kinetic terms for the additional 10D degrees of freedom obtained from the 11D metric, i.e., for the 10D vector and scalar fields. It is important to point out, however, that the presence of the metric determinant, together with the direct reduction of the curvature term, gives rise to dilaton factors in front of the various 10D terms, so that we cannot view the resulting action as a standard Einstein gravity theory unless we perform a rescaling of the metric. More concretely, the straightforward reduction of the Einstein–Hilbert term gives

$$e_{11} R_{11} \rightarrow e_{10} \, \mathrm{e}^{-2\phi} \left(R_{10} - \frac{1}{4} e^{2\phi} F^2 + 4(\partial\phi)^2 \right). \tag{10.21}$$

This has an apparent wrong sign for the kinetic term of the dilaton and at the same time a non-trivial scalar factor in front of the ten-dimensional Einstein–Hilbert term. To go to the Einstein frame, one performs a rescaling

$$g_{\mu\nu} \rightarrow e^{\frac{\phi}{2}} g_{\mu\nu}, \qquad (10.22)$$

so that (10.21) gets mapped to the more standard form

$$e_{10} \left(R_{10} - \frac{e^{\frac{3}{2}\phi}}{4} F^2 - \frac{1}{2}(\partial\phi)^2 \right). \qquad (10.23)$$

Once we reduced the theory to ten dimensions, we see that we now have two parameters: k_{11} and the vev of e^{ϕ}. The field ϕ is the first *modulus* we meet, i.e., a massless scalar field whose expectation value is related to some geometrical property of the internal spacetime and which affects the coupling constants of the effective theory. In fact, the scalar potential of type IIA supergravity is trivial,[2]

$$V(\phi) \equiv 0. \qquad (10.24)$$

The full bosonic sector of the type IIA supergravity action in the Einstein frame reads:

$$
S_{IIA} = \frac{1}{2\kappa_{10}^2} \int d^{10}x\, e \left(R - \frac{1}{2} \partial_\mu \phi \partial^\mu \phi - \frac{1}{12} e^{-\phi} H_{\mu\nu\rho} H^{\mu\nu\rho} \right)
$$
$$
- \frac{1}{2\kappa_{10}^2} \int d^{10}x\, e \left(\frac{1}{48} e^{\frac{\phi}{2}} G_{\mu\nu\rho\sigma} G^{\mu\nu\rho\sigma} - \frac{1}{4} e^{\frac{3}{2}\phi} F_{\mu\nu} F^{\mu\nu} \right) \qquad (10.25)
$$
$$
- \frac{1}{4\kappa_{10}^2} \int B_2 \wedge dC_3 \wedge dC_3 ,
$$

where

$$G_4 = dC_3 - A_1 \wedge H_3, \quad H_3 = dB_2, \quad F_2 = dA_1. \qquad (10.26)$$

In our simple reduction, the expectation value of ϕ is related to the radius of the S^1 we used to go from eleven to ten dimensions. If we interpret the IIA supergravity as a low energy limit of type IIA string theory, we can identify the vev of the dilaton with the string coupling constant as $\langle e^{\phi} \rangle \sim g_s$. It is then clear why taking a strong

---

[2] There exists a mass-like deformation of type IIA supergravity, called *massive type IIA supergravity*, which does contain a scalar potential for the dilaton [3].

coupling limit of type IIA strings, $g_s \to \infty$, implies to go from type IIA to M-theory. In our supergravity setup, which, however, cannot be fully trusted when one goes beyond the classical regime, this corresponds to send the radius of the S^1 to infinity and therefore to go back to an 11-dimensional background.

In order to complete this quick survey of ten-dimensional supergravity models, we just mention a couple of things. Minimal supergravity in ten dimensions (Type I) can be coupled to matter. In fact ten-dimensional ($\mathcal{N} = 1$) supersymmetry allows for another multiplet: a super-Maxwell multiplet, containing a vector and a spin 1/2 field:

$$\begin{matrix} \{ A_\mu , & \lambda^\pm \}. \\ 8 & 8 \end{matrix} \tag{10.27}$$

While at the classical supergravity level we could consider coupling an arbitrary number of such multiplets to the supergravity action, anomaly cancellation restricts the allowed possibilities to well-defined gauge groups and therefore to a well-defined and fixed number of vector multiplets. This is also reflected in the corresponding string theories with 10D, $\mathcal{N} = 1$ supersymmetry, namely, the type I string theory and the two heterotic string theories. Their low energy limits consist of type I supergravity coupled to $\mathcal{N} = 1$ super-Yang–Mills theory with gauge group SO(32) for the type I string theory and with gauge group SO(32) or $E_8 \times E_8$ for the two heterotic string theories.[3]

As we discussed before, we could also consider an $\mathcal{N} = 2$ theory with both supercharges of the same chirality. This is called *type IIB supergravity* [4,5], and its field content can be obtained by substituting the RR sector in Eq. (10.19) with the following:

$$\begin{matrix} \{ C_0 , & \lambda^{-(2)} , & C_{\mu\nu} , & C_{\mu\nu\rho\sigma} , & \psi_\mu^{+(2)} \} & \text{IIB RR sector} \\ 1 & 8 & 28 & 35 & 56 & = 64_B + 64_F . \end{matrix} \tag{10.28}$$

We point out that the rank-4 tensor field appearing in the spectrum has only 35 degrees of freedom on shell, because its field strength is a five-form in ten dimensions, and hence one can impose a self-duality constraint on it,

$$F_5 = \star F_5 , \quad F_{\mu_1 \ldots \mu_5} = \frac{1}{5!} \epsilon_{\mu_1 \ldots \mu_{10}} F^{\mu_6 \ldots \mu_{10}} . \tag{10.29}$$

---

[3] Despite the presence of gauge interactions and half-maximal supersymmetry, these theories do not have a scalar potential, which can be traced back to the gauge group being unrelated to the R-symmetry group or any scalar field isometries and to the fact that there are no scalars in the vector multiplets.

This constraint creates some problems when one wants to construct a Lorentz covariant action because, as we can easily see, the standard kinetic term would be identically zero

$$F_5 \wedge \star F_5 = F_5 \wedge F_5 = -F_5 \wedge F_5 = 0, \tag{10.30}$$

where we first used the constraint and then swapped the order of the two five-forms. A covariant and supersymmetric Lagrangian with a single scalar auxiliary field, which is a pure gauge of a new symmetry of the action and a singlet under supersymmetry, is provided in [6]. As a mnemonic tool, one can use the following bosonic Lagrangian, where the self-duality equation is imposed on the resulting equations of motion,

$$
\begin{aligned}
S_{IIA} = {} & \frac{1}{2\kappa_{10}^2} \int d^{10}x \, e \left(R - \frac{1}{2} \partial_\mu \phi \partial^\mu \phi - \frac{1}{12} e^{-\phi} H_{\mu\nu\rho} H^{\mu\nu\rho} \right) \\
& - \frac{1}{2\kappa_{10}^2} \int d^{10}x \, e \left(\frac{1}{2} e^{2\phi} \partial_\mu C_0 \partial^\mu C_0 + \frac{1}{2\cdot 5!} F_{\mu_1 \ldots \mu_5} F^{\mu_1 \ldots \mu_5} \right) \\
& - \frac{1}{2\kappa_{10}^2} \int d^{10}x \, e \, \frac{1}{12} e^{\phi} \left(F_{\mu\nu\rho} - C_0 H_{\mu\nu\rho} \right) \left(F^{\mu\nu\rho} - C_0 H^{\mu\nu\rho} \right) \\
& - \frac{1}{4\kappa_{10}^2} \int C_4 \wedge H_3 \wedge F_3 \,,
\end{aligned}
\tag{10.31}
$$

where

$$H = dB_2, \quad F_3 = dC_2 \quad F_5 = dC_4 - \frac{1}{2} C_2 \wedge H_3 + \frac{1}{2} B_2 \wedge F_3. \tag{10.32}$$

This theory arises as the low energy limit of type IIB string theory.

10.4   Dimensional Reduction and the Origin of Gauged Supergravities

In the same way as type IIA supergravity arises by reduction of 11-dimensional supergravity on a circle, one could try to produce many different models in four dimensions by compactifying ten- or eleven-dimensional supergravities on six- or seven-dimensional spaces with various geometries. Obviously, the matter content and the number of supersymmetries will depend on the vacuum expectation value of the metric on the internal space. However, following what happened in the IIA case, these reductions (when they preserve some supersymmetry) will usually generate a number of massless scalar fields associated with the shape and volume of the internal space, just like the IIA dilaton was associated with the volume (the radius) of the circle on which one compactifies M-theory. As discussed in detail in Sect. 7.4.1, these moduli fields are a serious phenomenological problem of the resulting effective theories, e.g., because they produce long range fifth forces, modifying the behavior of Newtonian gravity in an unacceptable way. A simple and efficient way out is given by assuming that also the other higher-rank tensor fields appearing in the ten- and eleven-dimensional theories acquire a non-trivial expectation value (the fluxes). As we will sketch in the following, this results in the deformation of the four-dimensional supergravity models by the gauging process and hence leads to gauged supergravity models, which then do have a scalar potential that can make (at least some of) the moduli sufficiently massive. The remarkable aspect of this type of reduction is that the fluxes can be treated perturbatively and produce closed computable expressions for the lower dimensional couplings and potentials.

We will give here an overview of the main features of flux compactifications leading to gauged supergravities. For this reason we will focus on the simple case of 11-dimensional supergravity reduced on the seven-dimensional torus, $\mathbb{T}^7$, and discuss only its bosonic sector. The standard reduction, where one truncates the spectrum to the massless modes, leads to $\mathcal{N} = 8$ ungauged supergravity in four dimensions. In detail, the reduction of the metric and three-form tensor field produces the following massless spectrum: As we already discussed previously,

$g_{\mu\nu}$	$g_{\mu I}$	g_{IJ}	
1 graviton		28	
$C_{\mu\nu\rho}$	$C_{\mu IJ}$	$C_{\mu\nu I}$, C_{IJK}	
0 dof	7+21 vectors	7	35
		= 63 scalars + 7 tensors	

massless tensors can be dualized to massless scalar fields, and hence, after such duality, one gets a total of 70 scalar fields, which is the standard scalar content of the $\mathcal{N} = 8$ supergravity multiplet.

An extremely important aspect of 4D supergravity is the U-duality group emerging as a generalization of the standard electric–magnetic duality (see Chap. 9). For $\mathcal{N} = 8$ supergravity, this is $E_{7(7)} \subset \mathrm{Sp}(56,\mathbb{R})$ and implies that there is an underlying symplectic action on 56 vector fields, 28 of which may appear simultaneously in the action. Using the appropriate language, we can then consider the vectors coming from the $\mathbb{T}^7$ compactification as the 28 electric ones: $A_\mu^\Lambda = \{g_\mu^I, C_{\mu IJ}\}$. However, one should also be able to identify the higher dimensional origin of the 28 dual vector fields, $A_{\mu\Lambda}$. Twenty-one of them can be readily identified by the reduction of the dual form of $G_4 = dC_3$. This dual form is schematically obtained as $G_7 \sim \star G_4$, so that its Bianchi identity reproduces the equation of motion for G_4: $dG_7 = G_4 \wedge G_4$. The seven-form field strength then is the curvature of a six-form potential that, upon reduction, produces

$$C_{\mu IJKLM} = \epsilon_{IJKLMNP} A_\mu^{NP}.$$ (10.33)

We also note that the seven scalars dual to the tensor fields above are also obtained from the same potential as C_{IJKLMN}. The remaining seven dual vector fields should have an origin as dual metric fields, $\tilde{g}_{\mu I}$, but it is not obvious how to achieve this yet, though we expect them to correspond to non-geometric deformations. Anyway, at the level of the 4D action, as we already discussed in Chap. 8, both descriptions are equally valid, and as long as we don't have gaugings we can (almost) freely dualize the curvatures and the vectors.

For what concerns the scalar fields, we get a σ-model, but no scalar potential, as expected for a standard Einstein–Maxwell extended supergravity.

The duality group is broken, and the 4D model gets deformed to a gauged supergravity by introduction of "fluxes" for the four-form, i.e., an expectation value for this field on the internal space:

$$\mathcal{G}_{IJKL} \equiv \langle G_{IJKL} \rangle \neq 0.$$ (10.34)

We now discuss how this is achieved, briefly noting that a similar situation occurs when introducing a flux for the dual seven-form field, or a combination of the two.

A simple sketch of the reduction of the kinetic term of the four-form explains how both the gauging and the scalar potential are generated. Without fluxes, the reduction of the four-form kinetic term produces the expected kinetic terms for the scalars and vectors:

$$\underbrace{g^{IK} g^{JL}}_{\text{gauge kin. function}} \partial_{[\mu} C_{\nu]IJ} \partial^{[\mu} C^{\nu]}{}_{KL} + \underbrace{g^{IA} g^{JB} g^{KC}}_{\text{scalar } \sigma\text{-model}} \partial_\mu C_{IJK} \partial^\mu C_{ABC} + \dots$$ (10.35)

In the presence of a non-trivial flux, a new term emerges at the quadratic level, containing only the scalar fields coming from the metric. This is an obvious scalar potential term, and it is of the second order in the coupling constants dictated by the fluxes $\mathcal{G}$:

$$V(g_{IJ}) = g^{IA} g^{JB} g^{KC} g^{LD} \mathcal{G}_{IJKL} \mathcal{G}_{ABCD}.$$ (10.36)

Another important coupling that emerges is linear in the flux,

$$\partial_{[\mu} C_{\nu]IJ} g^{\mu A} g^{\nu B} g^{IC} g^{JD} \mathcal{G}_{ABCD},$$ (10.37)

which reconstructs the $\mathcal{O}(\mathcal{G})$ couplings giving origin to non-Abelian covariant field strengths

$$\partial_{[\mu} C_{\nu]IJ} + g_\mu^A g_\nu^B \mathcal{G}_{ABIJ}.$$ (10.38)

From this same expression, it is also clear that $\mathscr{G}_{IJKL}$ play the role of structure constants. The gauging of the theory becomes even more evident if we look at the kinetic terms of the scalar fields. Also these kinetic terms get modified to new expressions involving $\mathscr{O}(\mathscr{G})$ couplings reconstructing covariant derivatives:

$$\partial_\mu C_{IJK} + \mathscr{G}_{IJKL}\, g_\mu^L = \widehat{\partial}_\mu C_{IJK}. \tag{10.39}$$

We therefore see that fluxes not only define the gauge couplings and the scalar potential, but they also tell us the form of the embedding tensor, because they tell us which vectors participate in the gauging and specify the couplings between the scalar and vector fields defining the covariant derivatives.

Since the fluxes now play the role of gauge structure constants, one could expect that they obey standard Jacobi identities. Although it may not be evident from the example above, it can be shown that there is a one- to-one correspondence between the consistency conditions on the gauge structure constants in four dimensions and the Bianchi identities of the form fluxes in ten or 11 dimensions. There is, however, an important subtlety that we want to emphasize here. When dealing with flux compactifications (and especially in the case of non-trivial geometric fluxes, i.e., globally defined torsion terms), the tensor fields appearing in the geometric reduction transform under gauge transformations of the vector fields and vice versa. This implies that generically the standard Jacobi identities coming from the gauge algebra of the vector fields do not close. In fact, the gauge algebra is now really a free differential algebra involving also the tensor fields, and therefore it is only the closure of this structure that imposes all the necessary consistency conditions on the structure constants (also to reproduce the higher dimensional Bianchi identities).

We conclude with some comments on the action of the duality group in relation to the situation described in the previous paragraph. Although the natural setup coming from a straightforward reduction of a flux background may result in a free differential algebra that includes tensor fields, one may always use the duality group to rotate the vector field basis so that no tensor fields are present in the final Lagrangian. Obviously, in order to be consistent, the embedding of the gauge group in the duality group will also be rotated accordingly, and the interpretation of the resulting algebra in terms of the original theory may not be straightforward anymore. Actually, this is a rough explanation of how non-geometric fluxes arise and why we should expect them if we believe that the U-duality group of the effective theory survives as a symmetry of the fully fledged higher dimensional fundamental theory (maybe with some restrictions).

10.5   Example: $D = 5$

As a further illustration of the new structures emerging in higher dimensional supergravity, we consider the case of five dimensions, where supergravity theories have numerous applications, e.g., in the context of the AdS/CFT correspondence, the study of black holes and domain walls, or phenomenological scenarios such as the

Randall–Sundrum or Hořava–Witten-type models. As we will see, the form of the R-symmetry group provides useful constraints on the scalar manifold geometries, which, for $\mathcal{N} \geq 4$, even fixes the target spaces completely once type and number of multiplets in the theory are specified.

In five dimensions, one cannot impose Majorana or Weyl conditions,[4] and the possible numbers of real supercharges are 8, 16, 24, 32. One might call this $\mathcal{N} = 1, 2, 3, 4$ supersymmetry, but in analogy with the counting in 4D, one usually refers to these possible 5D supersymmetries as $\mathcal{N} = 2, 4, 6, 8$, respectively, as we will also do in these lecture notes. As explained in Appendix 10.A, in 5D, the parameter ϵ defined in (10.79) is $\epsilon = -1$, so that one may impose a symplectic Majorana condition for the supersymmetry generators ($i = 1, \ldots, \mathcal{N}$),

$$(Q_i)^* = \Omega_{ij} B Q_j. \tag{10.40}$$

This has the advantage of making the action of the R-symmetry group manifest. More concretely, the linear rotations of the Q_i that preserve the above symplectic Majorana condition as well as the 5D supersymmetry algebra,

$$\{Q_i, Q_j^T\} = \Omega_{ij}(C\Gamma^a)P_a, \tag{10.41}$$

form the R-symmetry groups $USp(\mathcal{N}) \equiv U(\mathcal{N}) \cap Sp(\mathcal{N}, \mathbb{R})$.

10.5.1 $\mathcal{N} = 2$ in 5D

For $\mathcal{N} = 2$ ungauged supergravity in 5D, there are three important supermultiplets:

- Supergravity multiplet: $(e_\mu^a, \psi_\mu^i, A_\mu)$ ($i = 1, 2$)
  Apart from the fünfbein, e_μ^a, it contains two gravitini, ψ_μ^i, and a vector field, A_μ, the "graviphoton."
- Vector multiplet: $(A_\mu, \lambda^i, \varphi)$
  The superpartners of the vector field, A_μ, are two gaugini, λ^i, and one real scalar field, φ.
- Hypermultiplet: $(\zeta^{1,2}, q^{1,2,3,4})$
  Here, $\zeta^{1,2}$ are two spin-1/2 fermions ("hyperini"), and the $q^{1,2,3,4}$ denote four real scalar fields.

---

[4] A simple way to show that there are no Majorana spinors in 5D is to use the Majorana representation for the first four gamma matrices, which are then manifestly real in this representation and can thus naturally act on real spinors. The fifth gamma matrix, however, is then $\pm \gamma_5$, which is manifestly imaginary due to the i in its definition. There is thus no set of five gamma matrices with the same reality properties.

In the above, the index i of the gravitini and the gaugini is a doublet index of the R-symmetry group $USp(2)_R \cong SU(2)_R$. The hyperini, by contrast, are inert under $SU(2)_R$, which is the reason that we do not use i to label these two fermions. The scalar φ is likewise $SU(2)_R$-inert, whereas the hyperscalars $q^{1,2,3,4}$ form two doublets under $SU(2)_R$. All spinors are symplectic Majorana spinors.

In order to write down the general Lagrangian for $\mathcal{N} = 2$ supergravity coupled to n_V vector multiplets and n_H hypermultiplets, it is useful to group these fields as follows:

$$(e_\mu^a, \psi_\mu^i, A_\mu) \oplus n_V \times (A_\mu, \lambda^i, \varphi) \oplus n_H \times (\zeta^{1,2}, q^{1,2,3,4})$$
$$= (e_\mu^a, \psi_\mu^i, A_\mu^I, \lambda^i, \zeta^A, \varphi^x, q^u) \quad (10.42)$$

where

$$I = 0, 1, \ldots, n_V \quad (10.43)$$

$$x = 1, \ldots, n_V \quad (10.44)$$

$$u = 1, \ldots, 4n_H \quad (10.45)$$

$$A = 1, \ldots, 2n_H. \quad (10.46)$$

Here we have combined the graviphoton and the n_V vector fields of the vector multiplets into an $(n_V + 1)$-plet of vectors, A_μ^I. The indices x and u are curved indices on the scalar manifolds, $\mathcal{M}_V$ and $\mathcal{M}_H$, of the vector scalars and the hyperscalars, respectively. The gaugini, λ^{ix}, transform as tangent vectors of $\mathcal{M}_V$, and one may use curved indices, x, to label them, just as we did, or, as one often also finds in the literature, one could use a flat tangent space index, $a = 1, \ldots, n_V$, instead of the curved tangent space index x. Both notations can be easily converted into one another by contraction with vielbein on $\mathcal{M}_V$. As the holonomy group of $\mathcal{M}_V$ has no particular restrictions imposed by the supersymmetry algebra (see below), the use of flat vs. curved indices x and a has no clear advantage. This is different for the hyperini, as we will explain below.

In terms of the above fields, the bosonic Lagrangian of $\mathcal{N} = 2$ matter-coupled ungauged supergravity in 5D can be written as follows [7, 8]:

$$e^{-1}\mathscr{L}_{\text{bos}} = \frac{1}{2}R - \frac{1}{4}\tilde{a}_{IJ}(\varphi)F_{\mu\nu}^I F^{\mu\nu J} - \frac{1}{2}g_{xy}(\varphi)\partial_\mu\varphi^x\partial^\mu\varphi^y$$
$$- \frac{1}{2}h_{uv}(q)\partial_\mu q^u\partial^\mu q^v + \frac{1}{6\sqrt{6}}C_{IJK}\epsilon^{\mu\nu\rho\sigma\lambda}F_{\mu\nu}^I F_{\rho\sigma}^J A_\lambda^K. \quad (10.47)$$

In this expression, C_{IJK} is a constant, completely symmetric tensor. As was shown by Sierra [8], the two sigma models associated with the hyperscalars and the vector scalars do not mix, i.e., the scalar manifold metric is block diagonal in these two sectors and the total scalar manifold decomposes into a direct product, $\mathcal{M}_V \times \mathcal{M}_H$, just as in four dimensions.

10.5.1.1 The Geometry of $\mathcal{M}_V$

The scalars, φ^x, of the vector multiplets are inert under the R-symmetry group $SU(2)_R$. The holonomy group of $\mathcal{M}_V$ therefore does not receive any constraints from the R-symmetry group, as we already mentioned above. However, as they are connected by supersymmetry to the vector fields (or rather to n_V of them), the scalar manifold $\mathcal{M}_V$ inherits part of the vector field structure. In fact, one finds that $\mathcal{M}_V$ is completely determined by the constants C_{IJK} that define the Chern–Simons term in the action. More precisely, the C_{IJK} define a cubic polynomial [7],

$$N(X^I) \equiv C_{IJK} X^I X^J X^K \tag{10.48}$$

on an auxiliary space, $\mathbb{R}^{n_V+1}$, spanned by real coordinates X^I ($I = 0, 1, \ldots, n_V$). On this auxiliary space, N then defines a (not necessarily positive definite) metric,

$$a_{IJ}(X) \equiv -\frac{1}{3} \frac{\partial}{\partial X^I} \frac{\partial}{\partial X^J} \log N(X). \tag{10.49}$$

The scalar manifold $\mathcal{M}_V$ is then given as a cubic hypersurface in the auxiliary $\mathbb{R}^{n_V+1}$:

$$\mathcal{M}_V = \{X^I \in \mathbb{R}^{n_V+1} | N(X) = 1\}. \tag{10.50}$$

$\mathcal{M}_V$ can be parameterized by n_V real coordinates, which are identified with the physical scalar fields, φ^x. The metric, g_{xy}, on $\mathcal{M}_V$ is given by the pull-back of the auxiliary metric a_{IJ}, and the "gauge kinetic function" $\tilde{a}_{IJ}$ is the restriction of a_{IJ} to the hypersurface $\mathcal{M}_V$:

$$g_{xy}(\varphi) = \frac{3}{2} \frac{\partial X^I}{\partial \varphi^x} \frac{\partial X^J}{\partial \varphi^y} a_{IJ} \Big|_{N(X)=1} \tag{10.51}$$

$$\tilde{a}_{IJ}(\varphi) = a_{IJ} \Big|_{N(X)=1}. \tag{10.52}$$

The true scalar manifold is then actually the subspace of (10.50) for which $g_{xy}(\varphi)$ and $\tilde{a}_{IJ}(\varphi)$ are positive definite. The above-described geometry of the scalar manifold of the vector multiplet scalars is called *very special (real) geometry*. Upon dimensional reduction to four dimensions, $\mathcal{M}_V$ becomes a special Kähler manifold of restricted type, namely, one for which the holomorphic prepotential is purely cubic. Special Kähler manifolds that arise in this way from 5D are called *very special Kähler geometries*, and the corresponding map is called the *R-map* [7].

10.5.1.2 The Geometry of $\mathcal{M}_H$

As the hyperscalars transform non-trivially under the R-symmetry group $SU(2)_R$, we expect the holonomy group of $\mathcal{M}_H$ to respect this structure and hence to contain $SU(2)$ as a factor. The largest group G such that $SU(2) \times G$ is still a subgroup of (the maximal holonomy group) $O(4n_H)$ is $Usp(2n_H)$. The holonomy group of $\mathcal{M}_H$ should thus be contained in $SU(2) \times USp(2n_H)$, with the $SU(2)$ part being non-trivial. Manifolds of this type are called *quaternionic Kähler*, and we discussed them in detail already in Chap. 8, because the hypermultiplet geometry in 5D is really identical to the scalar field geometry of 4D hypermultiplets. The restricted holonomy group also means that the tangent space group can be restricted to $SU(2) \times USp(2n_H)$. Just as in four dimensions, this allows a natural split of the flat tangent space index of $\mathcal{M}_H$ into an $SU(2)$ index $i = 1, 2$ and an $USp(2n_H)$ index $A = 1, \ldots, 2n_H$. These indices are to be identified with the R-symmetry group index i and the index A of the hyperini ζ^A.

10.5.2 $\mathcal{N} = 4$ in 5D

For 5D, $\mathcal{N} = 4$ supersymmetry, the R-symmetry group is $USp(4)_R$. $USp(4)_R$ is a double cover of $SO(5)$, which, by abuse of notation, we will also call $SO(5)_R$. The two relevant multiplets in ungauged supergravity are [9]:

- Supergravity multiplet: This multiplet consists of the graviton, four gravitini, six vector fields, four spin-1/2 fields, and one real scalar field. This scalar is necessarily $SO(5)_R$-inert and parameterizes the real line:

$$\mathcal{M}_{SG} \cong \mathbb{R} \cong SO(1, 1) \tag{10.53}$$

- Vector multiplet: This multiplet contains one vector, four spin-1/2 fields, and five real scalars in the **5** of $SO(5)_R$. The holonomy group of the scalar manifold of n_V such vector multiplets should thus contain $SO(5)$ as a factor. The largest remaining group factor that still allows the embedding into the maximal holonomy group $O(5n_V)$ is $SO(n_V)$, i.e., $Hol(\mathcal{M}_V) \subset SO(5) \times SO(n_V)$ with the $SO(5)$ part being non-trivial. According to Berger's classification [10], the only Riemannian manifold of dimension $5n_V$ with this property is

$$\mathcal{M}_V = \frac{SO(5, n_V)}{SO(5) \times SO(n_V)}. \tag{10.54}$$

10.5.3 $\mathcal{N} = 6$ in 5D

In this case the R-symmetry group is $USp(6)_R$. The only multiplet relevant for supergravity is the supergravity multiplet. The scalars in this multiplet transform non-trivially under the R-symmetry group, and the holonomy group of the scalar manifold should contain USp(6) as a factor, which, together with the dimension fixes it to be

$$\mathcal{M}_{SG} \cong \frac{SU^*(6)}{USp(6)}, \tag{10.55}$$

where $SU^*(6)$ is a particular real form of $SU(6)$.

10.5.4 $\mathcal{N} = 8$ in 5D

In the maximally supersymmetric case, the R-symmetry group is $USp(8)_R$, and the supergravity multiplet contains the graviton, 8 gravitini, 27 vector fields, 48 spin-1/2 fields, and 42 real scalar fields. The latter transform non-trivially under the R-symmetry group, and we expect the holonomy group to contain a $USp(8)$-factor. The only 42-dimensional space with this property is

$$\mathcal{M}_{SG} \cong \frac{E_{6(6)}}{USp(8)}, \tag{10.56}$$

where $E_{6(6)}$ denotes the real form of E_6 for which the difference of compact and non-compact generators is 6.

10.5.5 Gaugings and Tensor Fields

As we discussed in Sect. 8.1, in four spacetime dimensions, a massless vector field without gauge interactions can be equivalently described by its magnetic dual vector field. In the presence of gauge interactions, on the other hand, this electric–magnetic duality is broken, and one has to make sure that the gauging is performed in a suitable duality frame.[5]

In five dimensions, there is no duality between electric and magnetic vector fields, but if there are no gauge interactions, there is an analogous Poincaré duality between vector fields, $A = A_\mu dx^\mu$, and two-form fields, $B = \frac{1}{2} B_{\mu\nu} dx^\mu \wedge dx^\nu$. In the simplest version, this is just the statement that the 5D source-free Maxwell equations,

$$d \star F = 0, \qquad dF = 0 \tag{10.57}$$

---

[5] As explained in Sect. 9.5, the embedding tensor formalism allows one to formally maintain electric–magnetic duality at the expense of a redundant field content.

for the two-form field strength $F = dA$ read

$$dH = 0, \qquad d \star H = 0, \tag{10.58}$$

when expressed in terms the dual three-form field strength, $H := \star F$, which imply $H = dB$ and $d \star dB = 0$, i.e., the field equation of a massless two-form field, B.

If one now instead considers 5D theories *with* gauge interactions, the above duality between 5D vector and tensor fields no longer holds, and one would naively expect that a consistent gauging would require working exclusively with vector fields. In many cases, this is also what happens, but there are also important situations where some of the vector fields have to be converted to tensor fields in a specific way in order to perform the gauging.

To understand this, let us assume we start from an ungauged 5D supergravity theory in the standard form, as described in the above subsections, where all potential tensor fields are dualized to vector fields. Suppose further the theory has n vector fields and a global symmetry group, G_{global},[6] such that the n vector fields transform in an n-dimensional representation of that global symmetry group (this representation may be reducible or irreducible, depending on the theory).

If G_{global} has a subgroup, G, such that this n-dimensional representation of G_{global} becomes the adjoint representation of G,

$$\mathbf{n}(G_{\text{global}}) \to \text{adj}(G), \tag{10.59}$$

one can replace the Abelian field strengths, F^I ($I = 1, \ldots, n$), by the corresponding non-Abelian field strengths, $\mathscr{F}^I$, and the partial derivatives, ∂_μ, of charged matter fields by gauge covariant derivatives, $\widehat{\partial}_\mu$, and gauge the group G. Just as in 4D, this covariantization will break supersymmetry, which, however, can be restored by introducing suitable Yukawa interactions and scalar potentials into the Lagrangian as well as fermionic shifts to the supersymmetry transformation laws.

The above also holds true in the more general situation when the n-dimensional representation of G_{global} decomposes into the adjoint of G plus singlets of G,

$$\mathbf{n}(G_{\text{global}}) \to \text{adj}(G) \oplus \text{singlets}(G). \tag{10.60}$$

If G has no Abelian factor, the singlet fields will just remain Abelian vector fields, and they will have no gauge couplings to the matter fields, i.e., they will be "spectator vector fields" with respect to the gauging. In case G has an Abelian factor, on the other hand, the singlet vector fields will still remain Abelian, but they

---

[6] In five dimensions, G_{global} is typically (a subgroup of) the isometry group, $\text{Iso}(\mathscr{M}_{\text{scalar}})$, of the scalar manifold. If all scalars are inert under the respective R-symmetry group (as, e.g., for 5D, $\mathscr{N} = 2$ supergravity coupled to vector multiplets only, or for 5D, $\mathscr{N} = 4$ pure supergravity), the global symmetry group also has the R-symmetry group, $G_R = \text{Usp}(\mathscr{N})$, as an additional factor.

might have minimal couplings to the matter fields, i.e., they might contribute to the Abelian part of the gauge group G.

As a simple example, consider a theory with four vector fields in the fundamental representation of a global symmetry group SO(4). With respect to the obvious subgroup SO(3), the **4** of SO(4) decomposes into **3** $\oplus$ **1**. As the **3** is the adjoint of SO(3), one can use the three vector fields in the **3** to gauge SO(3), with the SO(3) singlet vector field remaining a spectator vector field. Gaugings of this standard type were investigated in 5D, $\mathcal{N} = 2$ supergravity in [11] and in 5D, $\mathcal{N} = 4$ supergravity in [9].

A more problematic situation arises, however, if the decomposition of the n-dimensional representation of G_{global} with respect to the subgroup G also contains non-singlets of G,

$$\mathbf{n}(G_{\text{global}}) \rightarrow \text{adj}(G) \oplus \text{singlets}(G) \oplus \text{non-singlets}(G). \tag{10.61}$$

In this case, the gauging of G cannot be performed in the usual way, because vector fields can only couple consistently to other vector fields if they sit in the adjoint of the gauge group.

Historically, the first example of this situation occurred in 5D, $\mathcal{N} = 8$ supergravity in the 1980s [12–14]. In the ungauged version, this theory has 27 vector fields transforming in the 27-dimensional irreducible representation of the global symmetry group, $G_{\text{global}} = E_{6(6)}$. $E_{6(6)}$ is the maximally non-compact real form of the exceptional group E_6 and forms the isometry group of the scalar manifold $\mathcal{M}_{\text{scalar}} = E_{6(6)}/\text{USp}(8)$ of this theory.

A particularly interesting subgroup of the global symmetry group $E_{6(6)}$ is the subgroup SO(6), under which the **27** of $E_{6(6)}$ transforms as

$$\mathbf{27} \rightarrow \mathbf{15} \oplus \mathbf{6} \oplus \mathbf{6}. \tag{10.62}$$

Here, the **15** is the adjoint representation of SO(6), whereas the **6** denotes the fundamental representation of SO(6), which is clearly a non-singlet representation. Without the **6** $\oplus$ **6**, one could gauge SO(6) with the 15 vector fields in the adjoint, but due to the presence of the non-singlets, this is not possible in the standard way.

This by itself would not be a big deal, as not every subgroup of a global symmetry group needs to be gaugeable, but in this particular case, there were very strong arguments in favor of the existence of a gauging with the gauge group SO(6). These arguments have to do with the compactification of type IIB supergravity on the maximally supersymmetric background solution $AdS_5 \times S^5$, which was expected to admit a consistent truncation to the lowest lying Kaluza–Klein modes that should be identical to 5D, $\mathcal{N} = 8$ supergravity with gauge group SO(6) (the isometry group of the five-sphere).

The resolution of this problem came from a closer inspection of the Kaluza–Klein spectrum of this compactification [15, 16], which, apart from the 15 vector fields, also revealed the presence of 12 tensor fields, which are not equivalent to vector fields in an AdS_5 background (they transform in a different representation of the AdS_5 isometry group as the vector fields).

This suggests that a consistent gauging of SO(6) might require treating the fields in the $\mathbf{6} \oplus \mathbf{6}$ as antisymmetric tensor fields and not as vector fields. In fact, as tensor fields, they could be treated as a special type of matter fields, so that their derivatives could be covariantized by introducing minimal couplings to the 15 gauge fields using the six-dimensional representation matrices of SO(6) (see Eq. (10.64)). This approach turned out to be correct and led to the successful construction [12–14] of 5D, $\mathcal{N} = 8$ gauged supergravity with gauge group SO(6), which, many years later, also played a central role in the AdS/CFT correspondence.

The necessity of converting non-singlet vector fields to tensor fields in order to perform certain gaugings also was found in 5D, $\mathcal{N} = 4$ [17–19] and $\mathcal{N} = 2$ theories [20–22]. In fact, the $\mathcal{N} = 2$ cases allow one to isolate the contribution to the scalar potential that arises due to the presence of charged tensor fields from those contributions that come from the gauging of the R-symmetry group or the presence of non-Abelian gauge fields [20], or from charged hypermultiplets [21]. Interestingly, one finds that this scalar potential contribution is positive semidefinite, i.e., it cannot by itself lead to AdS vacua [20]. In fact, the $\mathcal{N} = 2$ theories with charged tensor fields were the first extended supergravity theories in which a meta-stable *de Sitter* vacuum could be constructed [23, 24]. This implies that the presence of tensor fields in 5D supergravity theories with non-singlet representations outside the adjoint representation of the gauge group is not a consequence of an AdS vacuum structure (although it is consistent with it), but follows from more general considerations.

Another noteworthy feature of the 5D gaugings with charged tensor fields is that the corresponding Lagrangians contain a first-order kinetic term for the tensor fields of the schematic form

$$\frac{1}{g}\Omega_{MN}\varepsilon^{\mu\nu\rho\sigma\kappa} B^M_{\mu\nu}\widehat{\partial}_\rho B^N_{\sigma\kappa}, \tag{10.63}$$

where g denotes the gauge coupling, $\Omega_{MN} = -\Omega_{NM}$ is an antisymmetric constant tensor, and $M, N, \ldots = 1, \ldots, m$ label the m tensor fields. The gauge covariant derivative, $\widehat{\partial}_\mu$, describes the minimal coupling of the n gauge fields, A^I_μ ($I, J, \ldots = 1, \ldots, n$), in the adjoint representation of the gauge group to the tensor fields and takes the form

$$\widehat{\partial}_{[\rho} B^N_{\sigma\kappa]} = \partial_{[\rho} B^N_{\sigma\kappa]} + g A^I_{[\rho} \Lambda^N_{IM} B^M_{\sigma\kappa]}, \tag{10.64}$$

where Λ^N_{IM} denotes the representation matrices of the tensor fields with respect to the gauge group.

Using this first-order form and taking into account also another, mass-like term for the tensor fields not shown here, it is in principle possible to integrate out half of the tensor fields so as to arrive at a Lagrangian with second-order kinetic term for the remaining (massive) tensor fields (see [25] for a detailed discussion). Doing this explicitly, however, yields fairly complicated expressions in general so that it is usually easier to stick with the above first-order form. For more detailed discussions of 5D gauged supergravity with tensor fields, we refer to the original literature.

10.A   Appendix: Clifford Algebras and Spinors in Arbitrary D

As described in Sect. 10.1, the balance between bosonic and fermionic degrees of freedom in supersymmetric field theories in general is implemented differently for different spacetime dimension, D, because the degrees of freedom of the corresponding fields have a different overall scaling with D. This is further complicated by the strong D-dependence of the possible chirality and reality conditions one can impose on Clifford algebra representations so as to generate minimal spinor representations of the Lorentz group. The D-dependence of these chirality and reality conditions also leads to D-dependent R-symmetry groups, which in turn contribute to a rich variety of possible scalar manifold geometries in the respective spacetime dimensions. It is the purpose of this subsection to classify the representations of Clifford algebras, the minimal spinor representations of the corresponding Lorentz groups as well as the resulting R-symmetry groups. This generalizes the discussion of spinors in four dimensions given in Chap. 1.

Starting point is the Clifford algebra, Cliff$(1, D − 1)$, in D Lorentzian dimensions,

$$\{\Gamma_a, \Gamma_b\} = 2\eta_{ab} \qquad (a, b, \ldots = 0, 1, \ldots, D − 1) \tag{10.65}$$

$$\eta_{ab} = \text{diag}(-1, +1, \ldots, +1). \tag{10.66}$$

Just as in 4D, the relation (10.65) implies that

$$\widehat{\rho}(M_{ab}) \equiv \Sigma_{ab} \equiv \frac{1}{4}[\Gamma_a, \Gamma_b] = \frac{1}{2}\Gamma_{ab} \tag{10.67}$$

form a representation of the Lorentz algebra. The exponentials $\exp\left[\frac{\omega^{ab}\Sigma_{ab}}{2}\right]$ with ω_{ab} being finite rotation angles or boost parameters then form a double-valued representation of the Lorentz group $SO_0(1, D − 1)$.

10.A.1   Irreducible Representations of Cliff$(1, D − 1)$

The structure of the irreducible representations of Cliff$(1, D−1)$ is slightly different for even and odd dimensions:

10.A.1.1   Even Dimensions
Up to equivalence, there is exactly one non-trivial irreducible representation (irrep) of Cliff$(1, D − 1)$ (see, e.g., [26]). It has complex dimension $2^{D/2}$, i.e., the Γ_a are

complex $(2^{D/2} \times 2^{D/2})$-matrices, generalizing the (4×4)-matrices in 4D. Explicit forms of these representations can be built up by successive tensor products of the irreps of lower-dimensional Clifford algebras, starting with the case $D = 2$ (see, e.g., [27]), but we do not need them for this book.

10.A.1.2 Odd Dimensions

If D is odd, irreps of Cliff$(1, D-1)$ can be obtained from an irrep of Cliff$(1, D-2)$ by defining the analogue of the γ_5 matrix in 4D:

$$\Gamma_* \equiv (-i)^{\frac{D+1}{2}} \Gamma_0 \Gamma_1 \dots \Gamma_{D-2}. \tag{10.68}$$

This matrix satisfies

$$(\Gamma_*)^2 = \mathbb{1} \tag{10.69}$$

$$\{\Gamma_*, \Gamma_a\} = 0 \qquad \forall a = 0, \dots, D-2 \tag{10.70}$$

so that either of

$$\Gamma_{D-1}^{(\pm)} \equiv \pm\Gamma_* \tag{10.71}$$

can be used as the remaining gamma matrix to promote $\{\Gamma_0, \dots, \Gamma_{D-2}\}$ to a representation of Cliff$(1, D-1)$. One thus obtains *two* inequivalent representations of Cliff$(1, D-1)$ for odd D, one for each sign in (10.71).

10.A.2 Irreducible Spinor Representations of $SO_0(1, D-1)$

Thus far, we have discussed the irreps of Cliff$(1, D-1)$ and described how these induce double-valued spinor representations of the corresponding Lorentz groups $SO_0(1, D-1)$. Just as in four dimensions, however, the spinor representations of the Lorentz group so-obtained are in general *not* irreducible, even though they descend from irreducible representations of Cliff$(1, D-1)$. In order to obtain an irreducible spinor representation of $SO_0(1, D-1)$, one in general has to impose additional constraints, which may be of the following type:

1. Chirality condition
2. Reality condition
3. Chirality and a reality condition

The possibilities to impose one of the above are strongly dimension dependent, as we will now describe.

10.A.2.1 Chirality Conditions

For even D, we can always impose the following chirality condition to define a left-
or right-handed *Weyl spinor*:

$$\Gamma_* \psi_{\substack{L \\ R}} = \pm \psi_{\substack{L \\ R}} \tag{10.72}$$

Note that that this condition is Lorentz covariant because of $[\Sigma_{ab}, \Gamma_*] = 0$.

For odd D, on the other hand, there is no non-trivial analogue of Γ_*, because

$$\underbrace{\Gamma_0 \Gamma_1 \dots \Gamma_{D-2}}_{\sim \Gamma_{D-1}^{(\pm)}} \Gamma_{D-1}^{(\pm)} \sim (\Gamma_{D-1}^{(\pm)})^2 \sim \mathbb{1}. \tag{10.73}$$

Thus a non-trivial chirality condition can only be imposed in *even* D.

10.A.2.2 Reality Conditions

It is again useful to distinguish between even and odd dimensions:

Even Dimensions

As discussed above, for even D there is only one equivalence class of irreps of
Cliff$(1, D - 1)$ generated by matrices Γ_a. Hence, the complex conjugate matrices
$\pm \Gamma_a^*$, which also satisfy the Clifford algebra, must be equivalent to the matrices Γ_a,
i.e., there has to be a matrix, B, such that

$$\Gamma_a^* = \eta B \Gamma_a B^{-1} \tag{10.74}$$

for both signs $\eta = \pm 1$.

Odd Dimensions

If D is odd, we can obviously find a matrix, B, that also satisfies (10.74) for the first
$(D - 1)$ gamma matrices with $\eta = \pm 1$. What is non-trivial, however, is to extend
(10.74) also to the remaining gamma matrix $\Gamma_{D-1}^{(\pm)} = \pm \Gamma_*$ (cf. Eq. (10.68)), i.e., to
have

$$(\Gamma_*)^* = \eta B \Gamma_* B^{-1}. \tag{10.75}$$

Indeed, using the definition (10.68) and (10.74) for $\Gamma_0, \dots, \Gamma_{D-2}$, one easily shows

$$(\Gamma_*)^* = (-1)^{\frac{D+1}{2}} B \Gamma_* B^{-1}, \tag{10.76}$$

which is consistent with (10.75) only for one sign:

$$\eta = (-1)^{\frac{D+1}{2}} = \begin{cases} -1 \text{ for D=5 mod 4} \\ +1 \text{ for D=3 mod 4} \end{cases} \tag{10.77}$$

Obviously, the defining Eqs. (10.74) and (10.75) define B only up to an arbitrary rescaling. We may thus choose the overall scaling such that

$$|\det B| = 1 \quad \text{(choice)} \tag{10.78}$$

With this normalization, one has (cf. Exercise 10.1)

$$B^* B = \epsilon \mathbb{1} \tag{10.79}$$

$$\epsilon = \pm 1. \tag{10.80}$$

The important point now is that this parameter ϵ is not arbitrary, but is instead fixed by the values of η and D. Concretely, for $D = 2n$ or $D = 2n + 1$, one finds[7]

$$\epsilon = -\eta \sqrt{2} \cos\left[\frac{\pi}{4}(1 + \eta 2n)\right], \tag{10.83}$$

and one arrives at the possible values for ϵ and η shown in Table 10.2.

The Majorana Condition

What makes the possible values of ϵ so important is that it determines whether one can impose a Majorana condition on a spinor, which, in terms of B, reads

$$\psi^* = \alpha B \psi \quad \text{(Majorana condition)}, \tag{10.84}$$

where α is an arbitrary phase. This condition is consistent with Lorentz invariance, because $\Gamma_{ab}^* = B \Gamma_{ab} B^{-1}$. A Majorana spinor thus furnishes a complete representation of the Lorentz algebra and has only half as many degrees of freedom as an unconstrained complex Dirac spinor. The consistency of (10.84) with $\psi^{**} = \psi$, however, imposes the consistency condition

$$\epsilon = +1 \quad \text{(for Majorana condition)}, \tag{10.85}$$

---

[7] This can be proven, e.g., with the help of the charge conjugation matrix C. In a friendly representation (i.e., for $\Gamma_a \Gamma_a^\dagger = \mathbb{1}$ (no sum) and symmetric or anti-symmetric Γ_a), $C \equiv B^T \Gamma_0$ satisfies, because of (10.74) and (10.79),

$$\Gamma_a^T = -\eta C \Gamma_a C^{-1} \tag{10.81}$$

$$C^T = -\eta \epsilon C. \tag{10.82}$$

The matrices $(C\Gamma_{a_1 \dots a_p})$ then have a definite symmetry under transposition. This symmetry depends on p, ϵ, and η. On the other hand, the set of all matrices $\Gamma_{a_1 \dots a_p}$ plus the unit matrix form a complete basis of all complex $(2^{[D/2]} \times 2^{[D/2]})$-matrices. As the number of linearly independent antisymmetric and symmetric of such matrices is fixed to be $2^{[D/2]}(2^{[D/2]} - 1)/2$ and $2^{[D/2]}(2^{[D/2]} + 1)/2$, respectively, one can determine the possible values of ϵ as a function of D and η (which, for odd dimensions, is itself fixed by D). (cf., e.g., [26]).

Table 10.2   The possible values for η and ϵ together with the resulting minimal spinor types, the minimal number of real supercharges, and the general form of the R-symmetry groups (M = Majorana, SM = Symplectic Majorana, W = Weyl, MW = Majorana–Weyl, SMW = Symplectic Majorana–Weyl)

D	η	ϵ	Min. spinor type	Min. # of real supercharges	R-symmetry group
2	+1	+1	MW	1	$SO(N_L) \times SO(N_R)$
	−1	+1			
3	+1	+1	M	2	$SO(N)$
4	+1	+1	M or W	4	$U(N)$
	−1	−1			
5	−1	−1	SM	8	$Usp(2N)$
6	+1	−1	SMW	8	$Usp(2N_L) \times Usp(2N_R)$
	−1	−1			
7	+1	−1	SM	16	$Usp(2N)$
8	+1	−1		16	$U(N)$
	−1	+1	M or W		
9	−1	+1	M	16	$SO(N)$
10	+1	+1	MW	16	$SO(N_L) \times SO(N_R)$
	−1	+1			
11	+1	+1	M	32	$SO(N)$
12	+1	+1	M or W	64	$U(N)$
	−1	−1			
$\cdots$	$\cdots$	$\cdots$	$\cdots$	$\cdots$	$\cdots$

limiting the possibility of Majorana spinors to certain dimensions, as indicated in Table 10.2.

Symplectic Majorana Spinors

If $\epsilon = -1$, one can impose a symplectic Majorana condition. To this end, one needs an even number of Dirac spinors ψ_i, $(i, j, \ldots = 1, \ldots, 2N)$ and an antisymmetric real matrix Ω_{ij} with $\Omega^2 = -\mathbb{1}_{2N}$ and imposes

$$(\psi_i)^* = \Omega_{ij} B \psi_j. \tag{10.86}$$

As one needs at least two Dirac spinors to impose the symplectic Majorana condition, it does not lead to a reduction of the minimal number of degrees of freedom relative to a single Dirac spinor. The symplectic Majorana condition is, however, convenient, because it makes the action of the R-symmetry group (which in these dimensions involve symplectic groups; see Table 10.2) manifest.

10.A.3 Majorana and Weyl Condition

In some dimensions, the Majorana and the Weyl condition can be imposed simultaneously. This reduces the number of independent degrees of freedom to one quarter relative to an unconstrained Dirac spinor. Imposing (we set the phase $\alpha = 1$ for simplicity)

$$\psi^* = B\psi \tag{10.87}$$

$$\Gamma_*\psi = \pm\psi, \tag{10.88}$$

at the same time, obviously requires the consistency condition

$$(\Gamma_*)^* = B\Gamma_*B^{-1} \tag{10.89}$$

which is possible only if $D = 4n - 2$. But as there are no Majorana spinors in $D = 6, 14, \ldots$, Majorana–Weyl spinors can only exist for

$$D = 2 \bmod 8 \qquad \text{(Condition for Majorana–Weyl spinors)}. \tag{10.90}$$

Note, in particular, that in 4D, one can have Majorana spinors *or* Weyl spinors, but not Majorana–Weyl spinors.

Analogously, in dimensions in which $\epsilon = -1$ allows a symplectic Majorana condition, one can sometimes also simultaneously impose a Weyl condition, and the corresponding spinors are then called symplectic Majorana–Weyl spinors. These are the dimensions $D = 6 \bmod 8$

The minimal amount of supersymmetry in each spacetime dimension is generated by a spinor operator that corresponds to the minimal spinor representation of the Lorentz group in the respective spacetime dimension. Extended supersymmetries then correspond to multiples of such minimal spinors. The R-symmetry group of the corresponding supersymmetry algebra has to respect these reality and chirality conditions and thus depends on the minimal spinor type as shown in Table 10.2. If the scalar fields of a given type of multiplet transform non-trivially under the R-symmetry group (or a factor thereof), the holonomy group of the scalar manifold typically contains this group (factor) as a factor. Especially for large amounts of supersymmetry, this already strongly constrains the possible scalar manifolds, as we described in detail for the theories in 4D and 5D.

For more than 32 real supercharges, one always has states with helicity $|h| > 2$ in the supermultiplets, which, for Lorentzian signature, limits supersymmetric field theories to $D \leq 11$.

We finally note that in spacetimes with non-Lorentzian signature, the possible reality and chirality conditions for a given D are in general different. This is in particular true for the Euclidean signature of the compactification manifolds in string compactifications, so that the possible spinor type on these manifolds cannot be read off from Table 10.2.

Exercises

10.1. Using (10.74) and Schur's Lemma, show that (10.79) holds for some $\epsilon \in$ $\mathbb{C}$. Using the complex conjugate of (10.79) and the choice (10.78), show that this implies (10.80).

References

1. E. Cremmer, B. Julia, J. Scherk, Supergravity theory in 11 dimensions. Phys. Lett. **B76**, 409–412 (1978)
2. F. Giani, M. Pernici, N = 2 supergravity in ten-dimensions. Phys. Rev. D **30**, 325–333 (1984)
3. L.J. Romans, Massive N = 2a supergravity in ten-dimensions. Phys. Lett. B **169**, 374 (1986)
4. J.H. Schwarz, Covariant field equations of chiral N = 2 D = 10 supergravity. Nucl. Phys. B **226**, 269 (1983)
5. P.S. Howe, P.C. West, The complete N = 2, D = 10 supergravity. Nucl. Phys. B **238**, 181–220 (1984)
6. G. Dall'Agata, K. Lechner, M. Tonin, D = 10, N = IIB supergravity: lorentz invariant actions and duality. JHEP **9807**, 017 (1998) hep-th/9806140
7. M. Günaydin, G. Sierra, P.K. Townsend, The geometry of N = 2 Maxwell-Einstein supergravity and Jordan algebras. Nucl. Phys. B **242**, 244–268 (1984)
8. G. Sierra, N = 2 Maxwell matter Einstein supergravities in D = 5, D = 4 and D = 3. Phys. Lett. B **157**, 379–382 (1985)
9. M. Awada, P.K. Townsend, N = 4 Maxwell-Einstein supergravity in five-dimensions and its SU(2) gauging. Nucl. Phys. B **255**, 617–632 (1985)
10. M. Berger, Sur les groupes d'holonomie homogènes des variétés a connexion affines et des variétés riemanniennes. Bull. Soc. Math. France **83**, 279–330 (1953)
11. M. Günaydin, G. Sierra, P.K. Townsend, Gauging the d = 5 Maxwell-Einstein supergravity theories: more on Jordan algebras. Nucl. Phys. B **253**, 573 (1985)
12. M.Günaydin, L.J. Romans, N.P. Warner, Gauged N = 8 supergravity in five-dimensions. Phys. Lett. B **154**, 268–274 (1985)
13. M. Günaydin, L.J. Romans, N.P. Warner, Compact and noncompact gauged supergravity theories in five-dimensions. Nucl. Phys. B **272**, 598–646 (1986)
14. M. Pernici, K. Pilch, P. van Nieuwenhuizen, Gauged N = 8 D = 5 supergravity. Nucl. Phys. B **259**, 460 (1985)
15. M. Günaydin, N. Marcus, The spectrum of the s**5 compactification of the chiral N = 2, D = 10 supergravity and the unitary supermultiplets of U(2, 2/4). Class. Quant. Grav. **2**, L11 (1985)
16. H.J. Kim, L.J. Romans, P. van Nieuwenhuizen, The mass spectrum of chiral N = 2 D = 10 supergravity on S**5. Phys. Rev. D **32**, 389 (1985)
17. L.J. Romans, Gauged N = 4 supergravities in five-dimensions and their magnetovac backgrounds. Nucl. Phys. B **267**, 433–447 (1986)
18. G. Dall'Agata, C. Herrmann, M. Zagermann, General matter coupled N = 4 gauged supergravity in five-dimensions. Nucl. Phys. B **612**, 123–150 (2001) [arXiv:hep-th/0103106 [hep-th]]
19. J. Schon, M. Weidner, Gauged N = 4 supergravities. JHEP **05**, 034 (2006) [arXiv:hep-th/0602024 [hep-th]]
20. M. Günaydin, M. Zagermann, The Gauging of five-dimensional, N = 2 Maxwell-Einstein supergravity theories coupled to tensor multiplets. Nucl. Phys. B **572**, 131–150 (2000) [arXiv:hep-th/9912027 [hep-th]]
21. A. Ceresole, G. Dall'Agata, General matter coupled N = 2, D = 5 gauged supergravity. Nucl. Phys. B **585**, 143–170 (2000) [arXiv:hep-th/0004111 [hep-th]]

22. E. Bergshoeff, S. Cucu, T. de Wit, J. Gheerardyn, S. Vandoren, A. Van Proeyen, N = 2 supergravity in five-dimensions revisited. Class. Quant. Grav. **21**, 3015–3042 (2004) [arXiv:hep-th/0403045 [hep-th]]
23. M. Günaydin, M. Zagermann, The Vacua of 5-D, N = 2 gauged Yang-Mills/Einstein tensor supergravity: Abelian case. Phys. Rev. D **62**, 044028 (2000) [arXiv:hep-th/0002228 [hep-th]]
24. B. Cosemans, G. Smet, Stable de Sitter vacua in N = 2, D = 5 supergravity. Class. Quant. Grav. **22**, 2359–2380 (2005) [arXiv:hep-th/0502202 [hep-th]]
25. P.K. Townsend, K. Pilch, P. van Nieuwenhuizen, Selfduality in odd dimensions. Phys. Lett. B **136**, 38 (1984)
26. P.C. West, Supergravity, brane dynamics and string duality [arXiv:hep-th/9811101 [hep-th]]
27. A. Van Proeyen, Tools for supersymmetry. hep-th/9910030

Index

© Springer-Verlag GmbH Germany, part of Springer Nature 2021
G. Dall'Agata, M. Zagermann, *Supergravity*, Lecture Notes in Physics 991,
https://doi.org/10.1007/978-3-662-63980-1

Printed in the United States
by Baker & Taylor Publisher Services